GENETIC RESPONSE OF FOREST SYSTEMS TO CHANGING ENVIRONMENTAL CONDITIONS

FORESTRY SCIENCES

Volume 70

The titles published in this series are listed at the end of this volume.

Genetic Response of Forest Systems to Changing Environmental Conditions

Edited by

Gerhard Müller-Starck

Technical University of Munich,
Faculty of Forest Science, Section of Forest Genetics, Germany

and

Roland Schubert

Technical University of Munich,
Faculty of Forest Science, Section of Forest Genetics, Germany

KLUWER ACADEMIC PUBLISHERS

DORDRECHT / BOSTON / LONDON

A C.I.P. Catalogue record for this book is available from the Library of Congress.

ISBN 1-4020-0236-X

Published by Kluwer Academic Publishers,
P.O. Box 17, 3300 AA Dordrecht, The Netherlands.

Sold and distributed in North, Central and South America
by Kluwer Academic Publishers,
101 Philip Drive, Norwell, MA 02061, U.S.A.

In all other countries, sold and distributed
by Kluwer Academic Publishers,
P.O. Box 322, 3300 AH Dordrecht, The Netherlands.

Printed on acid-free paper

Norway spruce in response to long term exposure to industrial pollution in the Lower Spiš region, Slovakia. (Photo: G. Müller-Starck). Underlayed DNA sequence refers exemplarily to a gene which is involved in stress response (Schubert, R., Müller-Starck, G., Ernst, D., Sandermann, H. and K.P. Häger (1997): The molecular structure and evolutionary relationships of a 16.9 kDa heat shock protein from Norway spruce (*Picea abies* (L.) Karst.). Forest Genetics 4(3), 131-138).

Printed in the Netherlands.

Preface

Forest ecosystems and various systems of plantations are increasingly exposed to changing environmental conditions following industrialisation, rapid population growth and subsequent anthropogenic pressure. In addition to the manifold ecological and economical benefits of forest systems, the global forest cover has become a highly significant tool in the fixation of carbon dioxide, which represents the predominant greenhouse gas associated with global warming. Changing environmental conditions substantially affect genetic variation and its dynamics in forest systems, infer genetic resources and challenge adaptive abilities.

In response to such challenges, a IUFRO conference was held 1999 in Freising/Munich, Germany, which was particularly addressed to Division 7 (Forest Health) and 2 (Physiology and Genetics). Moreover, this conference combined initiatives of previous IUFRO meetings in Vancouver 1996, Beijing 1998 and Edinburgh 1998.

The present book was initiated by the 1999'IUFRO conference. Part 1 focuses on the response to stress in terms of case studies which address physiological and genetic characters as well as various metric traits. Part 2 presents a choice of studies which refer to diversity and geographic variation of various species and site conditions, respectively. Part 3 deals with the characterisation of genetic resources and compiles a variety of studies which address reproduction and migration as well as management aspects. Part 4 presents a set of studies which focus on forest tree breeding with respect to uncertain climatic futures. Generally, conceptional contributions are integrated in the respective chapters.

The editors gratefully acknowledge the work of the chairmen of eight sessions of the conference: K. Percy (Fredericton, Canada), F. Scholz (Großhansdorf, Germany), D. Karnosky (Houghton, USA), X. Shen (Beijing, China), H. Hattemer (Göttingen, Germany), G. Vendramin (Florence, Italy), L. Paule (Zvolen, Slovakia) and A. Kremer (Pierroton, France). The review process selected a total of 29 contributions which are included in the present book. The editors greatly appreciate the work of 38 referees from Europe, North America, South America and Asia.

We would like to express our sincere thanks to M. Fischer, B. Kain, E. Röschter and M. Thole (Freising, Germany) for their excellent help in the technical preparation of the book manuscript. N. Gibson and M. Kelly (Dordrecht, The Netherlands) kindly supported the editorial work.

G. Müller-Starck, R. Schubert

Contents

PART 3: GENETIC RESOURCES, REPRODUCTION, MANAGEMENT

PART 4: TREE BREEDING FOR UNCERTAIN FUTURE ENVIRONMENTS

Contributors

AGÚNDEZ, M.D. — Forest Research Centre, INIA, Carr. Coruña Km. 7,5, E-28040 Madrid, Spain

ALFARO, R.I. — Pacific Forestry Centre, Canadian Forest Service, 506 W. Burnside Rd., Victoria, BC, Canada, V8Z 1M5

ALIA, R. — Forest Research Centre, CIFOR-INIA, P.O. Box 8111, E-28080 Madrid, Spain

ANZIDEI, M. — Consiglio Nazionale delle Ricerche, Istituto Miglioramento Genetico delle Piante Forestali, Via Atto Vannucci, 13, I-50135 Firenze, Italy

BANDOU, E. — INRA Station de Recherches Forestières Guyane, Campus agronomique, BP 709, 97387 Kourou cedex, French Guiana

BREYNE, P. — Department of Plant Genetics, Flanders Interuniversity Institute for Biotechnology, University of Gent, B-9000 Gent, Begium

BUENO, M.A. — Forest Research Centre, INIA, Carr. Coruña Km. 7,5, E-28040 Madrid, Spain

CARON, H. — INRA Station de Recherches Forestières, Laboratoire de Génétique des Arbres Forestiers, BP 45, F-33611 Gazinet cedex, France

ÇENGEL, B.N. — Forest Trees and Seeds Improvement Research Directorate, TR-06560 Gazi, Ankara, Turkey

CHEN, X. — Beijing Forestry University, 100083, China

CHIRON, H. — GSF – National Research Center for Environment and Health, Institute of Biochemical Plant Pathology, D-85764 Neuherberg, Germany

CSAIKL, U.M. — Österreichisches Forschungszentrum Seibersdorf (ARCS), A-2444 Seibersdorf, Austria

DEGEN, B. — Silvolab Guyane, INRA-Station de Recherches Forestières, Campus agronomique, B.P. 709, 97387 Kourou cedex, French Guiana

DEMESURE, B. Conservatoire Génétique des Arbres Forestiers, Office National des Forêts, Campus INRA, F-45160 Ardon, France

DROUET, D. Biology Department, University of Orléans BP6749, F-45067 Orléans cedex 2, France

DUTECH, C. CIRAD-Forêt, Campus agronomique, BP 709, 97387 Kourou cedex, French Guiana

ERNST, D. GSF – National Research Center for Environment and Health, Institute of Biochemical Plant Pathology, D-85764 Neuherberg, Germany

FINKELDEY, R. Institut für Forstgenetik und Forstpflanzenzüchtung, Universität Göttingen, Büsgenweg 2, D-37077 Göttingen, Germany

FROMM, M. Landesanstalt für forstl. Saat- und Pflanzenzucht, Forstamtsplatz 1, D-93317 Teisendorf, Germany

GERBER, S. INRA, Laboratoire de Génétique et d´Amélioration des Arbres Forestiers, B.P. 45, F-33611 Gazinet cedex, France

GIL, L. Unit of Anatomy, Physiology and Genetics, ETSIM, Ciudad Universitaria, E-28040 Madrid, Spain

GLÖSSL, J. Zentrum für Angewandte Genetik, Universität für Bodenkultur Wien, Muthgasse 18, A-1190 Vienna, Austria

GÓMEZ, A. Dept. Breeding and Biotechnology, INIA, Carr. Coruña Km. 7,5, E-28040 Madrid, Spain

GONZÁLEZ-MARTINEZ, S.C. Dept. of Breeding and Biotechnology, CIFOR-INIA, P.O. Box 8111, E-28080 Madrid, Spain

GREGORIUS, H.-R. Institut für Forstgenetik und Forstpflanzenzüchtung, Universität Göttingen, Büsgenweg 2, D-37077 Göttingen, Germany

GUESNET, D. Conservatoire Génétique des Arbres Forestiers, Office National des Forêts, Campus INRA, F-45160 Ardon, France

HAMANN, A.	University of British Columbia, Department of Forest Science, Forest Science Center, Vancouver, B. C., Canada, V6T 1Z4
HARTZ, R.	Institut für Forstbotanik und Forstzoologie, Technische Universität Dresden, Pienner Strasse 7, D-01737 Tharandt, Germany
HATTEMER, H.H.	Institut für Forstgenetik und Forstpflanzenzüchtung, Universität Göttingen, Büsgenweg 2, D-37077 Göttingen, Germany
HEINZE, B.	Institute of Forest Genetics, Federal Forest Research Centre, Hauptstrasse 7, A-1140 Vienna, Austria
HELLER, W.	GSF-National Research Center for Environment and Health, Institute of Biochemical Plant Pathology, D-85764 Neuherberg, Germany
HERTEL, H.	Federal Research Centre for Forestry and Forest Products (BFH), Institute for Forest Genetics and Forest Tree Breeding, Eberswalder Chaussee 3, D-15377 Waldsieversdorf, Germany
HERZOG, S.	Technische Universität Dresden, Wildökologie, Pienner Strasse 8, D-01737 Tharandt, Germany
HUSSENDÖRFER, E.	Forest Research Station Baden-Württemberg, Department of Botany and Site Ecology, Wonnhaldestr. 4, D-79100 Freiburg, Germany
KARNOSKY, D.F.	School of Forestry and Wood Products, Michigan Technological University, 1400 Townsend Drive, Houghton, MI, 49931-1295, USA
KAYA, Z.	Department of Biological Sciences, Middle East Technical University, TR-06531 Ankara, Turkey
KING, J.N.	Ministry of Forests, Research Branch, P.O. Box 9519, Stn. Prov. Govt., Victoria, British Columbia, Canada, V8W 9C2
KÖNIG, A.O.	Federal Research Centre for Forestry and Forest Products (BFH), Institute for Forest Genetics

and Forest Tree Breeding, Sieker Landstrasse 2, D-22927 Grosshansdorf, Germany

KONNERT, M.	Bavarian Institute for Forest Seeding and Planting, Forstamtsplatz 1, D-83317 Teisendorf, Germany
KORMUTAK, A.	Institute of Plant Genetics and Biotechnology, Slovak Academy of Sciences, Akademicka 2, P.O. Box 39A, SK-95007 Nitra, Slovak Republic
KOSHY, P.	University of British Columbia, Department of Forest Science, Forest Science Center, Vancouver, B. C., Canada, V6T 1Z4
KRABEL, D.	Institut für Forstbotanik und Forstzoologie, Technische Universität Dresden, Pienner Strasse 7, D-01737 Tharandt, Germany
KREMER, A.	INRA Station de Recherches Forestières, Laboratoire de Génétique des Arbres Forestiers, BP 45, F-33611 Gazinet cedex, France
LEXER, C.	Zentrum für Angewandte Genetik, Universität für Bodenkultur Wien, Muthgasse 18, A-1190 Vienna, Austria
LIESEBACH, M.	Federal Research Centre for Forestry and Forest Products (BFH), Institute for Forest Genetics and Forest Tree Breeding, Eberswalder Chaussee 3, D-15377 Waldsieversdorf, Germany
MACALKA-KAMPFER, S.	Zentrum für Angewandte Genetik, Universität für Bodenkultur Wien, Muthgasse 18, A-1190 Vienna, Austria
MADAGHIELE, A.	Consiglio Nazionale delle Ricerche, Istituto Miglioramento Genetico delle Piante Forestali, Via Atto Vannucci, 13, I-50135 Firenze, Italy
MAGGIA, L.	CIRAD-Forêt, Campus agronomique, BP 709, 97387 Kourou cedex, French Guiana
MEJNARTOWICZ, L.	Institute of Dendrology, Polish Academy of Sciences, PL-63-035 Kórnik, Poland

MONTAGU VAN, M. — Department of Plant Genetics, Flanders Interuniversity Institue for Biotechnology, University of Gent, B-9000 Gent, Belgium

MOSSELER, A. — Canadian Forest Service, Atlantic Forestry Centre, P.O. Box 4000, Fredericton, New Brunswick E3B 5P7, Canada

MÜLLER-STARCK, G. — Technical University of Munich, Section of Forest Genetics, Am Hochanger 13, D-85354 Freising, Germany

NAMKOONG, G. — University of British Columbia, Department of Forest Science, Forest Science Center, Vancouver, B. C., Canada, V6T 1Z4

NOTIVOL, E. — Unit of Forest Resources. SIA, Diputacion General de Aragon, Apdo. 727, E-50080 Zaragoza, Spain

ODDOU-MURATORIO, S. — Conservatoire Génétique des Arbres Forestiers, Office National des Forêts, Campus INRA, F-45160 Ardon, France

ÖZDEMIR, E. — Conservatoire Génétique des Arbres Forestiers, Office National des Forêts, Campus INRA, F-45160 Ardon, France

PAUL, M. — Department of Genetics and Breeding, Saxon State Institute for Forestry, Bonnewitzer Straße 34, D-01827 Graupa, Germany

PAULE, L. — Lesnicka Fakulta VSLD, SK – 96053 Zvolen, Tschechien

PETIT, R.J. — Station de Recherches Forestières, Laboratoire de Génétique des Arbres Forestiers, INRA, Pierroton, BP 45, F-33611 Gazinet cedex, France

RAHMAN, M.H. — Department of Renewable Resources, University of Alberta, Edmonton, Alberta T6G 2H1, Canada

RAJORA, O.P. — Department of Biology, Life Sciences Centre, Dalhousie University, Halifax, Nova Scotia B3H 4J1, Canada

RIEGEL, R.	Technical University of Munich, Section of Forest Genetics, Am Hochanger 13, D-85354 Freising, Germany
SALVADOR, L.	University of Huelva, Ctra. Palos de la Frontera, E-21071 Rábida-Palos de la Frontera, Spain
SANDERMANN, H.	National Research Center for Environment and Health, Institute of Biochemical Plant Pathology, D-85764 Neuherberg, Germany
SCHOLZ, F.	Federal Research Centre for Forestry and Forest Products (BFH), Institute for Forest Genetics and Forest Tree Breeding, Sieker Landstrasse 2, D-22927 Grosshansdorf, Germany
SCHOPPA, F.N.	Institut für Forstgenetik und Forstpflanzenzüchtung, Universität Göttingen, Büsgenweg 2, D-37077 Göttingen
SCHUBERT, R.	Technical University of Munich, Section of Forest Genetics, Am Hochanger 13, D-85354 Freising, Germany
SHEN, X.	Beijing Forestry University, 100083, China
SIREGAR, I.Z.	Laboratorium Silvikultur, Fakultas Kehutanan, Institut Pertanian Bogor, PO Box 168, Kampus Darmaga, Bogor 16001, Indonesia
SOUSA, V.A.	Centro Nacional de Pesquisa de Florestas (CNP). Empresa Brasileira de Pesquisa Agropecuária (EMBRAPA), Caixa Postal 319; CEP 83.411-000 – Colombo (PR), Brazil
STEINKELLNER, H.	Zentrum für Angewandte Genetik, Universität für Bodenkultur Wien, Muthgasse 18, A-1190 Vienna, Austria
TOLUN, A.	Department of Biological Sciences, Middle East Technical University, TR-06531 Ankara, Turkey
TREUTTER, D.	Institute of Fruit Science, Technical University of Munich, D-85354 Freising-Weihenstephan, Germany

UJVÁRI-JÁRMAY, E. Forest Research Institute, ERTI, Experiment Station, H-3232 Mátrafüred, Hegyalja ú. 14, Hungary

VELIOGLU, E. Forest Trees and Seeds Improvement Research Directorate, TR-06560 Gazi, Ankara, Turkey

VENDRAMIN, G.G. Consiglio Nazionale delle Ricerche, Istituto Miglioramento Genetico delle Piante Forestali, Via Atto Vannucci, 13, I-50135 Firenze, Italy

VOOKOVÁ, B. Institute of Plant Genetics and Biotechnology, Slovak Academy of Sciences, Akademicka 2, P.O.Box 39A, SK-95007 Nitra, Slovak Republic

WANG, K. Institut für Forstgenetik und Forstpflanzenzüchtung der Universität Göttingen, Büsgenweg 2, D-37077 Göttingen, Germany

WOLF, H. Division Forest Genetics and Tree Breeding, Saxon State Institute for Forestry, Bonnewitzer Straße 34, OT Graupa, D-01796 Pirna, Germany

ZIEGENHAGEN, B. Federal Research Centre for Forestry and Forest Products, Institute for Forest Genetics and Forest Tree Breeding, Sieker Landstrasse 2, D-22927 Grosshansdorf, Germany

ZIEHE, M. Institut für Forstgenetik und Forstpflanzenzüchtung, Universität Göttingen, Büsgenweg 2, D-37077 Göttingen, Germany

PART 1: VERIFICATION OF RESPONSE TO STRESS

Stress responses in Scots pine (*Pinus sylvestris* L.). Cloning and characterisation of an ozone-inducible pinosylvin methyltransferase

H. CHIRON[1,2], A. DROUET[2], D. TREUTTER3, W. HELLER[1], D. ERNST[1] AND H. SANDERMANN[1]

[1]*GSF – National Research Center for Environment and Health, Institute of Biochemical Plant Pathology, D-85764 Neuherberg, Germany;* [2]*Biology Department, University of Orléans, F-45067 Orléans cedex 2, France;* [3]*Institute of Fruit Science, Technical University of Munich, D-85354 Freising-Weihenstephan, Germany*

Key words: cDNA, methyltransferase, ozone, promoter, Scots pine, stilbene

Abstract: Pinosylvin *O*-methyltransferase (PMT) catalyses the transfer of the methyl group of *S*-adenosyl-L-methionine to the 3-hydroxyl group of the aglycon pinosylvin. Based on RT-PCR using oligonucleotide primers of a highly conserved region of *O*-methyltransferases (OMT)s a full-length *PMT* cDNA clone was isolated from an ozone-induced cDNA library. The deduced protein sequence matched that of a tryptic peptide sequence of the purified PMT protein and showed the typical highly conserved regions of OMTs. A search of protein databases revealed highest matches to caffeic acid 3-OMTs (58.5%), caffeoyl-CoA 3-OMTs (21%), pine bifunctional OMTs (39%), flavonoid OMTs (36%) and 3-OH phenol OMTs (51%). PMT expressed in *E. coli* corresponded to that of purified PMT (40 kDa) from pine cell cultures. The recombinant enzyme catalysed the methylation of pinosylvin, caffeic acid, caffeoyl-CoA and quercetin. Several other substances, e. g. astringenin, resveratrol, 5-OH-ferulic acid, catechol and luteolin were also methylated, while protocatechuic acid, chlorogenic acid, eriodictyol and kaempferol were not transformed. Treatment of 7-year-old Scots pine trees with ozone markedly increased the *PMT* mRNA level in needles.

Introduction

Stilbenes have so far been detected in relatively few plant families where they contribute to the resistance of woody tissues to degradation, and they act as phytoalexins (Gorham 1995). Phytoalexins are low-molecular-weight antimicrobial compounds which acccumulate in plants upon biotic or abiotic stressors. In Scots pine trees, the stilbenes pinosylvin (PS) and pinosylvin 3-*O*-methylether (PSM) are known to be major constitutive phenolic constituents of heartwood and root tissues (Billek, 1964). However, as major phytoalexins in Scots pine both compounds accumulate in response to fungal attack (Gehlert *et al.*. 1990; Schanz *et al.*. 1992; Bonello *et al.*. 1993; Lieutier *et al.*. 1996), upon

Müller-Starck, G. and Schubert, R. (eds.). Genetic Response of Forest Systems to Changing Environmental Conditions. Vol. 70 (For. Sci.), 1-9

elicitor treatment in cell suspension cultures (Lange *et al.*. 1994), and in needles of seedlings and young trees upon ozone treatment (Rosemann *et al.*. 1991; Langebartels *et al.*.1997).

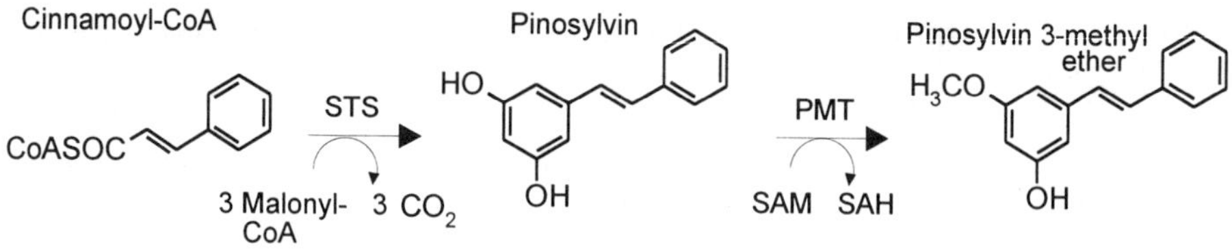

Figure 1. Biosynthesis of stilbenes. Abbreviations: STS, stilbene synthase; PMT, pinosylvin 3-*O*-methyltransferase; SAM, *S*-adenosyl-L-methionine; SAH, *S*-adenosyl-L-homocysteine.

PMT catalyses the transfer of the methyl group of *S*-adenosyl-L-methionine to the 3-hydroxyl group of PS (Figure 1), and plant OMTs have generally been assumed to possess strict substrate and position specificities for various methyl acceptor molecules (Poulton 1981; Ibrahim *et al.*. 1987). Various plant OMTs have been cloned and sequenced, encoding proteins of 38-43 kDa. Most of these are involved in lignin and flavonoid methylation steps (Ibrahim *et al.*. 1998). These enzymes share a high amino acid sequence similarity and five highly conserved regions have been identified (Ibrahim *et al.*. 1998).

To unravel stilbene biosynthesis in more detail we have cloned a PMT from Scots pine (Chiron *et al.*. 2000a,b). Here we present the state of the art of our research and are focussing on an active PMT protein expressed in *E. coli*, the substrate specificity of the recombinant protein, and an ozone-induced PMT transcript accumulation in needles of Scots pine saplings.

Materials and methods

We have screened an ozone-induced cDNA library of Scots pine (Wegener *et al.*. 1997). Plasmid DNA was isolated and PCR reactions were carried out using a primer, designed from a highly conserved region in OMTs (5'-(C/T)TIGTIGA(C/T)GTIGGIGGIGG-3'). PCR fragments were sequenced. The 5' region was obtained with a T7 primer and a 5'-GCATCGGCCACCACATGCGCCATGTC-3' primer derived from the PCR sequence. For the isolation of a full length cDNA PMT clone a linear PCR reaction betweeen the start and stop codon was carried out (Chiron *et al.*. 2000a).

DNA fragments were ligated into the pGEM-T vector and transferred to competent *E. coli* JM 109 cells. Cells were grown in standard LB medium and IPTG was added to induce expression of the fusion protein. Cells were collected by centrifugation and lysed by sonication. After centrifugation supernatants were used for SDS-PAGE and PMT activity measurements (Chiron *et al.*. 2000a).

Total RNA (2 µg) was isolated from ozone-treated or control needles and reverse transcribed with Superscript RT. One fourth of the reverse transcription volume was used for PCR with *Taq* polymerase. Amplification was performed during 35 cycles of 1 min denaturation at 94°C, 1 min annealing at 58°C and 2 min elongation at 72°C. PCR products were analyzed by electrophoresis and ethidium bromide staining (Chiron *et al.*. 2000a,b).

Results and discussion

Isolation and sequence analysis of a Pinus sylvestris PMT cDNA clone

We have isolated a full-length PMT cDNA clone from an ozone-induced cDNA library of Scots pine. The length of the Scots pine cDNA coding region (1092 bp), encoding a 40 kDa protein, was in good agreement with the size of OMTs (38-43 kDa) (Ibrahim *et al.*. 1998). Analysis of the deduced protein sequence revealed the presence of a 7 amino acid peptide (I,V,Y,S)LDTYEG sequenced from the 40 kDa protein of purified PMT (Figure 2) (Chiron *et al.*. 2000a). The deduced amino acid sequence contained the five conserved regions of plant OMTs (Ibrahim 1997). Regions 1 and 5 were perfectly conserved, whereas the other regions differed in a single amino acid. OMTs have been separated in different classes and within the class in several groups (Ibrahim *et al.*. 1998). According to their abundance this nomenclature may be given as follows: class 1: caffeic acid OMTs, class 2: caffeoyl-CoA OMTs, class 3: bifunctional caffeic acid/caffeoyl-CoA OMTs, class 4: flavonoid OMTs, class 5: other phenol OMTs.

AEHSTMLMNK***ILDTYEG***FKEVQE**LVDVGGGVG**STLNLIVSKYPHIS**GINFD**M**PHV**VADAP 240
HYPAVK**HVGGDMF**DSVPSGQAIFMKWILHDWSDDHCLRLLKNCHKALPEK**GKVI**VVDTIL 300
PVAAETSPYARQGFHIDLLMLAYNP**GGKERT**EQEFRDLAKEVGFAGGVKPVCCVNGHWVM 360

Figure 2. Part of the deduced amino acid sequence the *PMT*-cDNA clone. Only the regions of the conserved OMT signatures, which are in bold, are given. The amino acid sequence identical to those found in the tryptic peptide is in bold and italics. Data were taken from Chiron *et al.*. (2000a).

Comparative sequence analysis showed that the deduced protein sequence shared an average of 21-59 % identity with the above mentioned OMTs (Table 1). PMT shared highest sequence identity to caffeic acid OMTs. However, it has been suggested that caffeic acid OMT is not active *in vivo* for methylation of caffeic acid and that an alternative pathway via caffeoyl-CoA OMT might be responsible for the synthesis if ferulic acid moieties (Boudet 1998; Martz *et al.*. 1998). Therefore identity percentages of the deduced amino acid sequence with the already known OMT classes (Ibrahim *et al.*. 1998) suggest that the clone might be qualified as a member of class 4 or 5 OMTs for 3-OH phenols.

Table 1. Comparison of *PMT* deduced protein sequence with plant OMTs available in the EMBL databank. Identity percentages have been obtained by alignments performed by the PC/GENE program and average values are reported here. According to Ibrahim *et al.*. (1998), OMTs have been separated in five classes. Class 1: caffeic acid OMTs (D49710, Hayakawa *et al.*. 1996; X74452, unpublished); class 2: caffeoyl- CoA OMTs (M69184, Schmitt *et al.*. 1991; Z54233, Busam *et al.*. 1997); class 3: bifunctionnal caffeic acid/caffeoyl-CoA OMTs from pine species (U39301, Li *et al.*. 1997; U70873, unpublished); class 4: flavonoid OMTs (L10211, Maxwell *et al.*. 1993; X77467, Gregersen *et al.*. 1994; U16794, Gauthier *et al.*. 1996; U69554, Wu *et al.*. 1997); class 5: other phenol OMTs (X71430, Pellegrini *et al.*. 1993; D29809, Takeshita *et al.*. 1995).

Class		1	2	3	4	5
%	within the class	76	88	97	32	45
identity	between the class and PMT	58.5	21	39	36	51.5

Characterisation of the cDNA clone

To confirm that the cDNA clone did encode an OMT, PMT was cloned into an expression vector. A strongly expressed protein band of 40 kDa was observed after induction of *E. coli* cells, following SDS-PAGE (Chiron *et al.*. 2000 a). The recombinant enzyme had a similar substrate specificity as the purified enzyme from elicited Scots pine cell cultures. It primarily methylated the 3-OH group of stilbenes. The K_M-value of the *E.coli* expressed PMT towards pinosylvin (14 µM) was 3-fold lower compared to that of the enzyme isolated from Scots pine cells.

Differences in substrate specificities between the enzyme isolated from its natural source and *in vitro* expressed enzyme are not surprising, since post-translational protein modifications, important for *in vivo* enzyme activities of eukaryotes, are not carried out in *E. coli* cells. Such differences have also been described for *in vivo* and *in vitro* aspen OMT activities (Meng and Campbell 1998). It might also be possible that unidentified components, still present in the partially purified PMT enzyme extracts, contribute to modified substrate specificities, as has been found for a caffeoyl-CoA 3-*O*-methyltransferase from *Vitis vinifera* (Busam *et al.*. 1997). *In vivo* PMT enzyme acts as a dimer. It has been shown that *O*-methyltransferase cDNAs of *Thalictrum tuberosum* expressed in baculoviruses as different homodimers with a single amino acid exchange or as heterodimers, resulted in enzyme activities with a broad and different range of substrate specificities (Frick and Kutchan 1999). Denaturated *in vitro* expressed PMT migrated as a single 40 kDa protein band. The supposed homodimer formation still has to be shown.

In addition to pinosylvin, caffeic acid, caffeoyl-CoA and quercetin were good methyl acceptors for the *PMT* gene product (K_M = 28, 26 and 26 μM, respectively). The use of caffeoyl-CoA as a polyphenolic substrate by the *in vitro* expressed protein was unexpected, because main cellular PMT activity was separated from the main caffeoyl-CoA OMT activity by Matrix Red A chromatography (Chiron *et al.*. 2000a). As shown in Table 2 a very broad substrate specificity was found for the recombinant protein. The types of compounds methylated varied widely in their structural complexity. The smallest substrate was catechol consisting of a single aromatic ring, whereas a chromone residue is attached to a catechol structure in flavone. All methylations depended on a *meta*-standing OH-group. The expressed protein could methylate several stilbenes and flavonols with varying efficiencies. The most efficient methylation was achieved with the simple catechol. A similar broad substrate specificity for *in vitro* expressed OMTs has also been found in *Chrysosplenium americanum* and *Thalictrum tuberosum* (Gauthier *et al.*. 1998; Frick and Kutchan 1999). Therefore, the broad range of substrates used by the *in vitro* expressed Scots pine PMT supports the idea of an up-regulation of defence-related enzymes common to several biosynthetic pathways.

Table 2. Substrate specificity of Scots pine pinosylvin methyltransferase expressed in *E. coli.* (Substrates were used at 100 μM; enzymatic activities were determined by a radioactive assay of crude protein extract and are given relative to caffeic acid (100%); products were identified by TLC, [a]identified as ferulic acid after hydrolysis; products in italics were not identified). Data were taken from Chiron *et al.*. (2000a).

Substrate	Relative enzyme activity	Methylated product
Pinosylvin	31	Pinosylvin 3-methyl ether
Resveratrol	16	*Resveratrol 3-methyl ether*
Astringenin	130	*Astringenin 3,3'-dimethyl ether*
Caffeic acid	100	Ferulic acid
Caffeoyl-CoA	67	Feruloyl-CoA[a]
5-OH-Ferulic acid	42	*Sinapic acid*
Quercetin	84	Isorhamnetin
Luteolin	38	*Chrysoeriol*
Esculetin	166	Scopoletin
Catechol	273	*Guaiacol*

Ozone-induced PMT transcript accumulation

PMT transcript accumulation in one- to two-year-old needles of seven-year-old saplings in response to two different concentrations of ozone was examined by RT-PCR. Specific primers according to the *PMT*-cDNA sequence were designed. Ozone fumigation was performed in the GSF phytotron walk-in chambers (Thiel *et al.*. 1996; Chiron *et al.*. 2000b). Figure 3 indicates that the *PMT* mRNA level increased significantly in pine needles 48 hours after the beginning of a 10 h ozone exposure per day during two days. Without ozone, a very weak signal was also detected. When ozone concentration increased from 0.15 to 0.3 $\mu l.l^{-1}$, *PMT* transcripts were increased about five-fold. Similarely an ozone-induced *STS* mRNA increase has been found in Scots pine seedlings (Zinser *et al.*. 1998; Chiron *et al.*. 2000b). This indicates that in Scots pine needles ozone-induced stilbene biosynthesis is regulated at the transcriptional level. In addition an increase in STS and PMT activity, as well as stilbene accumulation upon ozone has also been reported (Rosemann *et al.*. 1991; Chiron *et al.*. 2000b). Thus a coordinated activation program of stilbene biosynthesis is induced in Scots pine by ozone. This is similar to the induction of phenylpropanoid biosynthesis in parsley induced by elicitor or UV radiation (Chappell and Hahlbrock 1984), as well as by ozone (Eckey-Kaltenbach *et al.*. 1994).

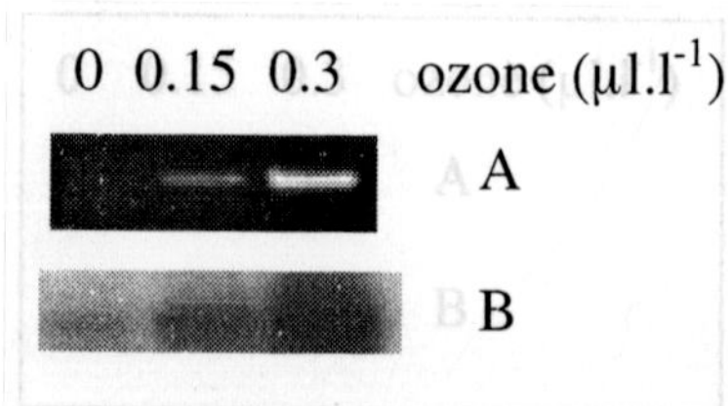

Figure 3. RT-PCR analysis of PMT gene transcripts. RT-PCR was carried out on RNA extracted from needles harvested 48 h after the beginning of ozone treatment (0, 0.15, 0.3 $\mu l.l^{-1}$ 10 h/day) on seven-year-old Scots pine saplings. Each treatment has been done on nine trees, and needles were harvested randomly. Controls were RT-PCR reactions omitting RNA and reverse transcriptase, respectively. RT-PCR products were separated on a 1% agarose gel. A: Ethidium bromide stained RT-PCR products of *PMT* transcripts. B: After transfer of total RNA to a nylon membrane, the identity of RT-PCR fragments was confirmed by hybridisation with a ^{32}P-labeled *PMT* cDNA probe.

Acknowledgements

This work was supported by a European Community grant (FAIR-BM-955721) attributed to H.C., as well as the Deutsche Forschungsgemeinschaft (SFB 607).

References

Billek, G. 1964: Stilbene im Pflanzenreich. pp. 115-152 in: Zechmeister, L. (Ed.) Progress in the Chemistry of Organic Natural Products, volume 22. Springer, Berlin

Boudet, A.M. 1998: A new view of lignification. Trends Plant Sci. 3: 67-71

Bonello, P., Heller, W. and Sandermann, H. 1993: Ozone effects on root-disease susceptibility and defence responses in mycorrhizal and non-mycorrhizal seedlings of Scots pine (*Pinus sylvestris* L.). New Phytol. 124: 653-663

Busam, G., Junghanns, K.T., Kneusel, R.E., Kassemeyer, H.-H. and Matern, U. 1997: Characterization and expression of caffeoyl-coenzymeA 3-*O*-methyltransferase proposed for the induced resistance response of *Vitis vinifera* L. Plant Physiol. 115: 1039-1048

Chappell, J. and Hahlbrock, K. 1984: Transcription of defence genes in response to UV-light or fungal elicitor. Nature 311: 1117-1125

Chiron, H., Drouet, A., Claudot, A.-C., Eckerskorn, C., Trost, M., Heller, W., Ernst, D. and Sandermann, H. 2000a: Molecular cloning and functional expression of a stress-induced multifunctional *O*-methyltransferase with pinosylvin methyltransferase activity from Scots pine (*Pinus sylvestris* L.). Plant Mol. Biol. 44: 733-745

Chiron, H., Drouet, A., Lieutier, F., Payer, H.-D., Ernst, D. and Sandermann, H. 2000b: Gene induction of stilbene biosynthesis in Scots pine in response to ozone treatment, wounding, and fungal infection. Plant Physiol. 124: 865-872

Eckey-Kaltenbach, H., Ernst, D., Heller, W. and Sandermann, H. 1994: Biochemical plant responses to ozone. IV. Cross-induction of defensive pathways in parsley (*Petroselinum crispum* L.) plants. Plant Physiol. 104: 67-74

Frick, S. and Kutchan, T.M. 1999: Molecular cloning and functional expression of *O*-methyltransferases common to isoquinoline alkaloid and phenylpropanoid pathways. Plant J. 17: 329-339

Gauthier, A., Gulick, P.J. and Ibrahim, R.K. 1996: cDNA cloning and characterization of a 3'/5'-*O*-methyltransferase for partially methylated flavonols from *Chrysosplenium americanum*. Plant Mol. Biol. 32: 1163-1169

Gauthier, A., Gulick, P.J. and Ibrahim, R.K. 1998: Characterization of two cDNA clones which encode *O*-methyltransferases for the methylation of both flavonoid and phenylpropanoid compounds. Arch. Biochem. Biophys. 351: 243-249

Gehlert, R., Schöppner, A. and Kindl, H. 1990: Stilbene synthase from seedlings of *Pinus sylvestris*: purification and induction in response to fungal infection. Mol. Plant Microb. Interact. 3: 444-449

Gorham, J. 1995: The Biochemistry of the Stilbenoids. Chapman & Hall, London

Gregersen, P.L., Christiensen, A.B., Sommer-Knudsen, J. and Collinge, D.B. 1994: A putative *O*-methyltransferase from barley is induced by fungal pathogens and UV light. Plant Mol. Biol. 26: 1797-1806

Hayakawa, T., Nato, K., Kawai, S., Katayama, Y. and Morohoshi, N. 1996: Molecular cloning and tissue-specific expression of two genes that encode caffeic acid -*O*-methyltransferase from *Populus kitakamiensis*. Plant Sci. 113: 157-165

Ibrahim, R.K. 1997: Plant *O*-methyltransferase signatures. Trends Plant Sci. 2: 249-250

Ibrahim, R.K., Bruneau, A. and Bantignies, B. 1998: Plant *O*-methyltransferases: molecular analysis, common signature and classification. Plant Mol. Biol. 36: 1-10

Ibrahim, R.K., De Luca, V., Khouri, H.E., Latchinian, L., Brisson, L. and Charest, P.M. 1987: Enzymology and compartmentation of polymethylated flavonol glucosides in *Chrysosplenium americanum*. Phytochemistry 26: 1237-1245

Lange, B.M., Trost, M., Heller, W. and Sandermann, H. 1994: Elicitor-induced formation of free and cell-wall-bound stilbenes in cell-suspension cultures of Scots pine (*Pinus sylvestris* L.). Planta 194: 143-148

Langebartels, C., Ernst, D., Heller, W., Lütz, C., Payer, H.-D. and Sandermann, H. 1997: Ozone responses of trees: results from controlled chamber exposure at the GSF phytotron. pp. 163-200 in: Sandermann, H., Wellburn, A.R. and Heath, R.L. (Eds.) Forest Decline and Ozone – A Comparison of controlled chamber and field experiments. Springer, Berlin

Li, L., Popko, J.L., Zhang, X.H., Osakabe, K., Tsai, C.J., Joshi, C.P. and Chiang, V.L. 1997: A novel multifunctional *O*-methyltransferase implicated in a dual methylation pathway associated with lignin biosynthesis in loblolly pine. Proc. Natl. Acad. Sci. USA 94: 5461-5466

Lieutier, F., Sauvard, D., Brignolas, F., Picron, V., Yart, A., Bastien, C. and Jay-Allemand, C. 1996: Changes in phenolic metabolites of Scots pine phloem induced by *Ophiostoma brunneo-ciliatum*, a bark beetle-associated fungus. Eur. J. For. Path. 26: 145-158

Martz, F., Maury, F., Pincon, G. and Legrand, M. 1998: cDNA cloning, substrate specificity and expression study of tobacco caffeoyl-CoA 3-*O*-methyltransferase, a lignin biosynthetic enzyme. Plant Mol. Biol. 36: 427-437

Maxwell, C.A., Harrison, M.J. and Dixon, R.A. 1993: Molecular characterization and expression of alfalfa isoliquiritigenin 2'-*O*-methyltransferase, an enzyme specifically involved in the biosynthesis of an inducer of *Rhizobium meliloti* nodulation genes. Plant J. 4: 971-981

Meng, H. and Campbell, W.H. 1998: Substrate profiles and expression of caffeoyl coenzyme A and caffeic acid *O*-methyltransferases in secondary xylem of aspen during seasonal development. Plant Mol. Biol. 38: 513-520

Pellegrini, L., Geoffroy, P., Fritig, B. and Legrand, M. 1993: Molecular cloning and expression of a new class of ortho-diphenol-*O*-methyltransferases induced in tobacco (*Nicotiana tabacum* L.) leaves by infection or elicitor treatment. Plant Physiol. 103: 509-517

Poulton, J.E. 1981: Transmethylation and demethylation reactions in the metabolism of secondary plant products. pp. 667-723 in: Conn, E.E. (Ed.) The Biochemistry of Plants, volume 7. Academic Press, London

Rosemann, D., Heller, W. and Sandermann H. 1991: Biochemical plant responses to ozone. II. Induction of stilbene biosynthesis in Scots pine (*Pinus sylvestris* L.) seedlings. Plant Physiol. 97: 1280-1286

Schanz, S., Schröder, G. and Schröder, J. 1992: Stilbene synthase from Scots pine (*Pinus sylvestris*). FEBS Lett. 313: 71-74

Schmitt, D., Pakusch, A.E. and Matern, U. 1991: Molecular cloning, induction and taxonomic distribution of caffeoyl-CoA 3-*O*-methyltransferase, an enzyme involved in disease resistance. J. Biol. Chem. 266: 17416-17423

Takeshita, N., Fugikawa, H., Mimura, H., Fitchen, J.H., Yamada, Y. and Sat, F. 1995: Molecular cloning and characterization of *S*-adenosyl-L-methionine:scoulerine 9-*O*-methyltransferase from cultured cells of *Coptis japonica*. Plant Cell Physiol. 36: 29-36

Thiel, S., Döhring, T., Köfferlein, M., Kosak, A., Martin, P. and Seidlitz, H.K. 1996: A phytotron for plant stress research: how far can artificial lighting compare to natural sunlight? J. Plant Physiol. 148: 456-463

Wegener, A., Gimbel, W., Werner, T., Hani, J., Ernst, D. and Sandermann, H. 1997: Molecular cloning of ozone-inducible protein from *Pinus sylvestris* L. with high sequence similarity to vertebrate 3-hydroxy-3-methylglutaryl-CoA-synthase. Biochem. Biophys. Acta 1350: 247-252

Wu, Q., Preisig, C.L. and VanEtten, H.D. 1997: Isolation of the cDNAs encoding (+)6a-hydroxymaackiain 3-*O*-methyltransferase, the terminal step for the synthesis of the phytoalexin pisatin in *Pisum sativum*. Plant Mol. Biol. 35: 551-560

Zinser, C., Ernst, D. and Sandermann, H. 1998: Induction of stilbene synthase and cinnamyl alcohol dehydrogenase mRNAs in Scots pine (*Pinus sylvestris* L.) seedlings. Planta 204: 169-176

Screening of Sitka Spruce genotypes for resistance to the White Pine Weevil in British Columbia

J. N. KING[1] AND R. I. ALFARO[2]

[1]Research Scientist, Ministry of Forests, Research Branch, Stn. Prov. Govt., Victoria, British Columbia, Canada, V8W 9C2; [2]Research Entomologist, Pacific Forestry Centre, Canadian Forest Service., Victoria, BC, Canada, V8Z 1M5

Key words: spruce, weevil, genetic resistance, screening

Abstract: Sitka spruce (*Picea sitchensis* (Bong.) Carr) in Coastal British Columbia (B.C.) and spruce plantations of interior B.C. (white spruce (*P. glauca* (Moench) Voss) and Engelmann spruce (*P.engelmanni* Parry)) have come under serious attack from the white pine weevil, *Pissodes strobi* Peck (Coleoptera: Curculionidae). This pest which attacks the current terminal leader causes serious growth loss and deformities. Genetic resistance to successful weevil attack was found in earlier Sitka provenance trials and interior spruce genetic trials. Since 1993 we have initiated a series of screening trials to confirm this resistance and to broaden the base of selections especially in the Sitka spruce breeding programme. To accelerate the screening process, and create a uniform weevil pressure, insect populations have been augmented at many of these trial sites. This artificial infestation has allowed quick and effective screening and has allowed us to now proceed with the construction of an *F-1* population to more fully understand the mechanisms behind this resistance. We hope that these selections will form the basis for successful establishment of Sitka spruce plantations in B.C. This paper describes our screening process and desribes some of the results from the earlier screening trials.

Introduction

The white pine weevil, *Pissodes strobi* Peck (Coleoptera: Curculionidae) is the most serious native pest of spruce regeneration in British Columbia, attacking Sitka (*Picea sitchensis* (Bong.) Carr), white (*P. glauca* (Moench) Voss), and Engelmann spruce (*P. engelmanni* Parry). The weevil has a one-year life cycle. The eggs are laid from late April to June in punctures made by the female in the bark, below the buds of the apical shoot (leader) grown in the previous season. The larvae burrow downward in the bark, feeding on the phloem and eventually killing the terminal leader. Destruction of the apical shoot reduces growth and causes deformities and

Müller-Starck, G. and Schubert, R. (eds.). Genetic Response of Forest Systems to Changing Environmental Condition. Vol. 70 (For. Sci.), 11-20

has had a devastating impact on spruce (*Picea* spp.) regeneration in western North America. Damage from this insect has forced forest managers to virtually eliminate the planting of Sitka spruce in most of coastal British Columbia (B.C.), Washington and Oregon. This is unacceptable given both the excellent growth and wood characteristics of Sitka spruce and its ultimate disappearance as an important part of the conifer diversity from coastal ecosystems. In the interior of B.C., reforestation with white spruce (B.C.'s most planted species) is now also under a threat of severe attack by *P. strobi*.

Although many control options for *P. strobi* have been investigated none have proven to be entirely effective (Cozens 1983, Alfaro *et al.* 1994, Alfaro and Omule 1990, McLean 1992, 1994, Rankin and Lewis 1994, Taylor and Cozens 1994). Selection and breeding of trees for insect resistance provides one of the more hopeful avenues for minimizing the damage caused by *P. strobi*. Previous work has identified spruce trees and seed sources as being putatively resistant to *P. strobi* attack in both Sitka (Alfaro and Ying 1990, Ying 1991, Ying and Ebata 1994, Ying 1997) and interior spruce (Kiss and Yanchuk 1991, Alfaro *et al.* 1996, King *et al.* 1997). Selective breeding efforts have begun for both Sitka and interior spruce that use this resistance information discovered in the natural populations of these species (King 1994, Kiss *et al.* 1994). There has been an especially important effort with Sitka spruce, which in the past decade has concentrated on screening populations where this resistance was noted from earlier provenance trials.

Screening in these earlier provenance trials was conducted by measuring attack rates in infestations that developed naturally (e.g. Ying 1991, Alfaro et al. 1996). However, natural population build up in a plantation is patchy and makes screening a slow and unpredictable process. Some susceptible individuals may appear resistant if they are located in areas of the plantation with low attack density (i.e., they simply escape attack), and conversely, weevils in high-density areas may overcome and attack potentially useful genotypes that have moderate resistance levels. To overcome these problems we have started augmenting natural weevil populations by artificially infesting our trial sites with breeding weevils. This has immediately exposed our trials to high weevil population pressures, greatly speeding up the screening of our populations. Using this method of creating artificial infestations by weevil augmentation we have been able to screen a series of clonal and open-pollinated trials established from promising natural populations. In all we have currently screened close to one thousand individual clones or families in trial sites throughout Vancouver Island. This report describes the overall process of screening for resistance and reviews in detail the results of the oldest screening tests, located at Jordan River and Cowichan Lake, on Vancouver

Island, B.C.. Besides allowing us to screen for resistant genotypes, this artificial infestation experiment has allowed us to follow in controlled conditions the population dynamics of weevil infestations.

Description of the screening programme in Sitka spruce

Screening populations

The older IUFRO provenance trials revealed two potential resistant sources: one from Big Qualicum on Eastern Vancouver Island and the other from Haney in the Fraser Valley (Alfaro and Ying 1990, Ying 1991, Ying and Ebata 1994, King 1994, Ying 1997). Candidate individuals for screening were of two types. The first were grafted individuals selected from the original provenance trials, or from natural trees from known resistant provenance sources. The second were open-pollinated progenies collected from the known resistant populations, or from populations from ecologically similar environments. Trial sites established to screen candidate genotypes typically are about 2 hectares and are established in a randomized complete block design with candidate resistant lines (clones or open-pollinated families). We attempted to put most of the trial sites in moderate to high weevil hazard areas in order to maintain weevil populations.

Augmentation of weevil populations

Weevil augmentation is typically made when the test plantations to be screened reach 4 to 5 years of age. The plantation is first examined for symptoms of previous weevil attack to determine attack history since establishment. Normally natural weevil populations have not yet developed significantly. The process of augmentation consists of releasing 2 to 4 weevils on each tree in the plantation. Weevils are normally released in the autumn by depositing the insects within the tree crown. At this time of the year, weevils will feed for a while and then move to their overwintering sites in the forest floor. The following spring, the weevils emerge from overwintering and proceed to attack the plantation. Attack rates are then monitored for up to 4 years by examining every tree for attack symptoms. These attack records form the basis for weevil resistance screening. In the older IUFRO provenance reliable information was not achieved for 15 or more years. In our augmentation process we can effectively screen trials in under 7 years.

Sources of weevils for augmentation

Care was taken to infest plantations using weevils of the local resident weevil population at the plantation site. Two separate studies of genetic variation in weevil populations, one using RAPD markers (Lewis 1995), and one using isozymes (Lewis *et al.*2000), separated Canadian weevil populations into four distinct groups.

In mid summer, naturally infested spruce regeneration on Northern Vancouver Island was searched for symptoms of infested apical shoots. Weevil infested leaders were then clipped and brought to the Pacific Forestry Centre, Victoria, to be placed in weevil rearing cages. As weevils emerged from the leaders they were collected and placed in rearing facilities containing lateral branches as weevil food. They were held until the autumn when they were released into the plantations (one plantation was augmented in the spring).

Jordan River and Cowichan Lake sites

The Jordan River and Cowichan Lake open-pollinated series were collected to sample a transect down East Vancouver Island and Western Puget Sound of Washington (Figure 1). One of the original resistant provenance sources is at Big Qualicum near Qualicum Beach on East Vancouver Island. This trial series screened 67 open pollinated (OP) families from 12 putatively resistant sources (including the original Big Qualicum source) as well as samples from similar ecological forest types. The assumption made was that the resistance found at Big Qualicum might be broadly based in these dryer hotter areas (of the East Vancouver Island rain shadow) perhaps brought about by natural selection to high weevil attack levels (King 1994). An additional 6 bulk control seedlots were added and were expected to be susceptible to the weevil.

These trials were established in the spring of 1992 with 20 replicates of each of the 73 seedlots (67 OP families and 6 controls) at Jordan River and 24 replicates at Cowichan Lake. The trials were established with one-year old container stock as single tree plots. Signs of weevil infestations at these sites were recorded by their 3rd year but the population was low and patchy. Additional weevils were introduced in both plantations. In October 1994, three weevils were placed on

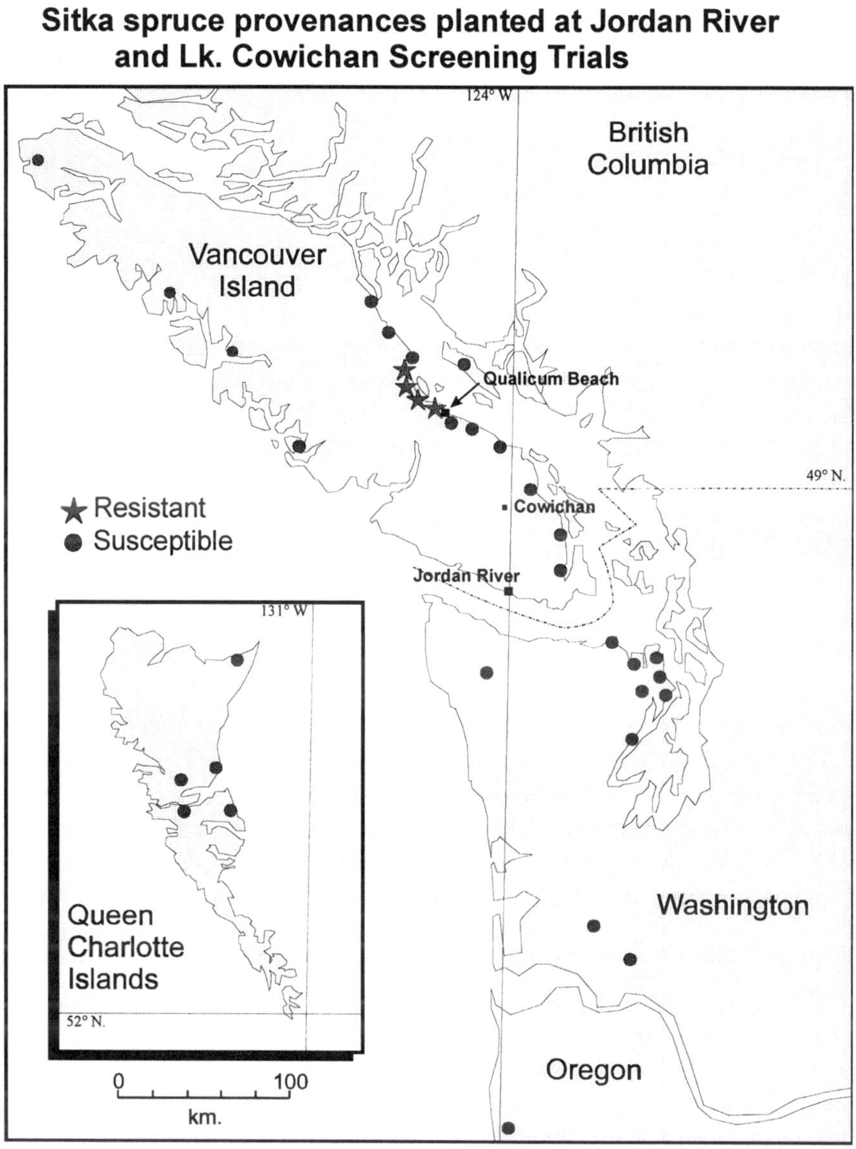

Figure 1. Location of parent trees for 75 Sitka spruce families at Jordan River and Cowichan Lake progeny trials on Vancouver Island, British Columbia, Canada. Stars indicate sources that demonstrated weevil resistance at both the Jordan River and Cowichan trials.

site. In a similar manner, two weevils were placed on each tree of 15 replicates at the Cowichan Lake trial in May of 1996.

Results from the Jordan River and Cowichan series

The mean annual weevil attack rates in the release areas, averaged over all families at each site, are indicated in Figure 2. Both sites showed extremely high rates of weevil attack immediately following weevil augmentation, this allowed separation of resistant and susceptible families after just one year. Weevil populations showed declines after the first year and this was especially dramatic at the Cowichan Lake site. These infestation trials are yielding important information on the population dynamics of the white pine weevil, the importance of its parasitoids in population control, the role of climate in maintaining weevil populations and the role of fungi in the weevil/host system. A detailed population dynamic study of the infestations at these sites is being undertaken. Understanding the dramatic population increase and then decline may help find factors that can control this pest.

Strong source level differences were noted in these trials which very much confirmed the results found from the original IUFRO provenance trials. However, our original hypothesis that the resistance found from the original sources might be widely based throughout the ecoclimatic region of East Vancouver Island was not borne out. We had originally suggested that natural selection due to high weevil pressure could have made populations from this area more resistant (King 1994). Results, however, show that only populations from the vicinity of the original Big Qualicum population demonstrated this resistance. Source effects (the 12 provenance sources) were among the strongest in the ANOVA models run, much stronger than family effects. Heritability estimates were 0.9 without source effects in the model, but only 0.2 with source effects (King *et al.* in prep.). Heritability estimates made with interior spruce, where strong provenance source effects are less apparent, are in the order of 0.5 (King *et al.* 1997). Another finding of the King *et al.* 1997 study was that although it had been noted that weevils prefer longer, more vigorous leaders, genetic resistance does not seem to come at the cost of reduced growth rate. In fact, the genetic correlation between weevil resistance and growth rate is positive in populations of white spruce ($r \cong 0.6$; King *et al.* 1997). The spruce families are performing similarly at the two sites with respect to weevil attack. The correlation between the percent weevil attack at Jordan River and Cowichan Lake (cumulative to 1997) was significant ($R^2 = 0.49$).

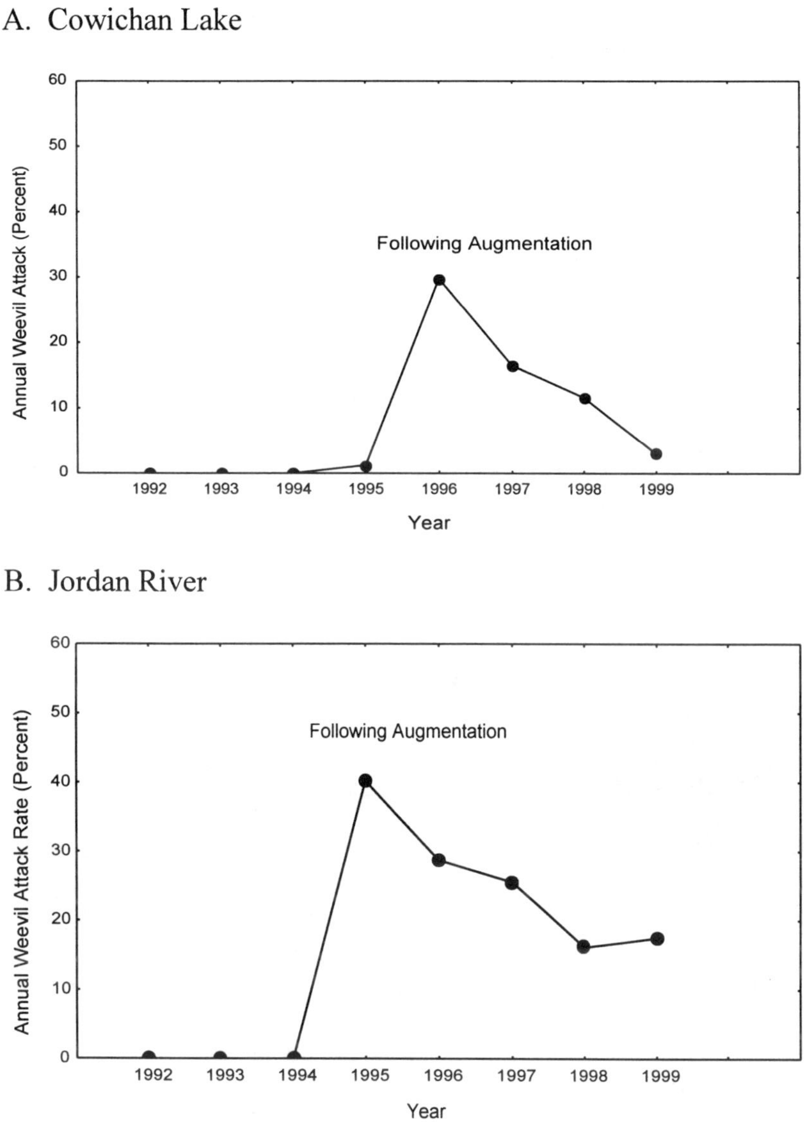

Figure 2. Mean annual weevil attack rates at the Cowichan Lake (A) and Jordan River (B) sites before and after weevil augmentation. Rates were based on the percentage of trees attacked in each release areas.

The strong well defined provenance source effect and bi-modal distribution between designated resistant and susceptible sources led us to believe that major gene resistance (MGR) could be behind some of this resistance (King 1994). We tested this by looking at the distribution of resistance within progeny families, and analysis of within-family variances has supported the evidence for allele segregation for MGR (King *et al.* in prep).

Implications for screening and weevil resistance breeding

Resistance screening

These results indicate that successful screening for weevil resistance using weevil population augmentation can be accomplished cheaply and effectively in only one or two years after weevil infestation. Test trees need to be of susceptible height, i.e., outplanted for approximately 3 - 4 years in coastal B.C. Infested plantations should be monitored for up to 3 years. After weevil augmentation, coastal sites can be planted to Douglas-fir (*Pseudotsuga menziesii*), Hemlock (*Tsuga* spp.) or Western redcedar (*Thuja plicata*), if damage appears severe. These species are non-hosts to *P. strobi*, hence the weevil infestation would die out gradually. This has now been done at the Jordan River trial (populations at the Cowichan site are collapsing on their own).

Resistance breeding

Genotypes from the Qualicum area of Vancouver Island, B.C., are emerging as very promising sources that can contribute to increased productivity of Sitka spruce in BC. The fact that these results have been repeated over a variety of sites, from our new screening trials to the older provenance trials, indicates the reliability of the information that has been gathered so far. Besides the Big Qualicum source, we know that another resistance source exists in the Fraser Valley near Haney B.C. With our extensive screening we hope to pick up other populations and individuals. Based on our screening trials we have so far selected approximately 60 individuals from these and other potential resistant source populations. With these individuals we are constructing an *F-1* breeding population in which we hope to search for the underlying mechanisms of resistance; understand how these resistant mechanisms are inherited (whether through major genes or polygenic inheritance); and construct populations that have strong and durable resistance to *Pissodes strobi*. We hope that the parents from our screening process and the resulting *F-1* population will help to restore Sitka spruce as a species of choice for many areas of Coastal British Columbia. We do not expect that genetic resistance will be the only component of a successful re-introduction programme, but we expect to use resistance as a component of an Integrated Pest Management System.

Acknowledgments

This work was financed by Forest Renewal BC Grant No. HQ 96251-RE, to Dr. Rene I. Alfaro, and by direct funding by the BC Ministry of Forests and the Canadian Forest Service. We thank the technical support of Mr. George Brown of the Pacific Forestry Centre and Mr Charles Cartwright of the B.C. Forest Service and Western Forest Products for providing the Jordan River site to screen for resistance. We thank Jill Peterson of the B.C. Forest Service for a careful review of this article.

References

Alfaro, R., Borden, J., Fraser, G. and Yanchuk, A. 1994: An integrated pest management system for the white pine weevil. pp. 226-238 *in*: The White Pine Weevil: Biology, Damage and Management. FRDA Report 226, Alfaro, R., Kiss, G., Fraser, G. (eds.). Proceedings of a symposium held January 19-21, 1994 in Richmond, British Columbia

Alfaro, R.I., He, F., Kiss, G, King, J. and Yanchuk, A. 1996: Resistance of white spruce to white pine weevil: development of a resistance index. Forest Ecology and Management 81: 51-62

Alfaro, R.I. and Omule, S. 1990: The effect of spacing on Sitka spruce weevil damage to Sitka spruce. Canadian Journal of Forest Research 20: 179-184

Alfaro, R.I and Ying, C. 1990: Levels of Sitka spruce weevil, *Pissodes strobi* (Peck), damage among Sitka spruce provenances and families near Sayward, British Columbia. The Canadian Entomologist 122: 607-615

Cozens, R.D. 1983: The spruce weevil, *Pissodes strobi* Peck (Coleoptera: Curculionidae): A review of its biology, damage and control techniques with reference to the Prince George Timber Supply Area. British Columbia Ministry of Forests, Forest Service Internal Report PM-PG-3.

King, J.N. 1994: Delivering durable resistant Sitka spruce for plantations. pp. 134-149 *in* Proc. of conf. on The White Pine Weevil: Biology, Damage and Management, Alfaro, R.,

King, J.N., Yanchuk, A., Kiss, G. and Alfaro, R. 1997: Genetic and phenotypic relationships between weevil resistance and height growth in spruce populations of British Columbia. Can. J. For. Res. 27: 732-739

King, J.N., Alfaro, A., Cartwright, C. and Ying, C.(in prep) Genetic resistance of Sitka spruce (*Picea sitchensis*) to terminal weevil *Pissodes strobi* . Forest Science (submitted).

Kiss, G.K. and Yanchuk, A. 1991: Preliminary evaluation of genetic variation in interior spruce in British Columbia. Canadian Journal of Forest Research 21(2): 230-234

Kiss, G.K, Yanchuk, A. and Alfaro, A. 1994: Recent advances in white pine weevil reasearch in British Columbia. *in* Proc. of conf. on The White Pine Weevil: Biology, Damage and Management, Alfaro, R., Kiss, G., Fraser, G. (eds.).Can. For. Serv. Pac. For. Cent. FRDA Rep. 226. Richmond B.C. Canada

Lewis, K. 1995:. Genetic variation among populations of *Pissodes strobi* (white pine weevil) reared from *Picea* and *Pinus* hosts as inferred from RAPD markers. M.Sc. Thesis, Faculty of Forestry, University of British Columbia, Vancouver. 232 pp.

Lewis, K, El-Kassaby, Y., Alfaro, R and Barnes, S. 2000: Population structure of the white pine weevil (Pissodes strobi). Annals Ent. Soc. Amer. 93(4):807-818

McLean, J. 1992: Overstory manipulation. Pp. 20-21 *in*: Spruce Weevil Symposium Proceedings, Terrace, B.C., March 12, 1992. British Columbia Ministry of Forests, Prince Rupert Region, Smithers, B.C.

McLean, J. 1994: Silvicultural control of the white pine weevil at the UBC Malcolm Knapp Research Forest. pp. 248-253 *in* Proc. of conf. on The White Pine Weevil: Biology, Damage and Management, Alfaro, R., Kiss, G., Fraser, G. (eds.).Can. For. Serv. Pac. For. Cent. FRDA Rep. 226. Richmond B.C. Canada.

Rankin, L. and Lewis, K. 1994: Effectiveness of leader clipping for control of the white pine weevil, *Pissodes strobi*, in the Cariboo Forest Region of British Columbia. pp. 262-269 *in* Proc. of conf. on The White Pine Weevil: Biology, Damage and Management, Alfaro, R., Kiss, G. and Fraser, G. (eds.).Can. For. Serv. Pac. For. Cent. FRDA Rep. 226. Richmond B.C. Canada.

Taylor, S. and Cozens, R. 1994: Limiting white pine weevil attacks by side and overstory shade in the Prince George Forest Region. Journal of the Entomological Society of British Columbia 91: 37-42.

Ying, C. 1991: Genetic resistance to the white pine weevil in Sitka spruce. British Columbia Ministry of Forest Research Note No. 196. Victoria, British Columbia, Canada. 17 pp.

Ying, C. 1997: Effects of site, provenance and site interaction in Sitka spruce in Coastal British Columbia. Forest Genetics 4: 99-112.

Ying, C. and Ebata, T. 1994: Provenance variation in weevil attack in Sitka spruce. pp. 98-109 *in*: The White Pine Weevil: Biology, Damage and Management. FRDA Report 226, Alfaro, R., Kiss, G., Fraser, G. (eds.). Proceedings of a symposium held January 19-21, 1994 in Richmond, British Columbia.

Genetic variation in two heavily polluted stands of Norway spruce (*Picea abies* [L.] Karst.) as indicated by nuclear and organelle DNA markers

R. RIEGEL[1], R. SCHUBERT[1], G. MÜLLER-STARCK[1], D. F. KARNOSKY[2] and L. PAULE [3]

[1]*Technical University of Munich, Section of Forest Genetics, D-85354 Freising, Germany;* [2]*School of Forestry and Wood Products, Michigan Technological University, Houghton, MI, 49931-1295, USA;* [3]*Faculty of Forestry, Technical University, SK-96053 Zvolen, Slovakia*

Key words: EST marker, cpDNA, microsatellites, isoenzyme gene marker, genetic variation, diversity, heterozygosity, air pollution, heavy metals, tolerance, *Picea abies*

Abstract: In two polluted neighbouring stands of Norway spruce, "tolerant" and "sensitive" adult trees were selected in pairs and genotyped by recently developed codominant EST markers as well as by chloroplast microsatellites. The results were compared with data previously obtained from isoenzyme gene markers. Larger allelic multiplicities are indicated in the two sensitive subsets as compared to the tolerant ones, i.e. a mean surplus of 11.5 % at 6 EST loci and of 27.2 % at 19 isoenzyme gene loci. Frequency distributions of genetic types deviate considerably between tolerant and sensitive subsets but statistical significance is indicated by only one of the EST loci tested. Another EST locus is the only one to reveal statistically significant deviations between the observed heterozygosities of the tolerant and the corresponding sensitive subset of the most heavily polluted stand (38 % vs. 15 %). Based on allelic distances, both categories of nuclear markers reveal a cluster of three subsets in contrast to one and the same sensitive subset which is part of the most heavily polluted stand. A different cluster is indicated by chloroplast markers: the two sensitive and the two tolerant subsets, respectively, reveal greater allelic similarities than the sensitive and tolerant subsets of each stand. Generally, the applied molecular markers appear to be more indicative for the genetic response of Norway spruce to the given environmental stress than isoenzyme gene markers. Different pollution intensities at the two locations are assumed to account for the observed deviations in the genetic response of the two stands.

Introduction

The region of Spiš, mainly the valley of the Hnilec river, belongs to the oldest industrial areas of Slovakia. Iron has been produced there since the 14th century (Cambel 1986). During the 18th century, the production of copper and zinc was initiated. At present, Hnilec valley is probably the most intensively polluted region of Slovakia. Recently, main sources of pollution have been

Müller-Starck, G. and Schubert, R. (eds.). Genetic Response of Forest Systems to Changing Environmental Conditions. Vol. 70 (For. Sci.), 21-34 © *2001 Kluwer Academic Publishers, Dordrecht, Boston, London*

smelting factories in Krompachy and Rudňany producing copper, mercury and manganese, which utilised brown coal containing arsenic. In addition, the region is subjected to long-distance transport of air pollutants (mainly sulphur dioxide) from the Silesian industrial complex around Ostrava and Katowice. High levels of arsenic, mercury, copper and sulphur have been therefore monitored in this region in the forest biomass, resulting from the industrial pollution mentioned (Maňkovská 2000).

This situation affected substantially the vitality of the local forests. Norway spruce is the main species of the Lower Spiš region. Currently, more than 60 % of the forest area of this region is covered by spruce stands which is more than twice as much as the average of Slovakia (27 %). At the same time, spruce belongs to the species which are known to be highly sensitive to air and soil pollution. Following the ICP Forest monitoring in year 2000, the average defoliation of Norway spruce in the Spiš region was 32.6 % and 21.6 % in the remaining part of Slovakia (Bucha *et al.* 2000).

Gene markers have been used to assess genetic responses of forest tree species to pollutants, such as *Picea abies* (Bergmann and Scholz 1989, Bergmann and Hosius 1996, Raddi *et al.* 1994), *Pinus sylvestris* (Mejnartowicz 1983, Prus-Głowacki and Nowak-Bzowy 1992), *Abies alba* (Konnert 1992), and *Fagus sylvatica* (Müller-Starck 1989, Müller-Starck and Ziehe 1991). Results of these studies indicate that pollution can affect the genetic structure of forest tree populations in various ways including impacts on the genetic multiplicity, diversity and observed heterozygosity. Corresponding changes in genotype and/or allele frequencies of tolerant and sensitive subsets of tree populations are common (for survey of phenomena see Müller-Starck and Schubert 2000).

In the past, the genetic consequences of environmental stress governed by pollution were studied by means of isoenzyme gene markers and thus a limited set of gene loci. Scotti *et al.* (1998) performed the first study on genetic variation in polluted Norway spruce populations by utilising molecular markers. These RAPD markers are targeted to anonymous genomic regions without distinguishing between homozygous and heterozygous genotypes due to their dominant inheritance. The present study is the first to employ recently developed co-dominant DNA markers which are linked to expressed genes (EST-marker) as well as DNA-markers which refer to the plastid genome, i.e. chloroplast microsatellites (cpSSRs).

These new tools are being tested to study impacts of severe and long lasting air pollution dominated by heavy metals on the genetic structures of Norway spruce populations. Subsets of tolerant and of sensitive trees are being genetically characterised and compared with each other. In addition to the nuclear and extra-nuclear molecular markers, isoenzyme gene markers are the focus. In the characterisation of the genetic response of Norway spruce populations to heavy pollution, our study particularly addresses two aspects: Firstly, whether or not the nuclear and cytoplasmatic markers reveal similar trends in genetic differentiation and, secondly, which variation parameters are most effective as indicators of differentiation between sensitive and tolerant subsets of trees.

Material and methods

Studied populations

Two neighbouring mixed stand complexes of Norway spruce [*Picea abies* (L.) Karst.] and silver fir [*Abies alba* Mill.], exhibiting different degrees of damage by industrial pollution, were selected in the Lower Spiš region near the village Novoveská Huta (48°53' N, 20°32' E): a more exposed, heavily damaged stand in the upper part of a slope (area I), and a better protected, moderately damaged stand near the bottom of a valley (area II). At both areas, pairs consisting of an apparently non-damaged tree and a neighbouring severely damaged tree were selected, both aged more than 80 years. The first category, i.e. trees obviously tolerant to the pollution load ("tolerant trees"), had dense crowns with green foliage. Damaged trees, putatively sensitive in response to pollution ("sensitive trees") were those trees which exhibited reduced needle retention levels, associated with discoloration of needles, dieback in the upper parts of the crowns and reduced growth increments over the last 10 years. At the same time, sensitive trees did not show any signs of bark beetle attack, fungal infection or mechanical damage by timber logging or extraction, in order to avoid involvement of other causes of decline than the pollution load. Additional analysis of some of the sampled trees revealed a higher damage of the needle epicuticular wax in the sensitive trees than in the tolerant ones, confirming the action of pollution (Blanka Maňkovská, pers. comm.).

The maximum distance between the two trees of a pair was 20 m. Sampling in pairs was performed in order to possibly randomise both the genealogical background in case of natural regeneration and the environmental heterogeneity (for methodical details see Gregorius 1989). During the winter seasons 1998/1999 and 1999/2000, twigs with dormant buds were sampled from 26 pairs in the heavily damaged stand and 26 pairs in the moderately damaged stand (Longauer *et al.* 2001).

Genotyping nuclear EST markers and chloroplast microsatellites

Single buds were used to extract total genomic DNA according to the method of Doyle and Doyle (1989).

The following six recently identified co-dominant EST marker loci were utilised with putatively targeted gene functions given in brackets: PA0002 [A-like cyclin], PA0034 [nonidentified gene harboring an imperfect repeat], PA0038 [halotolerance protein], PA0043 [belongs to the heat-shock protein 70 family], PA0055 [β-chain of the ATP synthase complex], PA0066 [60S ribosomal protein L13-2]. Genomic PCR, product separation and identification of individual marker alleles were done according to Schubert *et al.* (2001). At the level of the paternally inherited chloroplast DNA, microsatellite markers were employed (cpSSR) which have been originally developed and tested for various coniferous species by Vendramin *et al.* (1996). For the present study, three cpSSR markers (Pt26081, Pt63718 and Pt7193) were selected out of 20. These markers were proven to be more polymorphic than others in Norway spruce material and revealed substantial differences in product sizes, ensuring a simultan amplification by "multiplex-PCR" (Vendramin *et al.* 2000).

Isoenzyme gene markers

Genotypes were monitored at 19 polymorphic enzyme coding gene loci: FEST-A; GDH-A; GOT-A,-B,-C; G6PDH-A; IDH-A,-B; LAP-A,-B; MDH-A,-B,-C; PX-A; PGM-A; PGI-A,-B; SKDH-A,-B (for details see Longauer *et al.* 2001, Paule *et al.* 2001).

Quantification of genetic variation

Intrapopulation variation was studied by means of the observed proportion of heterozygotes (H_a), the conditional heterozygosity (H_c, Gregorius *et al.* 1986), the genetic multiplicity and diversity (Gregorius 1978, 1987) and the differentiation within populations (δ_T, Gregorius 1987). Fixation indices ($-1 \leq F_{IS} \leq +1$) were calculated according to the concept of Wright (1965). Interpopulational differentiation was quantified by means of a dendrogram based on the genetic distances (Gregorius 1974). Computation was performed by means of the programs GSED (E. Gillet; University of Göttingen, Germany) and Phylyp (J. Felsenstein, University of Washington, USA).

Differences among frequency distribution were tested statistically by utilizing the χ^2 homogenity test. Three hypotheses were tested, each of which implies statistically non-significant deviations between the genetic structures of the following subsets:

$H_{0\text{-}1}$: Sensitive subset I and sensitive subset II.
$H_{0\text{-}2}$: Tolerant subset I and tolerant subset II.
$H_{0\text{-}3}$: Pooled sensitive subsets and pooled tolerant subsets.

Only in case that $H_{0\text{-}1}$ and $H_{0\text{-}2}$ was not rejected, then the two sensitive and tolerant subsets, respectively, were pooled (H_{0-3}) in order to study the genetic response to stress under a larger heterogeneity of both, site conditions and genetic composition of the studied populations.

Results

Results are based on three different marker types, i.e. isoenzymes and EST-marker which identify genotypes of the nuclear genome, and haplotypes of the chloroplast genome. In the presentation of the results, diplotypes and haplotypes were considered separately.

Nuclear markers

In Table 1, genetic parameters are surveyed separately for each subset and each of the two kinds of nuclear markers. For each parameter, the two lowest values among the four subsets are underlined. Evidently, the number of alleles per locus deviates considerably among the subsets. Both kinds of markers indicate a reduction of the number of alleles per locus in the tolerant subsets as compared to the sensitive ones. In case of the isoenzyme gene markers, the average number of alleles is 22.8 % greater in the sensitive subset at area I and 31.6 % at area II. The corresponding values for the EST-markers are 12.0 % and 11.0 %, respectively. In terms of the number of genotypes, a similar range is evident: a surplus of 19.9 % and 22.8 % respectively for the isoenzyme gene markers, and 8.3 % and 21.9 % respectively for the EST markers. The correspondence of the values for alleles and genotypes indicate limited impacts of biases following small sample sizes (larger number of genotypes as compared to alleles). For demonstration of this correspondence, in Figure 1 the average number of alleles and of genotypes is contrasted for four subsets and two nuclear markers.

As can be further seen from Table 1, the isoenzyme gene markers indicate a similar but less pronounced trend with respect to the genetic differentiation. In case of the EST markers, deviations among subsets are very small. The same trends hold for the differentiation as derived from the genotypic structure. Generally, the frequency distribution of genetic types among the sensitive subsets tends to be more even than in the tolerant subsets so that deviations among both categories with respect to the number of genetic types per locus are partially compensated in case of the genetic diversity.

Trends with respect to the observed heterozygosity are not congruent in both experimental areas. Generally, deviations among the tolerant and sensitive subsets are small. It is evident that EST markers reveal much higher heterozygosities than isoenzyme gene markers (on average 0.40 vs. 0.11).

These deviations between both marker types are compensated if the observed heterozygosities are related to the maximum attainable ones (see H_C-values). This is a consequence of the marker dependent specific frequency distribution of genetic types (more pronounced trend for minor polymorphism in case of the isoenzyme gene markers as compared to the EST markers) in combination with marker dependent deviations from the corresponding Hardy-Weinberg proportions.

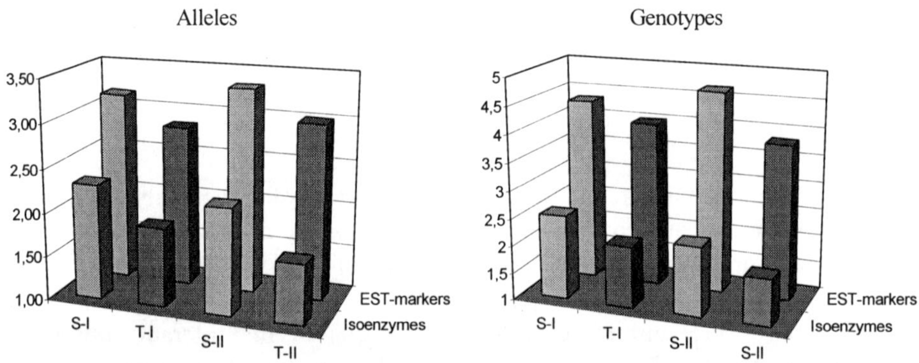

Figure 1. Number of alleles and genotypes per locus at 6 EST loci and 19 isoenzyme loci: S-I and T-I designate the sensitive and the tolerant subsets at area I, S-II and T-II at area II.

In all subsets, the mean F_{IS}-values (see Tab. 1) are very close to zero in case of the isoenzyme gene loci. A similar trend is indicated by the EST loci with one exception at area II (F_{IS}=0.11 for the tolerant subset). Generally, the fixation indices appear very homogeneous. A tendency of deviations from Hardy-Weinberg proportions in terms of excess of homozygotes can be denied.

The statistical testing of the genetic structures among tolerant and sensitive subsets resulted in the following:

Between the two tolerant subsets (H_{0-2}), statistical significant deviations are not indicated by any of the 25 nuclear loci tested. In case of the two sensitive subsets (H_{0-1}), allelic structures deviate significantly with respect to the markers PA0038 and GOT-C. If tolerant and sensitive subsets are compared, only the EST marker PA0038 reveals statistically significant deviations in case of the allelic frequencies. This effect is evident at the experimental area I, but not at the neighbouring area II. For both areas, the allelic frequencies at the locus PA0038 are surveyed in Figure 2. At area I, allele B is nearly 20 % more frequent in the sensitive subsets as compared to the tolerant one. This surplus is compensated by a deficiency with respect to allele C. In case of the experimental area II, the same two alleles reveal deviations among the sensitive and the tolerant subset but in opposite direction.

Table 1. Genetic variation in sensitive (S) and tolerant (T) subsets at two areas (I, II) in Spišská Nová Ves as indicated by 6 EST loci and 19 isoenzyme loci (smallest values in each area are underlined). Data refer to104 adult trees (4 × 26).

Parameter	Marker	Area I Sensitive (S-I)	Area I Tolerant (T-I)	Area II Sensitive (S-II)	Area II Tolerant (T-II)
Number of poly-morphic loci	EST	8	8	8	8
	Isoenzyme	17	<u>11</u>	15	<u>10</u>
Alleles per locus (all loci)	EST	3.17	<u>2.83</u>	3.33	<u>3</u>
	Isoenzyme	2.32	<u>1.89</u>	2.21	<u>1.68</u>
Genotypes per locus (all loci)	EST	4.33	<u>4.00</u>	4.67	<u>3.83</u>
	Isoenzyme	2.53	<u>2.11</u>	2.26	<u>1.84</u>
Differentiation ($\delta_{T\text{-allele}}$)	EST	<u>0.42</u>	0.42	0.43	<u>0.41</u>
	Isoenzyme	0.14	<u>0.11</u>	0.11	<u>0.09</u>
Differentiation ($\delta_{T\text{-genotype}}$)	EST	0.58	0.58	0.57	0.57
	Isoenzyme	0.22	<u>0.17</u>	0.19	<u>0.14</u>
Observed heterozygosity (H_o)	EST	<u>0.39</u>	0.42	0.42	<u>0.37</u>
	Isoenzyme	0.14	<u>0.11</u>	0.10	<u>0.08</u>
Conditional heterozygosity (Hc)	EST	0.69	0.69	0.65	<u>0.62</u>
	Isoenzyme	0.83	<u>0.80</u>	0.79	<u>0.70</u>
Fixation index (F_{IS})	EST	0.09	<u>-0.04</u>	<u>0.01</u>	0.11
	Isoenzyme	-0.03	<u>-0.04</u>	0.03	<u>0.02</u>

In Figure 3, observed heterozygosities are surveyed with respect to the markers PA0034, PA0043 and LAP-B, each of which reveals larger values among the sensitive subset as compared to the tolerant ones. The opposite trend is indicated by the markers PA0038 and PA0066. Only in case of the EST locus PA0066 at the experimental area I, the deviations among the sensitive and the tolerant subset are statistically significant. The corresponding heterozygosities of the sensitives are 15 % and of the tolerants 38 %.

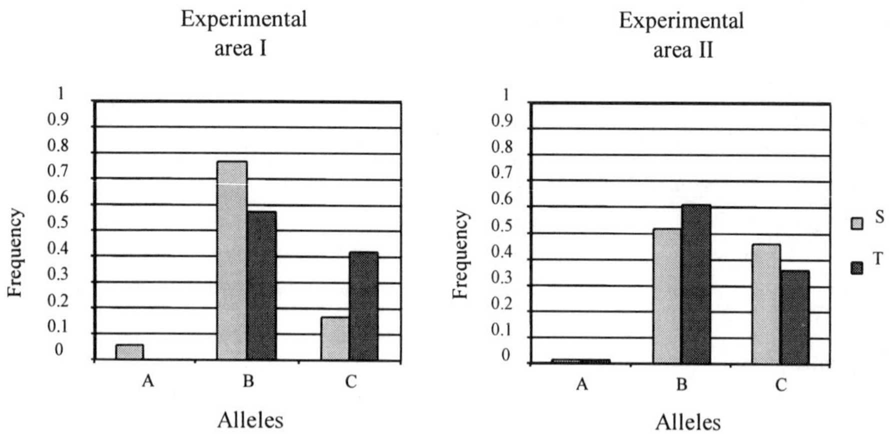

Figure 2. Allele frequencies at the EST locus PA0038 for sensitive (S) and tolerant (T) subsets at area I and II.

Figure 3. Observed heterozygosities (H_a) at 5 selected loci for sensitive subsets at area I and II

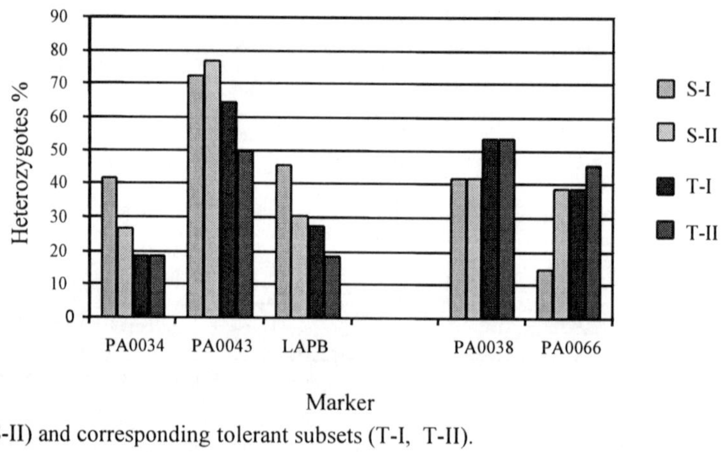

(S-I, S-II) and corresponding tolerant subsets (T-I, T-II).

Although the genetic distances between the four subsets are relatively small, both classes of nuclear markers reveal a distinct differentiation between the sensitive subset at the experimental area I and the remaining three subsets. Based on the allelic distances, dendrograms are plotted for all nuclear markers in Figure 4. The cluster of three populations as compared to the sensitive subset of the area I is evident.

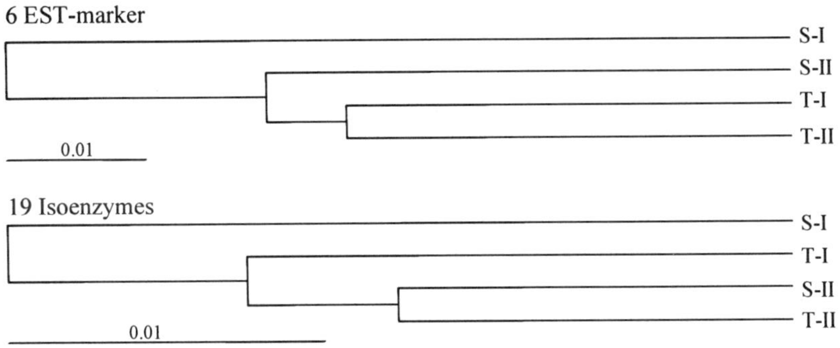

Figure 4. Dendrogram based on the allelic distances among sensitive and tolerant subsets at area I and II, respectively (S-I, T-I; S-II, T-II).

Chloroplast markers

In Table 2, the number of haplotypes and the values for genetic differentiation are compiled for three markers and four tree subsets. In contrast to nuclear markers, chloroplast markers display very small differences within and among sensitive and tolerant subsets. However, the combination of loci *Pt63718* and *Pt71936* reveal a remarkable trend with respect to both experimental areas, showing a much lower differentiation (δ_T) in the tolerant subsets.

Table 2. Genetic variation in sensitive (S) and tolerant (T) subsets of trees at two areas (I, II) in Spišská Nová Ves as indicated by 3 chloroplast microsatellites (cpSSR). Data refer to 104 adult trees (4 × 26).

	Area I				Area II			
	Sensitive (S-I)		Tolerant (T-I)		Sensitive (S-II)		Tolerant (T-II)	
Marker	Alleles/ haplotyp e	δ_T	Alleles/ haplotyp e	δ_T	Alleles/ haplotyp e	δ_T	Alleles/ haplotyp e	δ_T
Single locus								
Pt63718	4	0.55	5	0.35	4	0.44	4	0.40
Pt26081	2	0.47	4	0.62	3	0.55	3	0.51
Pt71936	3	0.54	3	0.46	5	0.68	3	0.56
Two-locus combination								
Pt63718+ Pt26081	7	0.75	8	0.72	7	0.74	8	0.71
Pt63718+ Pt71936	7	0.80	7	0.64	10	0.84	7	0.76
Pt26081+ Pt71936	5	0.68	7	0.75	7	0.75	6	0.78
Three-locus combination								
Pt63718+ Pt26081+ Pt71936	9	0.86	10	0.81	12	0.86	11	0.90

Statistically significant deviations among the frequencies of the 2-locus and 3-locus haplotypes of the four subsets are not indicated. However, the combination of loci *Pt63718* and *Pt71936* reveal a remarkable trend with respect to both experimental areas. As can be seen in Figure 5, 49 % of the studied tolerant trees at both experimental areas carry the haplotype "6-3" while the corresponding value for both sensitive subsets is only 29 %. Among the sensitives, the frequency of the haplotype "6-4" is 30 % while the corresponding value for the tolerant subsets is 21 %. The dominance of haplotyp "6-3" in the tolerant trees also explains the lower differentiation in this subset of trees (see Tab. 2).

In the dendrogram which is based on the genetic distances with respect to the haplotypes of cpSSR marker *Pt63718* and *Pt71936* (see Fig. 6), small genetic differences among the subsets are indicated. However, two clusters are grouping the sensitive subsets of trees separately from the tolerant ones. Evidently, the clusters are independent from the area were the samples come from.

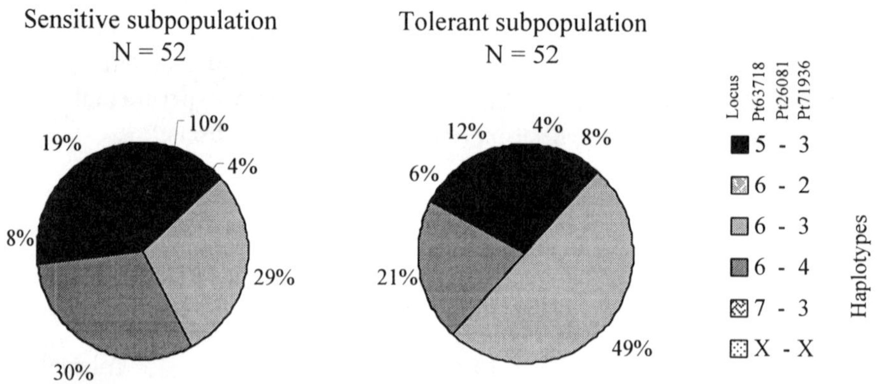

Figure 5. Frequencies of 2-locus haplotypes (cpSSR-loci *Pt63718* and *Pt71936*) in pooled tolerant and sensitive subsets from areas I and II. Haplotypes X-X represent a pool of different types each with frequencies lower than 5 %

Figure 6. Dendrogram based on the haplotype distances among sensitive and tolerant sub-sets at area I and II, respectively (S-I, T-I; S-II, T-II).

Discussion

The present investigation is a pilot study in order to test new molecular tools as indicators for the genetic response of Norway spruce population to heavy pollution. As a consequence of the limited size of the available sample (total sample sizes is 104 trees), allele frequencies were the focus rather than genotype frequencies. In spite of the small sample sizes of 26 trees per subset, the coincidence between the frequencies of alleles and genotypes is evident (see Fig.1).

A remarkable difference between sensitive and tolerant subsets refers to the number of alleles, respectively: the two sensitive subsets reveal on average 27.2 % more alleles at isoenzyme gene loci and 11.5 % at EST loci than the two tolerant subsets. Although information on the non-sampled trees in the studied stands is missing, this trend strongly suggest that pollution infers on genetic multiplicity and the risk of loosing genetic variation can not be denied.

Also with respect to genetic differentiation and heterozygosity, both kinds of codominant markers indicate smaller mean values for the tolerant subsets in most cases but the deviations are moderate. Remarkable differences are evident with respect to the observed heterozygosities at single loci (see Fig. 3) but the EST locus PA0066 is the only one to reveal statistical significant deviations between the tolerant and the corresponding sensitive subset at the most intensively polluted site. There is no evidence for a surplus of heterozygotes at certain isoenzyme gene loci as has been reported by Bergmann and Hosius (1996) for tolerant seedling populations of Norway spruce after exposure to heavy metal field pollution. Furthermore, fixation indices in our study do not confirm the excess of homozygosity at isoenzyme gene loci as has been reported in case of Scots pine trees which were classified as tolerant against heavy zinc pollution (Prus-Głowacki and Nowak-Bzowy 1992). On the other hand, considerabley higher heterozygosities were observed in subsets of *Picea abies* trees tolerant to air pollution (Prus-Głowacki and Godzik 1995).

In case of the frequency distribution of alleles, various deviations are evident between tolerant and sensitive subsets but the only locus which reveals statistically significant deviations is the EST locus PA0038. Again, these deviations are indicated for the experimental area I which is considerably more exposed to pollution than the neighboured area II.

The study of genetic (allelic) distances among the two sensitive and two tolerant subsets indicates that the sensitive subsets of the most heavily polluted experimental area I is clearly separated from the cluster of the remaining three subsets (see Fig. 4). Important is the fact that both codominant nuclear types of markers, addressing entirely different gene loci, verify this fact in surprising congruency. In contrast to these findings, the chloroplast microsatellites indicate a grouping of haplotypes of the two sensitive subsets separate from the two tolerant subsets. The check of combinations of haplotypes verifies the two locus haplotypes as most discriminative with respect to the differentiation of sensitive and tolerant subsets. Remarkable deviations in allele frequencies are evident in the combination of the cpSSR loci Pt 63718 and Pt 71936 (e.g. 49 % vs. 29 %, see Fig. 5).

These results are interpreted for the present that chloroplast markers can be considered as very efficient tools in the genetic differentiation between subsets which reveal particularly tolerant or sensitive response to heavy pollution. The present findings support the hypothesis that chloroplast markers can function well as diagnostic tools in the verifications of tolerant and sensitive responds to stress in addition to codominant nuclear markers. In contrast to this, the applied nuclear markers appeared to reflect the differences in the pollution scenario between the two experimental sites. This hypothesis is supported by the fact that the sensitive subset of the most heavily polluted spruce stand I is genetically clearly differentiated from the remaining subsets.

The sample sizes used in this study are not large enough to infer selective effects with respect to certain genotypes. Further investigations are required in order to verify the trends which are indicated in the present study under a broader scale of environmental conditions and by applying larger sample sizes. Anyway, the given results strongly suggest the indicative and possibly the diagnostic value of recently developed molecular markers for genetic monitoring in polluted forest ecosystems.

Acknowledgements

We greatly appreciate the help of D. Gömöry and R. Longauer, Zvolen, who collected material and supported our work substantially. This project is part of the co-operation within the Carpathian Mountains Project (USDA Forest Service, IUFRO-SPDC). Parts of this study were supported financially by SFB-607 of the Deutsche Forschungsgemeinschaft, Germany.

References

Bergmann, F. and Scholz, F. 1989: Selection effects of air pollution in Norway spruce (*Picea abies*) populations. In: Scholz, F., Gregorius, H-R. and Rudin, D. (eds.), Genetic Effects of Air Pollutants in Forest Tree Populations, Springer Verlag, Berlin-Heidelberg 143-160

Bergmann, F. and Hosius, B. 1996: Efects of heavy metal polluted soils on the genetic structure of Norway spruce seedling populations. Water-, Air- and Soil-Pollution. 89: 363-373

Bucha, T., Maňkovská, B., Minďáš, J., Molnárová, H., Pavlenda, P., Pajtík, J. and Raši, R. 2000: Zdravotný stav lesov Slovenska. [Health state of forests of Slovakia]. MP SR, Bratislava, LVÚ, Zvolen, 50+2 pp.

Cambel, S. 1986: Dejiny Slovenska I [History of Slovakia I]. Veda, Bratislava, 533 pp.

Doyle, J. and Doyle, J. 1989: Isolation of plant DNA from fresh tissues. Focus 1: 13-15

Gregorius, H.-R. 1974: Genetischer Abstand zwischen Populationen. I. Zur Konzeption der genetischen Abstandsmessung. Silvae Genetica 34: 22-27

Gregorius, H.-R. 1978: The concept of genetic diversity and its formal relationship to heterozygosity and genetic distance. Math. Bioscience 41: 253-271

Gregorius, H.-R. 1987: The relationship between the concepts of genetic diversity and differentiation. Theor. Appl. Genetics 74: 397-40

Gregorius, H.-R. 1989: The attribution of phenotypic variation to genetic or environmental variation in ecological studies. In: Scholz, F., Gregorius, H-R. and Rudin, D. (eds.). Genetic Aspects of Air Pollutants in Forest Tree Populations. Springer Verlag Berlin-Heidelberg 3-15

Gregorius, H.-R., Krauhausen, J. and Müller-Starck, G. 1986: Spatial and temporal genetic differentiation among the seed in a stand of *Fagus sylvatica* L. Heredity 57: 255-262

Konnert, M. 1992: Genetische Untersuchungen in geschädigten Weißtannenbeständen (*Abies alba* Mill.) Südwestdeutschlands. Mitt. d. Forstl. Ver. u. Forschungsanstalt Baden-Würtenberg. Freiburg. Heft 167, 119 pp.

Longauer, R., Gömöry, D., Paule, L., Karnosky, D.F., Maňkovská, B., Müller-Starck, G., Percy, K. and Szaro, R., 2001: Selection effects of air pollution to gene pools od Norway spruce, European silver fir and European beech. Environ. Pollut. (in press)

Maňkovská, B. 2000: Chemical composition of forest tree species, mosses and humus as environmental indicator in Slovakia. In: Lomský, B. and Šrámek, V. (eds.), Ore Mountains 2000. (in press)

Mejnartowicz, L. 1983: Changes in genetic structure of Scots pine (*Pinus silvestrys* L.) population affected by industrial emission of fluorid and sulphur dioxide. Genetica Polonica 24: 41-50

Müller-Starck, G. 1989: Genetic implications of environmental stress in adult forest stands of *Fagus sylvatica* L. In: Scholz, F., Gregorius, H-R. and Rudin, D. (eds.). Genetic Aspects of Air Pollutants in Forest Tree Populations. Springer Verlag Berlin-Heidelberg, 127-142

Müller-Starck, G. and Ziehe, M. 1991: Genetic variation in populations of *Fagus sylvatica* L., *Quercus robur* L., and *Quercus petraea* Liebl. in Germany. In: Müller-Starck, G. and Ziehe, M. (eds.). Genetic Variation in European Populations of Forest Trees. Sauerländer's Verlag, Frankfurt a. M. 125-140

Müller-Starck, G. and Schubert, R. 2000: Genetic markers as a tool for bioindication in forest ecosystems. In: Young, A., Boshier, D. and Boyle, T. (eds.): Forest Conservation Genetics: Principles and Practice. CABI Publ., Oxford 227-237

Paule, L., Gömöry, D., Karnosky, D. and Longauer, R. 2001: Genetic effects of industrial pollution in Norway spruce. In: Lomský, B. and Šrámek, V. (eds.), Ore Mountains 2000. (in press)

Prus-Głowacki, W. and Godzik, S. 1995: Genetic structure of *Picea abies* trees tolerant and sensitive to industrial pollution. Silvae Genetica 44: 62-65

Prus-Głowacki, W. and Nowak-Bzowy, R. 1992: Genetic structure of a naturally regenerating Scots pine population tolerant for high pollution near a zinc smelter. Water-, Air- and Soil-Pollution 62: 249-259

Raddi, S., Stefanini, F.M., Camussi, A. and Giannini, R. 1994: Forest decline index and genetic variability in *Picea abies* (L.) Karst. Forest Genetics 1: 33-40

Schubert, R., Müller-Starck, G. and Riegel, R. 2001: Development of EST-PCR markers and monitoring their intrapopulational genetic variation in *Picea abies* (L.) Karst. Theoretical and Applied Genetics. (in press)

Scotti, I., Soranzo, N., Ferrario, S. and Binelli, G. 1998: Genetic variation in Norway spruce as revealed by mapped PCR-based markers. In: Proc. Plant Biotechnology as a Tool for the Exploitation of Mountain lands. 363-369

Vendramin, G.G., Lelli, L., Rossi, P. and Morgante, M. 1996: A set of primers for the amplification of 20 chloroplast microsatellites in *Pinaceae*. Molecular Ecology 5: 595-598

Vendramin, G.G., Anzidei, M., Madaghiele, A., Sperisen, C. and Bucci, G. 2000: Chloroplast microsatellite analysis reveals the presence of population subdivision in Norway spruce (*Picea abies* K.). Genome 43: 68-78

Wright, S. 1965: The interpretation of population structure by F-statistics with special regard to systems of mating. Evolution 19: 395-420

Effects of extreme SO_2-air pollution in winter 1995/96 on vitality and growth of SO_2-tolerant Norway spruce (*Picea abies* [L.] KARST.) clones in the Ore mountains

H. WOLF

Division Forest Genetics and Tree Breeding, Saxon State Institute for Forestry, D-01796 Pirna, Germany

Key words: Norway spruce, *Picea abies*, air pollution, SO_2-tolerance, breeding, autovegetative propagation

Abstract: Between 1988 and 1990, 6 field tests were established in air polluted regions of the Ore mountains and the Thuringian forest in order to evaluate SO_2-tolerance, vitality, growth and tree form of about 450 Norway spruce clones selected as SO_2-tolerant in the field and propagated autovegetatively. 6 and 7 years respectively after establishment, the results indicate significant differences in SO_2-tolerance among the clone ramets investigated. On the other hand, clones with a high level of SO_2-tolerance show insufficient survival rates, a height growth below average and average tree form. Clonal mixtures originating from seedlings as well as seedling progenies planted as standards were damaged more intensively by SO_2 than SO_2-tolerant clones. In contrast to the tolerant clones, the clonal mixtures and the seedling progenies show survival rates, height growth and tree form above average. Regarding the parameter flushing, the clonal mixtures the tolerance characters of which are not defined show only slight differences compared to resistant clones. The reason for the differences observed in survival rate, height growth and tree form is the better ability of juvenile material to be propagated autovegtatively. The results presented show the difficulties to implement the results of breeding for SO_2-tolerance to forestry using autovegetatively propagated adult basic material SO_2-tolerant in the field.

Introduction

Since the 1950s, serious damages and dieback of Norway spruce (*Picea abies* [L.] Karst.) stands caused by SO_2 could be observed in the higher elevations and at the ridges of the mountainous areas in the south east of Germany along the border to Czech Republic and Poland (Anonymous 1973; Liebold and Drechsler 1991; Raben *et al.* 1996). In 1998, the forest area damaged by SO_2 air pollution comprised about 120,000 ha in the mountainous regions of Saxony, 23 % of which could be described as extremely damaged. Out of this area, about 10,400 ha of Norway spruce stands were damaged totally – 8,800 ha between 1962 and 1991, 1,600 ha in 1996 (Butter and Richter 1998).

Müller-Starck, G. and Schubert, R. (eds.). Genetic Response of Forest Systems to Changing Environmental Conditions. Vol. 70 (For. Sci.), 35-49

Forced by the increase of forest areas damaged by SO_2, two main strategies were implemented for the reforestation of the damaged areas by the forestry authorities and administration since the 1970s: Firstly, the immediate reforestation of cleared forest areas using SO_2-tolerant indigenous and exotic species of the genera *Abies, Picea, Pinus* and *Pseudotsuga* as well as *Alnus, Fagus, Populus* and *Quercus* among others (Anonymous 1973, 1978, 1985); Secondly, in the middle- and the long-term the plantation of forest reproductive material especially of Norway spruce improved in tolerance of air pollution and frost. In the long-term breeding programme, several activities were integrated for the *ex situ* conservation of genetic resources as well as hybrid breeding using European and Japanese larch, different varieties of Douglas fir and different clones and families of Aspen (Tzschacksch 1981, 1983a; Matschke *et al.* 1984; Kohlstock 1985; Braun *et al.* 1987; Weiser and Schachler 1988; Braun and Kohlstock 1990; Weiss *et al.* 1990). However, there was no political and economical approach in the German Democratic Republic (GDR) to reduce significantly the SO_2 emissions until the political change in 1989.

One major part of the Norway spruce breeding programme was the selection of about 1,500 vital, so-called field-tolerant, Norway spruce clones for the development of a clonal mixture with increased tolerance of SO_2 between 1973 and 1990. The selection was done in the extremely air polluted areas of the mountainous regions of south-eastern Germany, the northern part of Czech Republic and the south-western part of Poland (Tzschacksch 1983b; Anonymous 1991). In the field, selection criteria were the condition of the needle foliage after continous impact of SO_2-pollution and a low percentage of visible needle damage after frost. The age of the basic material selected ranged from 15 to 110 years. In the nursery, only clones with a minimum rooting rate of 35 % were considered for further testing. From 1986 to 1990, 13 field tests were established for the evaluation of vitality, growth, tolerance and form of more than 500 clones propagated autovegetatively. The field tests were located exclusively in air polluted areas of the Ore mountains and Thuringian forest (Anonymous 1991).

However, following a decision of the Politbuero of the Sozialistische Einheitspartei Deutschlands and the Council of Ministers of the GDR dated 26.02.1983, the autovegetative mass propagation of field-tolerant spruce clones was started at the same time as the selection and before the field testing of the clones in question. Between 1984 and 1987 about 200,000 – 300,000 twigs were rooted per year after the establishment of sufficient nursery capacities (Schachler *et al.* 1987).

In the following paper, results of an assessment of two field tests will be presented. The field tests are integrated in a network of 6 field tests established with 450 field-tolerant clones from 1988 to 1990. In winter 1995/1996, the two field tests to be described were exposed to extreme SO_2-pollution. After 6 and 7 years of growth respectively, survival, height growth and tree form of the tested clones were evaluated as well as needle damage and flushing of the tested clones after the heavy impact of SO_2. Following the presentation and discussion of the results, several conclusions are drawn on the suitability of the material investigated for reforestation of air-polluted regions.

Material and methods

Clones and progenies investigated

The Norway spruce clones investigated were selected in heavily air polluted mountainous regions of Czech Republic, Poland and Saxony, Germany (Tab. 1). The basic material selected in Saxony was located in the District Forest Offices (DFOs) Altenberg, Eibenstock, Marienberg and Olbernhau (northern part of the Ore mountains), Bad Gottleuba and Cunnersdorf (Elbsandstone mountains) as well as Loebau (Zittauer mountains). The clones from Czech Republic descended from the DFOs Janov, Klasterec and Litvinov in the southern part of the Ore mountains, the clones selected in Poland from spruces in the Iser and the Sudeten mountains (DFOs Izera and Orle). The age of the basic material tested corresponded to the age of all clones selected. The clones were autovegetatively propagated by rooting of secondary scions mainly in 1984 (Altenberg) and 1985 (Baerenfels) respectively. 10 clones selected in the DFO Olbernhau and one clone from the DFO Marienberg were included in both field tests. They will be called standard clones in the following.

Several clonal mixtures descended from 7 full sib families (controlled crosses between adult individuals selected as SO_2-tolerant in the field) and 3 open pollinated progenies were also included in the experiment. The age of the seedling progenies used for the autovegetative propagation by rooting was 5 years or younger. Finally, 5 year old seedlings of two open pollinated progenies were planted at the field test in Baerenfels (Tab. 1).

Field tests

The two field tests are situated in the eastern part of the Ore mountains. The ecological conditions demonstrated in Table 2 are characterized by a rough climate with strong continental influence.

Table 1. Tested clones and progenies of Norway spruce tolerant of SO_2 selected in Germany (D), Czech Republic (CR), Poland (P) and Ukraine (UK)

Country of provenance	Basic material	Reproductive material	Number of clones and progenies investigated per field test	
			Altenberg	Baerenfels
D	Clones	Cuttings	99	138
CR	Clones	Cuttings	52	6
P	Clones	Cuttings	-	54
D	Full sib families	Cuttings	7	1
CR	Seed stands	Cuttings	2	1
UK	Seed stands	Cuttings	-	1
D	Seed stands	Seedlings	-	2
Total			160	203

Table 2. Location and site conditions of the field tests Altenberg and Baerenfels

District Forest Office	Longitude	Latitude	Altitude In m a. s. l.	Mean annual temp. [°C]	Mean annual precipitation [mm]	Ground rock
Baerenfels	13° 41' E	50° 45' N	740	4.7	993	Gneis
Altenberg	13° 43' E	50° 46' N	780	5.4	993	Granit

Table 3. Range of SO_2-concentration in μg/m³ in the period October 1995 to March 1996 at the air pollution control stations Fichtelberg and Zinnwald (Anonymous 1995a,b,c; 1996a,b,c)

Air pollution control station	Range of SO_2-concentration in μg/m³ in the period October 1995 to March 1996		
	30-d mean	Maximum 24-h mean	Maximum 3-h mean
Fichtelberg	20 - 100	117 – 399	354 – 1,030
Zinnwald	42 - 100	117 – 326	244 – 616

Until 1991, this region was heavily polluted by SO_2 emitted by power stations burning brown coal and located in the Bohemian depression. From 1991 to 1998, the annual mean SO_2-concentration decreased significantly from 85 μg/m³ to values below 20 μg/m³ (Anonymous 1999). In the period October 1995 to March 1996 however, the mean monthly SO_2-concentration increased again to a damaging extent due to continuous transport of SO_2-polluted air masses from W/SW. The daily mean SO_2-concentration reached values between 326 μg/m³ at the air pollution control station Zinnwald located east of the field tests and 399 μg/m³ at the air pollution control station Fichtelberg located west of the field tests. The 3-h mean SO_2-concentration yielded in maximum values between 616 µg/m³ and 1,035 µg/m³ at the respective stations (Tab. 3) (Anonymous 1995a,b,c; 1996a,b,c).

The design of the field tests established in spring 1989 and 1990 respectively can be seen in Table 4.

Table 4. Design of the field tests Altenberg (established in 1989) and Baerenfels (established in 1990)

DFO	No. of progenies	Replicates	Progenies without replicates	Plants/ plot	Spacing	Area in ha
Altenberg	160	3	-	6	2.5 m x 1.3 m	0.94
Baerenfels	203	2	86	15	2.0 m x 1.5 m	1.44

Assessments and statistical evaluation

In late summer 1995, mortality, height and tree form were assessed using a three-stage evaluation scheme of the clones and progenies tested. To evaluate the tolerance of SO_2 and the capacity of regeneration after SO_2 impact, the parameters "flushing (bud break and elongation of the shoot)", "intensity of flushing", "needle defoliation rate" (four stage scheme) of every annual shoot and "number of remaining annual shoots" were ascertained in summer 1996. However, the parameters "intensity of flushing" and "number of remaining annual shoots" will not be considered since they did not show any variation among the clones and progenies investigated.

The statistical analysis (descriptive statistics, Spearman rank correlation, analysis of variance) was carried out with statistic software developed at the former Research and Implementation Centre Graupa (Weissleder 1990). The confidence level of significance was set at ≤ 0.05.

Results

Mortality

Six or seven years respectively after establishment, the mean mortality rate over all clones and progenies investigated ranged from 50 % in Baerenfels to 63 % in Altenberg (Tab. 5). Between 5 % (Baerenfels) and 15 % (Altenberg) of the clones investigated did not survive at all until 1995. Among them, several clones originating from Olbernhau (6 clones in Baerenfels; 19 clones in Altenberg) could be found.

At both field experiments, the mean mortality rate of the standard clones corresponded more or less to the mean loss rate of all clones investigated. However, clonal mixtures originating from seedlings or seedling progenies showed significantly lower mortality rates than the clone material autovegtatively propagated from adult trees (Tab. 5).

Table 5. Mortality of the clones and progenies tested 6/7 years after the establishment of the field tests Altenberg and Baerenfels

Clones and progenies tested	Mortality in %							
	Altenberg				Baerenfels			
	N	Mean	Min.	Max.	N	Mean	Min.	Max.
Clones	151	66	0	100	197	51	0	100
Standard clones	11	72	33	100	11	53	0	80
Clonal mixtures/ Seedling progenies	9	17	0	39	6	21	10	47
All progenies	160	63	0	100	203	50	0	100

SO_2-caused damages and flushing

In spring and summer 1996, redish needle discolouration followed by loss of needles of varying extent could be observed on all parts of plants which were not covered by snow during the winter period at the field tests Altenberg and Baerenfels. Parts of plants covered by snow had green needles without exception. The results of the evaluation of damage are presented at the example of the one year old needle shoot 1995.

In Altenberg, no or only slight needle loss could be observed on 56 % of the progenies investigated. 30 % of the progenies showed needle loss between 26 % and 75 %. 14 % of all plants had needle loss rates above 76 % (Tab. 6). However, there was a big variation of the needle loss rates among the clones and progenies tested. It is possible to demonstrate significant differences in SO_2-tolerance between the group of 20 clones most heavily damaged (called sensitive clones in the following) and the group of 20 clones only slightly damaged (called tolerant clones in the following).

Within the group of tolerant clones 12 clones originated from Olbernhau, 3 clones each from Klasterec and Janov as well as 2 clones from Litvinov can be found. The group of sensitive clones is composed of 14 clones from Olbernhau and two clones each from Janov, Klasterec and Litvinov.

Table 6. Needle loss after SO_2-impact in winter 1995/96

Clones and progenies tested	Number		Percentage of individuals [in %] in needle loss classes							
			0 – 25 %		26 – 50 %		51 – 75 %		76 - 100 %	
	A	B	A	B	A	B	A	B	A	B
Tolerant clones	20	20	91	72	8	16	1	5	0	7
Sensitive clones	20	20	17	20	19	14	15	11	50	54
Standard clones	9*	9*	46	51	25	13	11	7	18	28
Clonal mixtures/ Seedling progenies	9	6	75	47	15	19	7	6	3	28
Average	136 *	192*	56	45	21	19	9	10	14	27

A=Altenberg; B=Baerenfels; *Number of remaining clones and progenies in 1995

The clones originating from Olbernhau showed clearly the variation in SO_2-tolerance. Out of 95 clones selected, 12 clones can be described as tolerant of SO_2 and 14 as sensitive to SO_2. Since 19 clones from Olbernhau have already died in Altenberg until 1995, about one third of the field-tolerant clones selected in Olbernhau can be described as sensitive clones.

In Baerenfels, a tendency to heavier needle damage could be observed with exception of the sensitive clones (Tab. 6). The differences assessed between the groups of tolerant and sensitive clones were similar to those in Altenberg. At both field tests, the clonal mixtures originating from seedling progenies and the seedling progenies had heavier needle damage than the tolerant clones. However, the mean mortality rate of the clonal mixtures as well as the seedling progenies tested was significantly lower than the one of the tolerant clones (Tab. 6).

After the heavy SO_2-impact in winter 1995/96, the assessment of the flushing in 1996 showed similar response of the progenies investigated in Altenberg and Baerenfels. On average, between 79 % (Baerenfels) and 84 % (Altenberg) of the clones and progenies flushed normal, 14 % (Altenberg) and 20 % (Baerenfels) of the material observed did not flush at all. Only in few cases, elongating shoots died during the vegetation period (Fig. 1).

However, the differences in flushing between the group of the tolerant and sensitive clones are remarkable. Based on the data of all individuals assessed, the failure of flushing is significantly correlated to the extent of SO_2-damages observed (Altenberg: r_s=0.65; Baerenfels: r_s=0.67).

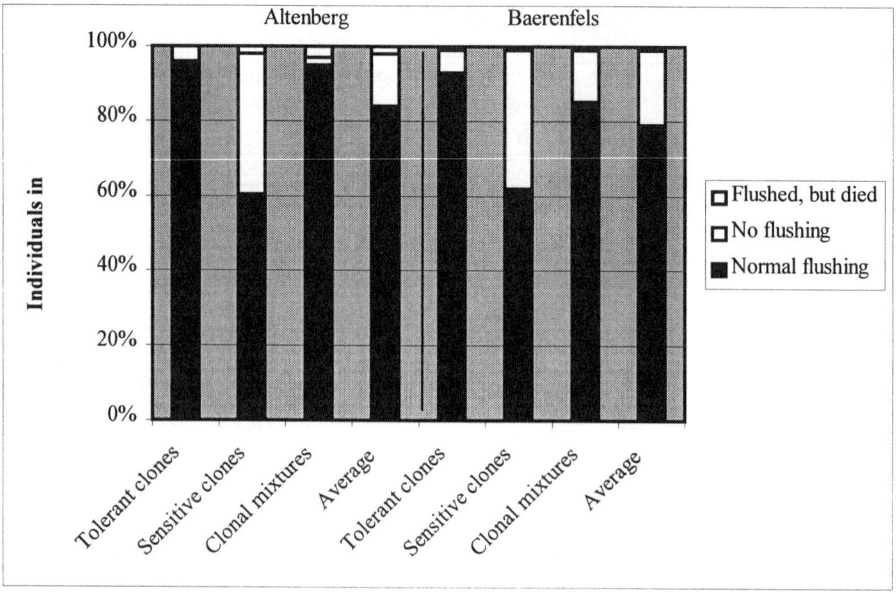

Figure 1. Flushing of the clones and progenies assessed at the field tests Altenberg and Baerenfels in 1996 after SO_2-impact in winter 1995/96

Height and tree form

The mean height over all clones and progenies tested did not show any differences between the field tests Altenberg and Baerenfels 6 or 7 years respectively after the establishment. In Altenberg however, the mean height of the respective clones and progenies varied significantly and ranged from 23 cm to 151 cm. Differences of the mean height could be seen between the group of the tolerant and sensitive clones on the one hand and the group of the clonal mixtures and the seedling progenies on the other hand. The clonal mixtures showed the best height growth with a mean height of 95 cm (Tab. 7). No significant correlation could be established between the height growth and the needle damage observed (r_s=0.21).

In Baerenfels, the height growth of the progenies tested was influenced by damage caused by game. Therefore, the variation of the height growth is not so clear as in Altenberg (Tab. 7). Significant differences could be observed among several clones and progenies but also between the replicates. Similar to the results of Altenberg, there is no correlation between the height growth and the needle damage (r_s=-0.08).

Table 7. Height of the clones and progenies assessed in 1995

Clones and progenies tested	Number		Height in 1995 [in cm]					
			Average		Minimum		Maximum	
	A	B	A	B	A	B	A	B
Tolerant clones	20	20	66	51	23	35	123	92
Sensitive clones	20	20	56	57	26	43	100	91
Clonal mixtures/ Seedling progenies	9	6	95	67	64	58	151	72
Standard clones	9*	9*	77	55	23	30	129	72
Average	136*	192*	56	57	23	26	151	108

A=Altenberg; B=Baerenfels; * Remaining clones and progenies in 1995

On both field tests, tree form varied clearly among the tolerant clones, sensitive clones and the clonal mixtures/seedling progenies. The best form showed the clonal mixtures/seedling progenies with a portion of 97 % and 96 % respecitively of upright growing individuals (Fig. 2).

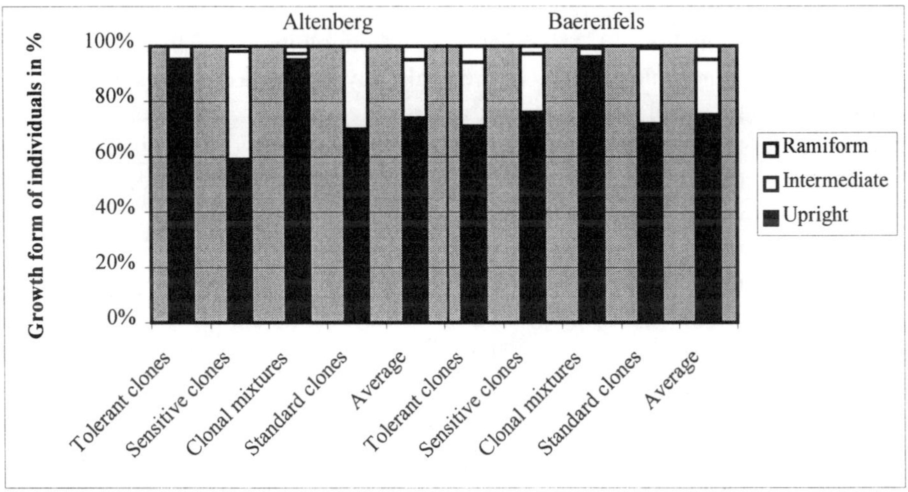

Figure 2. Tree form of the clones and progenies assessed in 1995

Discussion

The material investigated at the two field tests Altenberg and Baerenfels is very heterogeneous according to the origin and age of the basic material, the propagation methods used, the number of clones per field test as well as the number of clone mixtures and seedlings progenies investigated. No common standard progenies i.e. seedling progenies were used in both clonal tests with exception of 11 clones. Therefore, the results of both field tests can be compared only with difficulties.

As a result of the exposure to extremely high SO_2-concentrations in the winter 1995/96, the needles of the clones and progenies investigated at the field tests Altenberg and Baerenfels were damaged to a different extent. The needle damage, expressed by the percentage of needle loss, varied significantly among the Norway spruce clones selected as SO_2-tolerant in the field. A group of 20 clones with little damage, called tolerant clones, differed significantly from the average damage of the field tests as well as from the clonal mixtures and seedling progenies tested in Altenberg and Baerenfels. On the other hand, the group of 20 clones with the most serious damage, called sensitive clones, showed clearly more serious damage than the progenies mentioned above. Both groups of tolerant and sensitive clones consist of clones descended from basic material of different age ranging from 15 to 110 years. In every case, the selection criterion for SO_2-tolerance was a better vitality of the trees selected compared to neighbouring trees in areas already polluted by SO_2 since 10 to 20 years (Tzschacksch 1983b). According to Weiser and Schachler (1988), the realibility to select trees with high SO_2-tolerance improves with increasing age of the trees and with the period of time the selected trees were exposed to air pollution. However, the SO_2-tolerance observed in the field does not constitute tolerance to immission peaks (Wiese *et al.* 1996).

The significant differences of SO_2-tolerance among clones originating from one and the same stand confirm results on the genetic control of SO_2-tolerance (Rohmeder and von Schönborn 1965; Tzschacksch and Weiss 1972; Karnosky 1977; Tzschacksch 1982).

Despite the SO_2-tolerance of the material selected, the parameters mortality, growth and tree form have crucial importance for the successful introduction of improved material in forestry. In these parameters, significant differences can be observed between the groups of tolerant clones and the clonal mixtures or seedling progenies. The clonal mixtures originated from seedlings aged 3 to 5 years, and the seedling progenies show the best survival rates, the best height growth and the highest percentage of upright growing individuals.

The reason for these observations is clearly the different age of the selected basic material. Having clonal differences in mind, the increasing age of the basic material diminishes the ability of clones to be rooted, the frequency of rooted plants and the stem-root-proportion. Additionally, the period of time needed to overcome topophysis- and zyclophysis-effects is influenced negatively (Kleinschmit *et al.* 1973; Schachler *et al.* 1983, 1987; Weisgerber 1983; Schachler and Matschke 1984). On SO_2-polluted sites, the intensity of rooting has decisive importance, since SO_2 influences negatively not only assimilation or frost hardiness of plants but also root growth (Keller 1976, 1979, 1981).

The results presented show clearly the difficulties to transfer the results of breeding for SO_2-tolerance into forestry by autovegative propagation of adult basic material. However, the autovegetative propagation of juvenile basic material which has proved its tolerance of SO_2 in laboratory conditions is possible to a limited extent since laboratory results can not be tranferred necessarily to field conditions (Scholz 1983). Nevertheless, results of practical relevance can only be expected after longterm research in the field (Kleinschmit 1983; Weiser and Schachler 1988).

The genetic principles of tolerance against air pollution, the possibilities and limitations of breeding for tolerance of air pollution, the loss of genetic variation caused by air pollution and the danger to decrease genetic variation by the use of a limited number of clones were already discussed (a.o. Kleinschmit 1983; Scholz 1983, 1984, 1985; Weisgerber *et al.* 1985; Hattemer and Ziehe 1987; Ziehe and Hattemer 1987; Albrecht 1988; Bergmann and Scholz 1989; Gregorius 1989; Scholz and Bergmann 1994).

Since the 1980s, various measures were undertaken by the institutions in charge of forest tree breeding to conserve forest genetic resources of air polluted areas in order to counteract loss of genetic variation caused by forest decline or tree breeding (Braun *et al.* 1987; Braun and Kohlstock 1990; Weiss *et al.* 1990; Paul 1998).

However, only little attention was paid to the origin of the basic material investigated on SO_2-tolerance until today. Results of investigations on different damaged Norway spruce stands in the Bavarian forest based on biochemical-genetical analysis emphasize the crucial role especially of the degree of adaptedness for the survival of stands on a given site apart from the ability of adaptation (Ruetz *et al.* 1996). The investigations of Zimmermann (1931) and Nestler *et al.* (1957) show that a variety of more or less unsuitable Norway spruce provenances was planted in the Ore mountains since the begin of the 19th century. On the other hand, results of provenance trials and biochemical-genetical analysis show that especially populations of Norway spruce autochthonous to the Ore mountains have a high genetic variation and ability for adaptation to very different site conditions (Gaertner *et al.* 1996).

Conclusions

The results presented indicate significant variation in SO_2-tolerance among the material investigated. However, the use of the material selected for SO_2-tolerance by autovegetative propagation of adult basic material did not produce satisfactory improvement in survival rate, height growth or form. Therefore, this procedure can not be recommended for the transfer of adult basic material selected and proved as SO_2-tolerant into forestry, at least not until more suitable propagation methods are developed.

On the other hand, individuals of clonal mixtures descending from progenies of controlled crossed SO_2-tolerant clones have also a relatively high SO_2-tolerance compared to all material tested. Contrary to the tolerant clones of adult basic material, the clonal mixtures tested showed high survival rates combined with height growth above average and upright tree form.

The latter results indicate a possible way for the procurement of reproductive material with improved SO_2-tolerance in the case of continuing SO_2 pollution in the mountainous areas of Saxony. The procedure could be the crossing of SO_2-tolerant individuals, the testing of the progenies in the field, the selection of suitable progenies and the reproduction of tested progenies by crossing with subsequent autovegetative propagation of seedlings by cuttings or by *in vitro* methods. However, such an approach is never a substitute for political, legal and economical activities for the reduction of air pollutants of any nature.

Under the assumption of a further decrease of SO_2-pollution as observed in the past decade, the use of reproductive material of autochthonous Norway spruce seed stands, which show a high level of adaptability, is recommended for reforestation in the air polluted regions of Saxony. A basis for the selection of suitable seed stands could be the results of provenance trials in combination with a biochemical-genetical assessment of the seed stands in question.

References

Albrecht, J. 1988: Forstpflanzenzüchtung und genetische Variation. Allgemeine Forst- und Jagdzeitung 159: 84-87

Anonymous 1973: Richtlinie für die Bewirtschaftung des Rauchschadgebietes Oberes Erzgebirge und Sächsische Schweiz. Vereinigung volkseigener Betriebe, Forstwirtschaft Karl-Marx-Stadt (Chemnitz), not published

Anonymous 1978: Richtlinie für die Bewirtschaftung immissionsgeschädigter Fichtengebiete. Ministerium für Land-, Forst- und Nahrungsgüterwirtschaft der DDR, Ost-Berlin, not published

Anonymous 1985: Richtlinie für die Bewirtschaftung immissionsgeschädigter Fichtengebiete. Ministerium für Land-, Forst- und Nahrungsgüterwirtschaft der DDR, Ost-Berlin, not published

Anonymous 1991: Annual report 1990 (in German). Forschungs- und Überleitungszentrum, Graupa

Anonymous 1995a, b, c; 1996a, b, c: Monthly reports on the air pollution situation October 1995-March 1996 (in German). Sächsiches Landesamt für Umwelt und Geologie, Dresden

Anonymous 1999: Report on forest condition 1999 (in German). Sächsische Staatsministerium für Umwelt und Landwirtschaft, Dresden

Bergmann, F. and Scholz, F. 1989: Selection effects of air pollution in Norway spruce (*Picea abies*) populations. pp. 143-160 in: Scholz, F., Gregorius, H.-R., Rudin, d. (Eds.) Genetic Effects of Air Pollutants in Forest Tree Populations. Springer Verlag, Berlin, Heidelberg

Braun, H., Weiss, M. and Kohlstock, N. 1987: Erhaltung des genetischen Potentials der Fichte (*Picea excelsa*). Sozialistische Forstwirtschaft 37: 149-152

Braun, H. and Kohlstock, N. 1990: Aufgaben und Ergebnisse der Forstpflanzenzüchtung speziell für die Immissionsschadgebiete der Mittelgebirge der DDR. AFZ 45: 868-873

Butter, D. and Richter, H.-J. 1998: Beurteilung der im Immissionsschadgebiet des oberen Erzgebirges vorhandenen Bestände aus unterschiedlichen Baumarten und weiteres Vorgehen bei der Waldschadenssanierung. pp. 73-80 in: Forstpflanzenzüchtung für Immissionsschadgebiete. Schriftenreihe der Sächsischen Landesanstalt für Forsten N° 13/98, Graupa

Gaertner, G., Wolf, H. and Braun, H. 1996: Untersuchungen zur genetischen Struktur der autochthonen Fichten-Population Carlsfeld im Erzgebirge und ihrer Nachkommenschaften als Grundlage zur Beurteilung der Effektivität von Generhaltungsmaßnahmen. Silvae Genetica 45: 294-301

Gregorius, H.-R. 1989: The importance of genetic multiplicity for tolerance of atmospheric pollution. pp. 163-172 in: Scholz, F., Gregorius, H.-R., Rudin, d. (Eds.) Genetic Effects of Air Pollutants in Forest Tree Populations. Springer Verlag, Berlin, Heidelberg

Hattemer, H.H. and Ziehe, M. 1987: Genetische Grundlagen der Resistenz gegenüber Umweltschäden. Allg. Forst- und Jagdzeitschrift 158: 169-174

Karnosky, D.F. 1977: Evidence for genetic control of response to sulfur dioxide and ozone in Populus tremuloides. Can. J. For. Res. 7: 437-440

Keller, Th. 1976: Der Einfluß von Schwefeldioxid als Luftverunreinigung auf die Assimilation der Fichte. Berichte der Eidgenössischen Anstalt für das forstliche Versuchswesen, Birmensdorf N° 161: 3-8

Keller, Th. 1979: Der Einfluß langandauernder SO_2-Begasungen auf das Wurzelwachstum der Fichte. Schweiz. Zeitschrift für Forstwesen 130: 429-435

Keller, Th. 1981: Die Beeinflussung physiologischer Prozesse der Fichte durch eine Winterbegasung mit SO_2. Mitteilungen forstl. Bundesversuchsanstalt, Wien N° 137: 115-120

Kleinschmit, J. 1983: Möglichkeiten der Züchtung resistenter Waldbäume für die immissionsbelasteten Flächen. Der Forst- und Holzwirt 38: 196-199

Kleinschmit, J. Mueller, W., Schmidt, J. and Racz, J. 1973: Entwicklung der Stecklingsvermehrung bei Fichte zur Praxisreife (*Picea abies* [L.] Karst.). Silvae Genetica 22: 4-15

Kohlstock, N. 1985: Stand und Perspektiven der Rauchtoleranz-Züchtung und Möglichkeiten der Überführung. Beitraege für die Forstwirtschaft 19: 153-157

Liebold, E. and Drechsler, M. 1991: Schadenszustand und -entwicklung in den SO_2-geschädigten Fichtengebieten Sachsens. AFZ 46: 492-494

Matschke, J., Schoenborn, H.-J. and Schachler, G. 1984: Möglichkeiten und Probleme bei der Selektion und Vermehrung rauchharter Forstgehölze. Beitraege für die Forstwirtschaft 18: 58-65

Nestler, K., Mann and Riess 1957: Die Nebelfrostablagerungen im Fichtelberggebiet. Abschlußarbeit Fachschule für Forstwirtschaft, Schwarzburg, unpublished Master Thesis

Paul, M. 1998: Die Nutzung immissionsresistenter Fichten (*Picea abies* [L.] Karst.) für die forstliche Generhaltung. pp. 69-72 in: Forstpflanzenzüchtung für Immissionsschadgebiete. Schriftenreihe der Sächsischen Landesanstalt für Forsten N° 13/98, Graupa

Raben, G., Andreae, H. and Leube, F. 1996: Schadstoffbelastungen in sächsischen Waldökosystemen. AFZ/Der Wald 51: 1244-1248

Rohmeder, E. and Schönborn, A. von 1965: Der Einfluß von Umwelt und Erbgut auf die Widerstandsfähigkeit der Waldbäume gegenüber Luftverunreinigung durch Industrieabgase. Forstwirtschaftl. Cbl 84: 1-13

Ruetz, W. F., Konnert, M. and Behm, A. 1996: Sind Waldschäden auch eine Frage der Herkunft? AFZ/Der Wald 51: 759-761

Schachler, G., Matschke, J. and Kohlstock, N. 1983: Einbeziehung von autovegetativ vermehrten Fichten in die Waldbaukonzeption des mittleren Erzgebirges. Sozialistische Forstwirtschaft 33: 250-252

Schachler, G. and Matschke, J. 1984: Stand und Perspektiven zur autovegetativen Vermehrung von Fichte. Beitraege für die Forstwirtschaft 18: 19-24

Schachler, G., Matschke, J. and Kohlstock, N. 1987: Verfahren zur autovegetativen Vermehrung von *Picea abies* und *Fagus silvatica.* IFE-Berichte aus Forschung und Entwicklung 6, Institut fuer Forstwissenschaften Eberswalde

Scholz, F. 1983: Kann Züchtung auf Immissionsresistenz zur Lösung des Problems "Waldsterben" beitragen? AFZ 38: 281-283

Scholz, F. 1984: Drohen unsere Wälder durch Luftverunreinigungen genetisch zu verarmen? AFZ 39: 1258-1261

Scholz, F. 1985: Waldschäden durch Luftverunreinigungen als Herausforderung für Forstgenetik und Forstpflanzenzüchtung. Der Forst- und Holzwirt 40: 238-247

Scholz, F. and Bergmann, F. 1994: Genetic Effects of Environmental Pollution on Tree Populations. pp. 34-50 in: Zin-Suhkim and Hattemer, H. H. (Eds) Conservation and Manipulation of Genetic Resources in Forestry. Kwang Moonkag, Seoul

Tzschacksch, O. 1981: Stand und Perspektiven der forstlichen Rauchresistenzzüchtung in der DDR. Beitraege für die Forstwirtschaft 15: 134-137

Tzschacksch, O. 1982: Untersuchungen zur Erblichkeit der SO_2-Resistenz bei Kiefer (*Pinus sylvestris* L.) und Douglasie (*Pseudotsuga menziesii* [Mirb.] Franco) mit Schlußfolgerungen für die Forstwirtschaft. Beitraege für die Forstwirtschaft 16: 103-106

Tzschacksch, O. 1983a: Möglichkeiten der Begründung von Fichtenbeständen verminderter Anfälligkeiten gegenüber phytotoxischen Immission durch Hochpfropfungen. Sozialistische Forstwirtschaft 33: 339-341

Tzschacksch, O. 1983b: Immissionsresistente Fichten-Herkunftssorte - Immissionsresistente Fichten-Mehrklonsorte. Report of the Institute for Forestry, Eberswalde, not published

Tzschacksch, O. and Weiss, M. 1972: Die Variation der SO_2-Resistenz von Provenienzen der Baumart Fichte (*Picea abies* [L.] KARST.). Beitraege für die Forstwirtschaft 6: 21-23

Weiser, F. and Schachler, G. 1988: Aufbau sowie erste Ergebnisse zur Entwicklung und Nutzung eines Stecklings-Mutterquartiers mit Fichten-Klonen verminderter Anfälligkeit gegenüber SO_2. Beitraege für die Forstwirtschaft 22: 55-61

Weisgerber, H. 1983: Forstpflanzenzüchtung. Mitteilungen der Hessischen Landesforstverwaltung, Bd. 19

Weisgerber, H., Albrecht, J., Bohnens, J., Kechel, H.-G., Rau, H.-M., Schulzke, R. 1985: Bedrohung der genetischen Vielfalt unserer Wälder durch Immissionen - Gegenmaßnahmen der Forstpflanzenzüchtung. Der Forst- und Holzwirt 40: 235-238

Weiss, M., Paul, M. and Braun, H. 1990: Das genetische Potential der Fichte. Der Wald 40: 332-334

Weissleder, S. 1990: Anwenderdokumentation zum Programm "Datenverarbeitungssystem (DVS)". Forschungs- und Überleitungszentrum Graupa

Wiese, C., Slovik, St. and Kaiser, W. M. 1996: Wirkungen hoher SO_2-Konzentrationen auf Fichte, Stechfichte, Waldkiefer und Latsche. AFZ/Der Wald 51: 570-572

Ziehe, M. and Hattemer, H. H. 1987: Populationsgenetische Ansätze zur Resistenz gegenüber Umweltschäden. Allg. Forst- und Jagdzeitung 158: 217-222

Zimmermann, H. 1931: Fichtensamenbeschaffung in Sachsen. Tharandter Forstliches Jahrbuch 82: 821-864

Variation in adaptation and growth as indicated by provenance trial *Platycladus orientalis* (L.) Franco

X. SHEN AND X. CHEN

Beijing Forestry University, 100083, China

Key words: Platycladus orientalis, (L.) Franco, provenance trial, genetic variation, adaptation

Abstract: Platycladus orientalis (L.) Franco is an essential tree species for natural conservation and ornamental use in northern and central China. It grows under diverse range of soils with pH 5-8. In 1982 and 1986 provenance trials of whole range, including over 30 seed origins derived from 17 provinces were laid out at 30 experiment sites in total. Response to drought, frost stresses growth and growth rhythm was investigated. Significant differences in performance of seedlings and young trees derived from different origins were found at all sites. Seed origins for this species significantly differ for both adaptation and growth. The variation is closely correlated with altitude and longitude in particular. According to comprehensive investigation and evaluation, five zones and two sub-zones were divided for seed transfer and reforestation using statistic technique. Furthermore, provenance trial provides intraspecific variation patterns, which are extremely valuable information in consideration of the global change.

Introduction

Provenance trial has a long history since Duhamel du Monceau laid out his experiment in 1745. At the beginning of the 20th century provenance trials have increased rapidly. They have played an important role in understanding geographic variation, the selection of superior planting materials and the conservation of gene resources.

In China provenance trials were initiated in the mid 1950's. From the mid 1970's the number of tested species increased considerably, up to now more than 40 species have been tested (Hong 1995). The major reforestation conifer tree species *i.e. Cunninghamia lanceolata* Lamb, *Pinus massoniana* Lamb, *Pinus tabulaeformis* Carr., *P. koraiensis* Sieb. et Zucc. *P. sylvestris* var. *mongolica* Litv., *Pinus armandii* Franch. *Larix olgensis* Henry, *L. principis-rupprechtii* Mayr., *L. Gmelinii* (Rupr.) Rupr., *Platycladus orientalis* (L.) Franco and some exotic species were included. Based on the provenance trials, geographic variations in growth and adaptation have been determined and seed zones primarily delineated for some of them (National Technic Criteria 1988).

In recent years great attention and concern have been paid to rising air

Müller-Starck, G. and Schubert, R. (eds.). Genetic Response of Forest Systems to Changing Environmental Conditions. Vol. 70 (For. Sci.). 51-61 © 2001 Kluwer Academic Publishers, Dordrecht,Boston, London

temperature due to rapidly increasing CO_2 concentration (Jarvis 1998; Mohren *et al.* 1997). With climate change, rational maintenance of genetic diversity within species is of critical importance. Knowledge from the provenance trial is valuable for this purpose (Mátyás 1997). *Platycladus orientalis* is an essential tree species for natural conservation and ornamental use in northern and central China. It is mainly distributed in the middle and lower branches of the Yellow River and in the north region of Huai River. The annual average temperature in its range varies between 5-17°C and -6 to -16°C in January and from 18 to 28°C in July. Annual precipitation ranges between 200 and 1200 mm. It can be grown under a diverse range of soils with pH 5-8. It is a unique species for this purpose in areas with arid climate and infertile soils.

Materials and methods

The provenance trials for the whole range for *Platycladus orientalis* were conducted in 1982 and 1986. A total of 20-24 sites covering 14 provinces (autonomous regions) were laid out respectively. 136 seed samples from 106 origins covering 17 provinces (regions) were collected, and among them more than 30 seed origins were included for each site. The representation of ecological conditions was primarily taken into consideration when seed collection and site setup were organized. Seed origins and site locations mentioned in the text are shown in Figure 1. Their geographic and environmental data are presented in Table 2.

Response to drought and frost stresses, growth performance and growth rhythm, as well as the morphology of root systems for tested origins at the sites were investigated using conventional methods. Various stresses for testing seed germination accrued with polyethylene glycol (PEG). The anatomical structure of shoots was observed with electron microscope. Data accumulated from both field observations and laboratory experiment were analyzed with statistical technique, including variance, partial and canonical correlation, cluster and trend surface analyses. Concrete experiment designs are described in relevant sections.

Results

Performance of seeds and seedlings under critical drought stress

Seed germination percentage and germination speed were tested under gradient stress from 0, -2, -4, -6, –9 to -12 bar, prepared with PEG. With increasing stress the germination percentage decreased and the speed slowed down. Decrease was exceptionally rapid within a range of –2 to –4 bar. A few seed samples of the north origin gave germination even under –6 bar stress. No seed germinated under –12 bar. Transformation from 0 to – 2 bar was critical. For the north-west origins, *i.e.* Wulashan (Inner-Mongolia), Miyun (Beijing), Wenshui (Shanxi), the germination

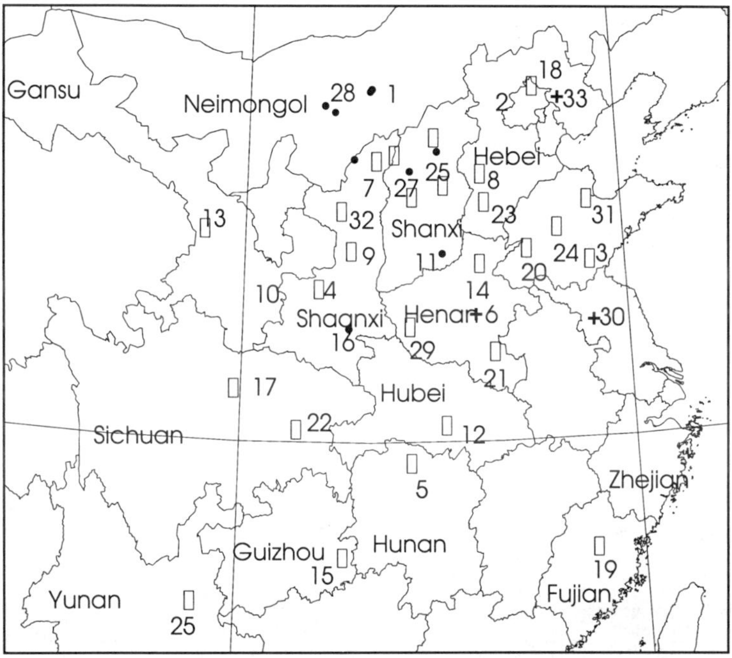

Figure 1. A sketch of seed origins and site locations mentioned in the text. The symbols •, □ and + indicate seed origins, site locations and origin and location respectively; The arabic numerals on the sketch show the place names, which are arranged in alphabet order. 1 Baotou, Neimongol; 2 BJFU, Beijing; 3 Cangshan, Shandong; 4 Chunhua, Shaanxi; 5 Cili, Hunan; 6 Dengfeng, Henan; 7 Fugu, Shaanxi; 8 Huolu, Hebei; 9 Huangling, Shaanxi; 10 Huixian, Gansu; 11 Jincheng, Shanxi; 12 Jingmen, Hubei; 13 Lanzhou, Gansu; 14 Linxian, Henan; 15 Liping, Guizhou; 16 Luonan, Shaanxi; 17 Mianyang, Sichuan; 18 Miyun, Beijing; 19 Nanping, Fujian; 20 Pingyin, Shandong; 21 Queshan, Henan; 22 Shehong, Sichuan; 23 Shexian, Hebei; 24 Taian, Shandong; 25 Taiyuan, Shanxi; 26 Wenshan, Yunnan; 27 Wenshui, Shanxi; 28 Wulashan, Neimongol; 29 Xichuan, Henan; 30 Xuzhou, Jiansu; 31 Yidu, Shandong; 32 Zhidan, Shaanxi; 33 Zunhua, Hebei

Table 1. Geographic and environmental data of seed origins and experimental site locations

Places	Lati.	Longi.	Annual temp (°C)	Mean temp. in July (°C)	Annual solar hrs	Annual. humid. %	Annual. Precip. mm
1 Baotou	40.67	110.27	4.9	-14.0	3075.1	54	370.5
2 BJFU	39.91	116.67	11.5	-4.6	2780.2	60	644.2
3 Cangshan	35.05	118.35	13.2	-1.5	2021.2	68	902.3
4 Chunhua	34.78	108.52	9.9	-4.3	2305.1	67	590.4
5 Cili	29.43	111.13	16.9	5.0	1621.8	78	1421.9
6 Dengfeng	34.50	113.05	14.3	0.4	2341.4	61	637.5
7 Fugu	39.12	110.85	9.1	-8.4	2887.9	51	459.2
8 Huolu	38.33	114.42	12.9	-.2.9	2737.8	62	549.9
9 Huangling	35.83	116.98	10.6	-3.2	2356.4	64	587.9
10 Huixian	33.60	106.00	12.1	-0.7	1744.2	73	765.6
11 Jincheng	35.47	112.83	10.7	-4.1	2387.6	63	679.7
12 Jingmen	30.40	112.83	16.1	3.4	1910.3	80	1114.6
13 Lanzhou	36.05	103.88	9.1	-6.9	2607.6	59	327.7
14 Linxian	36.00	113.80	12.7	-2.7	2526.1	65	693.1
15 Liping	26.23	109.15	15.6	4.5	1317.9	83	1321.9
16 Luonan	33.86	109.97	12.9	0.1	2153.7	65	715.0
17 Mianyang	31.47	104.48	16.3	5.2	1298.1	79	963.2
18 Miyun	40.38	116.83	10.8	-6.5	2788.0	58	661.3
19 Nanpi ng	27.00	118.50	19.3	9.1	1709.9	79	1663.9
20 Pingyin	36.25	116.50	13.6	-2.2	2737.3	-58	600.0
21 Queshan	32.70	114.02	15.0	-1.4	2082.2	72	946.7
22 Shehong	34.80	108.83	17.6	6.5	1354.7	79	1020.1
23 Shexian	36.57	113.67	12.4	-2.7	2706.8	60	572.9
24 Taian	36.80	117.08	12.8	-2.7	2633.7	65	722.6
25 Taiyuan	37.80	112.25	9.5	-6.6	2675.8	60	459.5
26 Wenshan	23.38	105.25	17.8	10.5	2029.8	77	962.2
27 Wenshui	37.53	112.15	10.1	-6.2	2741.8	62	491.1
28 Wulashan	40.47	110.27	4.9	-14.0	3075.1	54	370.5
29 Xichun	33.30	110.50	15.1	2.0	2018.5	69	881.6
30 Xuzhou	34.28	117.15	14.2	0.0	2317.0	69	848.1
31 Yidu	36.72	118.50	12.6	-3.4	2656.3	62	715.1
32 Zhidan	36.62	108.80	7.8	-8.0	1510.8	64	527.6
33Zunhua	40.20	117.95	10.4	-7.1	2827.0	59	803.6

percentage declined less than 30%, while 44% to 74% decreased for the south ones, *i.e* Queshan, Nanping, Jingmen, and the others from Shandong, Henan, Hebei provinces were inter-medium. Meanwhile the germination speed was extended for all origins with stress raising (see Table 2).

Table 2. Seed germination percentage (before slash) and speed in days (after slash) for 12 seed origins representative different ecological regions under different water stress, 20 °C and 16 hr illumination. For each treatment three repetitions of 50 seeds per germination disk

Seed origins	Water stress (bar)			
	0	-2	-4	-6
Wulashan	82.5/ 9.5	64.0/19.9	13.0 /25.8	1.0 / 27.2
Miyun	100.0/10.1	26.0/22.6	10.0/24.5	0 / -
Wenshui	84.0/11.0	86.0/20.2	14.0/25.3	4.0/ 25.0
Jincheng	94.0/10.4	45.0/23.9	17.0/21.8	0 /
Huangling	95.0/10.0	70.0/21.8	18.0/26.2	3.0/29.0
Luonan	96.0/ 9.5	76.0/18.8	15.0/24.2	1.0/26.0
Shexian	85.0/12.9	53.0/23.8	22.0/26.1	0/
Yidu	95.0/9.4	62.0/23.1	5.0/20.5	4.0/24.0
Queshan	98.0/9.4	42.0/21.9	8.0/27.0	0/
Jingmen	100.0/10.0	46.0/23.9	9.0/25.8	0/
Wenshan	100.0/ 9.9	56.0/24.2	4.0/28.0	0/
Nanping	100.0/ 10.1	26.0/22.6	10.0/24.5	0/

Subsequently, Performance of 2.5-year-old seedlings of different origins planted in pots under critical drought stress was investigated and more than 60 plants for each treatment were tested. The results showed: when soil water content dropped from 19.30% - 25.85% to 3.17% - 4.85%, for southern origins *i.e* Liping, Xuzhou, Dengfeng 82.4% to 100% of seedlings felled, while for the northern ones, *i.e.* Zhidan, Miyun 38.2% to 43.9% wilted (see Figure 2). The northern origins had a stronger ability to maintain water balance and were less susceptible to injure than southern ones.

In addition, the water potential of shoots at different drought stresses was studied using pressure-volume technique. The water potential of shoots for

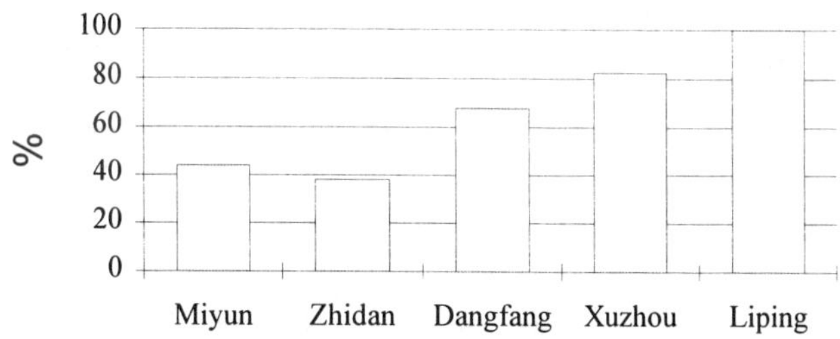

Figure 2 . Percentage of wilted seedlings from southern and northern origins under critical drought stress

northern origins decreased from -0.42 - 0.56 mpa to -4.26 - -6.10 mpa at a slow rate. A highly negative correlation was found between the rates and latitude (Dai *et al.* 1997). Clinal geographic variation pattern was detected.

Performance of 3-year-old saplings and 2-year-old seedlings of different origins exposed to natural frost and controlled low temperature

There were significant differences in frost tolerance for tested origins in the northern experiments. Percentage of wilted saplings derived from 39 origins at age three varied from 0 to 100% at the site of forest farm of Beijing Forestry University (BJFU), located in Beijing suburb. Saplings derived from Nanping (Fujian) and Mianyang (Sichuan) were completely damaged by frost, while the northern origins were barely injured. The geographic variation of frost damage is presented in Figure 3 (Nationwide A 1987).

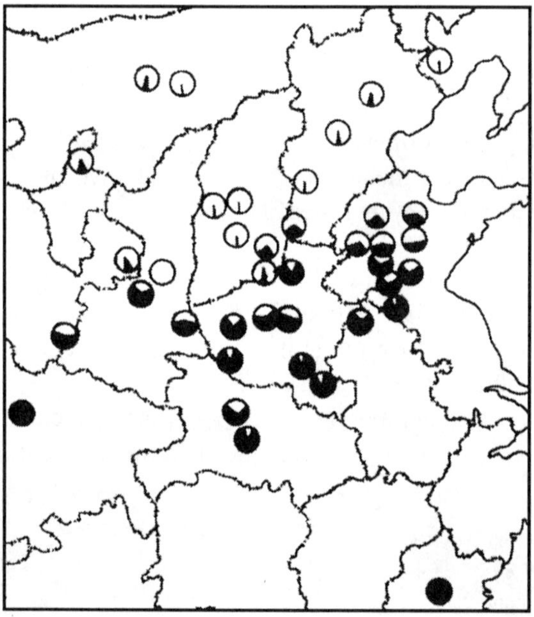

Figure 3. Geographic variation in frost damage of three-year–old saplings at forest farm of BJFU. Legend: Dark – damaged; White –undamaged.

Growth and recovery method was used to evaluate the variation in frost tolerance for seedlings derived from 15 origins. During the treatment, the temperature change was restricted to 5 degrees per hour. The performance of 6 representative origins is shown in Figure 4. Seedlings from northern, southern and middle regions performed quite differently when gradually exposed to 0 to -

40°C. Samples from the south, *i.e.* Liping and Queshan were killed even at -5°C, while those from the north, *i.e.* Baotou and Miyun could tolerate -35°C without injury. However, all the samples of the origins were killed at - 40° C., The frost resistance of all tested origins generally increased with ages, but the northern ones more notably. Seedlings from the northern regions entered the frost-hardening phase earlier and de-hardened later in comparison with those from southern regions.

Growth performance and its interaction of origins and planting sites

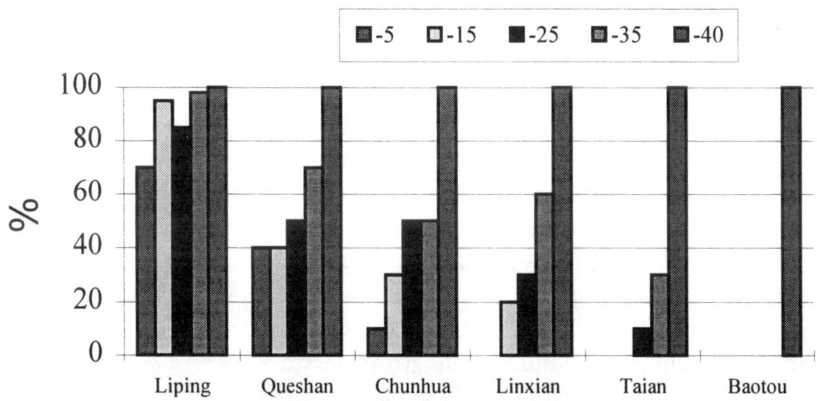

Figure 4. Percentage of killed seedlings from representative origins exposed down to -40 °C

At Pinying site, Shandong province the average tree height at age 7 from the tallest origins was about 3 times of the shortest, although the gap of tree height between the tallest, *i.e.* Wenshan and the shortest, *i.e.* Miyun was diminished in comparison with that at seedling stage. The dynamics of height growth is shown in Figure 5. Significant differences in growth of seedlings and saplings derived from different origins were found at all sites. The extremum (ratio of extremes) of height growth was over three. The variance attributed to the origins accounted for 37.0% to 88.3%. Partial correlation analysis showed that there was significant negative correlation between the seedling height and latitude of seed origins. The variation in latitudes was larger in comparison with that in longitude. With a few exceptions the southern origins grew faster than the northern and western ones, but they were sensitive to drought stress and frost.

Therefore, it must be cautions to move the seeds from south to north.

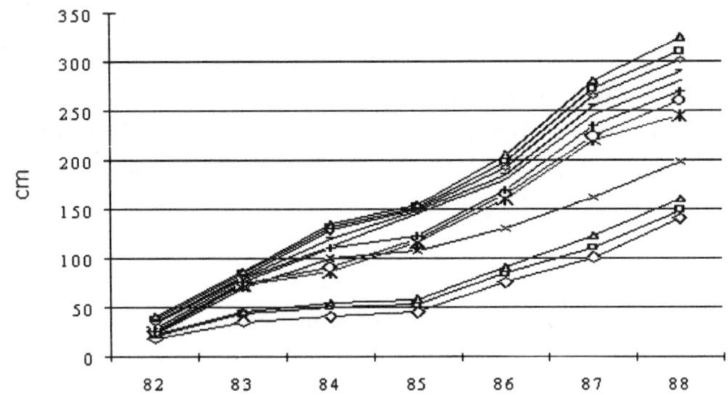

Figure 5: Dynamics of height growth of 7-year-old trees derived from 12 sources at Pinying, Shandong.

According to growth and its rhythm of seedlings, all seed origins were basically divided into three groups, using cluster analysis. The first one originating from northern and western areas of the range was characterized by delayed responses to climate changes and low growth rates. The second one from the south and east of the range grew faster and was sensitive to climate changes. The remaining origins were intermediate (Fu *et al.*. 1989).

The interaction between growth and sites was significant. It showed that the order of height growth for 7 tested origins at 6 sites, widely located in the range was strongly fluctuant (see Figure 6). The growth performance of certain origins at different sites was sharply different (Nationwide B 1987).

Difference in morphology of root systems and anatomic structure of shoots

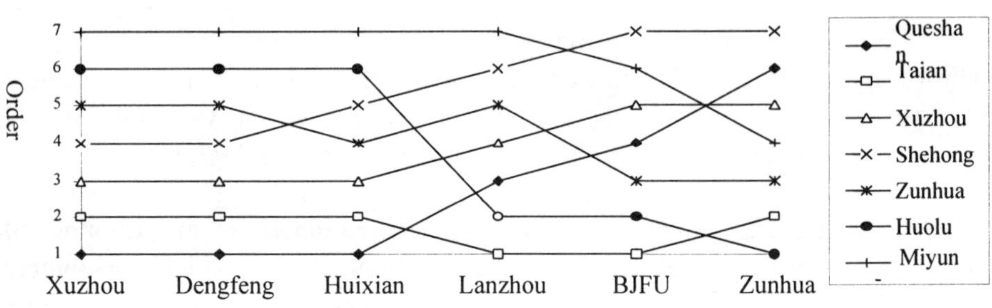

Figure 6. Interaction of height growth and experiment location

Root systems of one-year-old seedlings from tested origins were invested at four sites. The length of main roots from various origins at the sites was significantly different and varied between 22.6cm - 46.6cm, 42.8 - 93.2cm, 16.6cm - 31.3cm and 34.1cm - 62.2cm respectively. The extremum was 1.8 - 2.2. It positively related with latitude and negatively with longitude where the seedling originated. The extremum of the seedling height to main root was 4.2 - 7.6 The variation ratio was more evident among origins and the variance component accounted for 41.2% to 54.6%. On the contrary, it negatively correlated with latitude and positively with longitude. This means that the main root systems were more developed for the northern and western origins and the seedlings from the south were taller. The number of lateral root system was also significantly distinct among the origins. Lateral roots of seedlings from the south more thrived than those from the north and west. The extremum was 2 to 4. The total length and dry weight of root system were significantly different among tested origins. They were obviously correlated with geographical origins. Significant correlation, negative with latitude and positive with longitude was found in the study (Chen *et al.* 1990).

Anatomic structure of one-year-old seedlings from 6 origins in connection with drought-resistance was observed with scan electron microscope. *P. orientalis* is generally characterized by distinct structural features of xeromorphism: The numbers of stomata on lower epidermis of scale accounted for 75%-85% of total; Stomata chamber is identified with deep structure. Differences were detected in thickness of low epidermis, depth of stomata chamber and stomata density on lower epidermis of scale top in particular among 6 representative origins (Dai *et al.* 1999) (see Figure 7).

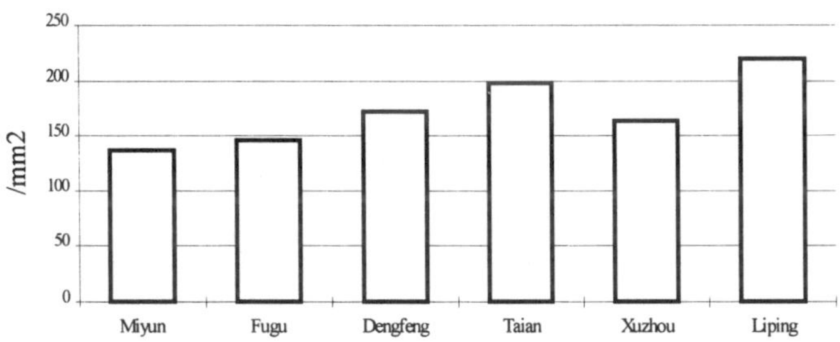

Figure 7. Variation in density of stomata number on lower epidermis of scale top

Conclusions

As mentioned above, the seed origins of the species significantly differed on both adaptation and growth performance and interaction between origins and experimental sites were detected as well. The variation in studied traits is closely correlated with altitude and longitude in particular. Five zones basically paralleling latitudes were divided, and in consideration of the difference in frost resistance of seedlings planted at the east and west sides of the middle zone, two sub-zones were separated. Thus, five zones i.e. North (I), Middle-North (II), Middle (III), Middle-South (IV), South (V) and two sub-zones i.e. North-East (VI) and South-West (VII) were delineated. (See Figure 8).

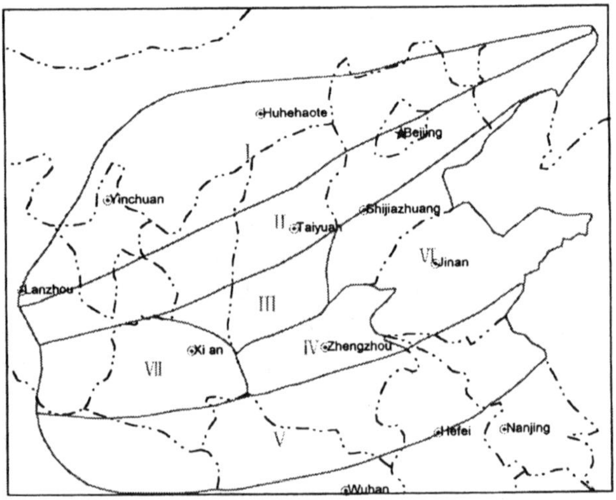

Figure 8. Seed zones of *P. orientalis* (L.) Franco delineated on the base of adaptation and growth characteristics

Based on performance of the origins in trials, the principles of seed transfer were worked out for commercial reforestation (National Technic Criteria 1988). The main points are listed as follows: 1) Seeds produced in north (I), middle-north (II) and south zones (V) are recommend for local reforestation of the same zones; 2) Seeds produced from the north and west of the range were basically not permitted to move to the south and east correspondingly; 3) Seeds from south (V), middle-south (IV) zones and south-west sub-zone(VII) were not allowed to transport to middle-north (II) zone; 4) Seeds of north-east sub-zone (VI) might be used for reforestation in eastern region of zone II, since seedlings derived from sub-zone VI were not suffered from frost and their growth slightly improved. 5) Increased growth could be expected, when seeds were moved from (VI) to middle zone (III) and from V to IV respectively. Meanwhile some superior origins for individual zones were found. Superior seed origins were suggested to preferentially utilize for local reforestation (Shen *et al.* 1993).

Obviously, provenance trial is of great ecological and economical importance. Furthermore, provenance trial provides valuable information about intraspecific variation patters as well as superior plant stock. In consideration of global change they are of extreme importance for the future. Our studies clearly show that the trial is a powerful, available tool for testing climate adaptation of tree species, although it is time consuming and costly. It is advisable to integrate programs applying traditional tree breeding and new technologies (Third 1999).

References

Chen Xiaoyang, Wang Dongyang and Wu Shuanzhu 1990: Studies on the genetic variations of seedling root characteristics in provenance trial of *Platycladus orientalis*. Journal of Beijing Forestry University 12 (2): 47-54

Dai Jianliang, Wang Fang and He Hulin. 1997: Response to water stress of different origins of *Platycladus orientalis.* Forestry Science and Technology of province Gansu (2)

Dai Jianliang, Dong Yuan, Chen Xiaoyang and Shen Xihuan. 1999: Anatomical structure and its relation to drought-resistance for provenance of *Platycladus orientalis*. Journal of Beijing Forestry University 11 (4): 26-31

Fu jianming and Shen Xihuan 1989: A Study of growth rhythm in provenance trial of *Platycladus orientalis*. Journal of Beijing Forestry University 26 (1): 26-31

Hong Jusheng 1995: Forest tree provenance trials in China. IN: Shen Xihuan (Ed.) Forest Tree Improvement in the Asia-Pacific Region. China Forestry Publishing House: 67-79

Jarvis P.G. 1998: European forests and global change. Cambridge University Press.

Mátyás, Csaba 1997: Climatic adaptation of trees: rediscovering provenance tests. P.M.A. Tigerstedt (Ed.). Adaptation in plant breeding: 49-58, Kluver Academic Publishes.

Mohren G.M. K. Kramer and S. Sabate (Eds.). 1997: Impacts of Global Change on Tree Physiology and Forest Ecosystems. Kluwer Academic Publishers

National Technic Criteria of the PRC. 1988: Seed zones of Chinese Forest Trees. UDC 581.527:634.0.2(510); GB 8822. 1-8822. China Forestry Publishing House: 13-88

Nationwide cooperative group for provenance trial of *Platycladus orientalis*. A 1987: Studies on variation of growth and overwintering characterristics of seedling of provenance trial of *Platycladus orientalis.* Journal of Beijing Forestry University 9 (3): 20-27

Nationwide cooperative group for provenance trial of *Platycladus orientalis*. B 1987: Studies on growth interaction between origins and sites for seedlings of *Platycladus orientalis*. Journal of Beijing Forestry University 9 (3): 28-35

Shen Xihuan and Cheng Xiaoyang. 1993: Progress in provenance studies of *Platycladus orientalis*. IN Tu Zhongyu, Shen Xihuan (Eds) Progress in genetic improvement of forest tree in China. Scientific and Technical Document Publishing House: 14-18

Third IUFRO-FAO Consultation on Forest Genetics and Tree Improvement. 1999: Contribution of genetics to the sustained management of global forest resources. Forest Genetics 6(3): 61-64

Influence of nursery environment and pollution on alders

L. MEJNARTOWICZ

Institute of Dendrology, Polish Academy of Sciences, PL-63-035 Kornik, Poland

Key words: *Alnus*, nursery, polluted areas, reforestation, selection, heavy metals

Abstract: For reforestation in polluted areas, especially high quality plant material should be used. It is a common practice to grow plants for one or two years under the cover of plastic tents. We have found that probably selection against specific families occurs in a polyethylene tent environment. Some progenies that were among the tallest when grown in the open nursery were below average when grown under the tent. Leaves of alders raised in a tent were more responsive to polluted environments than leaves of alders raised in an open nursery. Alder progenies differed substantially in leaf size. Progenies that had large leaves had more injured leaves and grew less than small-leaved alders. Speckled alder grew better than black alder in a polluted environment. Spontaneous hybrid alders growing in the polluted environment grew worse than two parental species, though they exhibited less leaf injury. After two years of growth of outplanted alder trees in the polluted zone, the content of copper, zinc and magnesium in the leaves was significantly higher than the amount found in the leaves of control trees. Nitrogen in the leaves did not differentiate significantly the control trees from those growing in polluted environment.

Introduction

Of the many deciduous plant species, alders appear very useful trees for reforestation in areas degraded by industrial emissions. In Europe five species of alder occur, only two of which are of importance for forestry and reclamation, namely the black alder (European alder) - *Alnus glutinosa* (L.) Gaertn., and the speckled alder (gray or white alder) - *Alnus incana* (L.) Moench. These are the only native forest trees in Europe, which owing to the symbiotic *Frankia alni*, fix atmospheric nitrogen and simultaneously act as a water and air filter (Peters and Peitzmeier 1989). Alders are also an important component of sustainable European meadow forests.

Alder trees can grow on wet soils, but also on moderately dry soils, providing wood for fiberboard and veneer, and can serve as a substitute for imported tropical hardwoods. Alder has long been used for reclamation of quarried sand pits and in regions with boggy soils resulting from a disturbance of the water regime.

Müller-Starck, G. and Schubert, R. (eds.). Genetic Response of Forest Systems to Changing Environmental Conditions. Vol. 70 (For. Sci.), 63-73

Studies are needed on the utility of alders for reforestation in areas polluted by emissions from copper smelters where the soil has accumulated heavy metal elements, especially Cu and Zn. On the other hand reports on the utility of alders in regions exposed to gaseous and particulate industrial emissions are not clearly reconcilable. Jamrich (1968), Niklefeld (1967) and also Wenzel (1968) and Perez *et al.* (1992) considered alders to be resistant to the action of sulfur dioxide, HF and other air pollutants. In contrast, Hobjørg (1975) when studying the influence of SO_2 on seedlings of speckled alder found that a 4-month exposure of seedlings to this gas resulted in a 40% decline in dry-weight increment. Similarly, Yee-Meiler (1978) found that SO_2 at a concentration of 0.1 – 0.2 ppm injured the leaves of black alder.

The primary purpose of this experiment was to see whether differences exist among black alder families in their growth response to emissions from a copper smelter, and also to examine which of black alder or speckled alder is more resistant. An evaluation of natural hybrids was of lesser importance given their rare natural occurrence.

Material and methods

The results presented concern comparisons of 33 half-sib progenies originating from three provenances of black alder. In all 7 progenies were collected from Babki, 12 from Czeszewo and 14 from Karczma Borowa populations. One mixed seed sample was gathered from five trees of speckled alder (N-05-525), and one half-sib family (N-05-518) was collected from a natural hybrid between *Alnus incana* and *Alnus glutinosa*. The generation of the hybrid tree is unknown. The hybrid individual had leaf characters intermediate between the black and speckled alder and closely resembles the hybrid obtained artificially from pollination of *A. incana* with *A. glutinosa* pollen (Mejnartowicz 1999).

During the nursery stage of the study, we decided to test the influence of covering the beds with a polyethylene tent on the growth of individual progenies. It was of interest to see whether the family differences in the second year were similar between the seedlings grown under the tent as between those grown in the open. Both series were otherwise treated identically. Details of the experiment are given in Mejnartowicz (1983).

Two-year seedlings (1+1) were lifted, and their height was measured. The weight of the aerial part and of the root was measured. Leaf samples were collected from each family and width, length and leaf area were measured, and their nitrogen content was estimated in order to enable later comparisons of the influence of industrial emissions from a copper smelter on the vegetative organs of these alder trees. Nitrogen contents were calculated as a percentage of dry weight (dw) and were determined with the standard Kjeldahl method (Grewling 1976).

For the field experiment, an area was selected in a region under the influence of emissions from the Głogów copper smelter. Soil on the experimental site was brown and formed from a loess formation underlain by sand. Due to excessive accumulation of heavy metals in agricultural produce, cultivation of this land was abandoned. The experiment was established in four replicates both for the material raised in the plastic tent and for the material grown in the second year in an open nursery.

Results and discussion

Tree height in nursery stage

After the end of each vegetative season, the height of seedlings was measured. The mean values for each progeny and differences between progenies were evaluated statistically and are presented in Table 1. After the first year of growth the differences between progenies were highly significant. The tallest seedlings were from the Czeszewo population with a mean value of 90 cm. Progeny of the spontaneous hybrid had a mean height of 81 cm, and thus was among the group of the tallest progenies. This progeny has segregated into trees with leaves typical of *Alnus glutinosa*, trees with leaves typical of *A. incana* and a small portion of trees with intermediate features. The progeny of speckled alder with a mean height of 58 cm was below the average height for black alder populations.

After the second year, black alder seedlings that continued to grow under the polyethylene cover were twice as large as those growing in the open nursery, having a mean height of 113 cm compared to 50 cm for the seedlings grown in the open. After transplantation to the nursery seedling growth was disturbed by the late spring frost as is manifest in the dying of apical shoots, which were however being regenerated already by mid-season.

The speckled alder cultivated only in the tent attained a mean height of 140 cm, substantially exceeding the mean value of 120cm for the black alders.

Some progenies of black alder that were among the tallest in the second year when grown in the open nursery were below average among those grown under the polyethylene. It is possible that some progenies have the potential to better utilise the specific microclimate of the polyethylene tent, whereas others grow better in the open nursery conditions.

The differences between progenies cultivated under the tent were statistically significant. In contrast to seedlings raised in the tent, differences in height between populations and progenies of 3-year-old seedlings from the open nursery were not statistically significant (Tab.1).

Table 1. Height [cm] of alder trees measured over four years. After the second year the trees were outplanted at a site polluted by a copper smelter.

	Grown in tent		Polluted area		Increm.	Nursery	Polluted area		Increm.
Year:	1	2	3	4	4th	2	3	4	4th
			Black alder population Babki						
Average	65	106	118	158	40	43	71	123	52
Minimum	50	78	94	135	29	41	70	118	48
Maximum	82	152	141	188	50	46	72	127	55
			Black alder population Czeszewo						
Average	64	120	130	163	33	49	83	129	46
Minimum	39	93	115	143	17	35	64	73	9
Maximum	90	148	148	211	63	70	95	165	75
			Black alder population Karczma Borowa						
Average	68	111	124	164	40	52	86	135	49
Minimum	40	77	111	135	19	29	56	101	25
Maximum	88	149	142	197	63	73	103	161	74
Grand mean	66	113	125	162	37	50	84	132	48
Ai x *Ag*	81	143	117	149	32	54	101	151	50
A. incana	58	140	139	217	78	-	-	-	-
F between:									
Families	**	**	**	n.s.	n.s.	n.s.	n.s.	n.s.	*
Populations	n.s.	n.s.	*	n.s.	n.s.	n.s.	n.s.	n.s.	

*: < 0.05, **: p < 0.01, n.s.: not significant

Growth and increment of alder trees in a polluted area

From the third year onward the trees grew under the influence of emissions from a copper smelter. The influence of the emitted SO_2 and heavy metals accumulated in the soil was manifested in the first year by inhibited growth and the death of apical shoots, which already in mid-season began regenerating. Differences between progenies that had grown under a tent for the first two years were significant. In contrast, the height of three-year-old black alder progenies that grew in the open nursery, is not a trait that statistically differentiates the progenies (Tab.1).

The growth increment in tree height under conditions of industrial emissions was distinctly different for the group of trees from the nursery and the group of trees from the tent. The mean increment of the seedlings from the tent was 10.6% of the planting stock height, which in absolute terms amounted to 12 cm.

The absolute increment in seedlings raised in the open nursery was 34 cm, nearly three times larger than the tent seedlings (Tab.1). If one can consider the growth increment in polluted conditions as a measure of tolerance to environmental pollution, then the smaller seedlings withstood the emissions of the copper smelter better than the larger ones, and the seedlings from the nursery were better than those from the tent. After one year of growth in the polluted site, the seedlings from the nursery attained, on average, 71% of the height of the planted seedlings. The highest increment was 196% for progeny N-05-510, which was the shortest at the time of planting, both in the group of plants from the open nursery (29 cm) and in the group from the tent (77 cm).

At the age of four years, the trees have passed transplanting shock and the results are likely more robust than comparisons made in the third year. The trees that had been grown in the tent were much taller than those that grew their second year in an open nursery, having an average height of 162 cm and 132 cm respectively. However, the measured annual increment in height for the fourth year demonstrated that the trees from the nursery have in fact grown faster (48 cm) than those from the tent (37cm). This is probably associated with the damage inflicted to the large root systems of the tent-grown seedlings. In terms of tree height there were no significant differences between progenies after the age four years, both for the group of plants originating from the tent and for those from the open nursery (Tab.1).

The hybrid progeny in the emission zone faired comparatively worse, exhibiting shoot death and slow shoot regeneration, such that after an age of 3 years the plants were shorter than at the time of outplanting in the experimental area. Trees of speckled alder easily regenerated any loss of leader or have sustained no damage. The height increment of speckled alder in the fourth year was 78cm, whereas black alder was only 37cm (Tab.1).

Shoots and root weight

Tree height increment under conditions of industrial emissions was distinctly different among the progenies. It is rather rare to find information about growth in more than one environment, as plants are typically destructively harvested to obtain growth data. After the end of the second growing season, 40 seedlings were selected randomly from the open nursery and 26 of those that had grown for two years under the tent. Tent-grown seedlings had a significantly higher dry weight on average (43.4g) compared to seedlings grown in the open nursery which averaged 27.6 g. The root bundle of seedlings grown under the tent had a relatively low number of roots, frequently with one, thick, root growing at a slant or horizontally with a small number of fine rootlets. The shoot/root weight ratio for these plants was 1.25 (Tab.2).

The seedlings that in the second vegetative season grew in the open nursery have had a much better formed root system. It consisted for a large number of small thin rootlets growing compactly together. Their mean weight (31g) was lower than that of the roots in plants grown in the tent, and which exceeded the weight of the aerial part of the plant. The shoot/root ratio was 0.88 (Tab.2). This character may prove of importance later when evaluating tolerance of these alders to soil and air pollution in the vicinity of a copper smelter.

At the hybrid seedlings the shoot/root ratio was identical as in black alder. These hybrids grew in the open nursery during the second year.

Table 2. Root and shoot weight [g] of alder trees grown under the tent and in a nursery.

Weight [g]	Alders grown 2-years under tent			Alders grown in their second year in nursery		
	Roots	Shoots	Shoot/Root	Roots	Shoots	Shoot/Root
			Black alders			
Average	36.5	43.4	1.25	30.9	27.6	0.88
Minimum	5.0	4.0	0.78	5.0	4.5	0.57
Maximum	118.0	118.0	1.87	63.3	65.0	1.29
				Alnus incana x *Alnus glutinosa*		
Average	-	-	-	22.2	19.3	0.88
Minimum				8.0	7.0	0.64
Maximum				62.7	56.0	1.05

Influence of emissions on alder leaves

Emissions from a copper smelter influenced not only the physiology, but also the morphology of alder leaves present in the emission zone. The leaves of trees raised in a tent responded much more to the pollutants than the leaves of alders raised in an open nursery. Under stress conditions, leaves were significantly reduced in size except for the petiole, which remained the same. Leaf width was reduced from 8.1 to 6.2 cm, length from 10 to 6.9 cm, leaf area from 68.43 cm^2 to 36. 9 cm^2 (reduced by 46%) and the number of vein pairs from 7.7 to 7.0. Alders raised in an open nursery did not change significantly. Their leaf width was even slightly higher (6.2 to 6.7cm) and their length not significantly different from 7.9 to 7.5 cm. The petiole length increased substantially from 1.5 to 2.2 cm, attaining a size similar to those observed in alders raised under the tent. Leaf area, though not significantly, increased from 42.62 cm^2 to 44.98 cm^2.

Alder progenies, regardless of whether they were raised in a tent or in an open nursery, differed substantially in leaf size. Besides "large-leaved" forms there were also "small-leaved" forms. To check whether the size of leaves has any link with the mortality of trees in the emissions zone and with leaf injury and reduction in tree growth, the traits in question were compared to values measured before outplanting in the emission zone. The highest percentage of mortality was found in small-leaved alders from the open nursery. Both forms of alders, when grown in plastic tents did not differ significantly in mortality in the emission zone. So, it is likely that tree mortality is more associated with the conditions under which the plants were grown in the nursery stage than with the size of their leaves.

Analysing the relationship between the size of leaves and their viability under growth conditions in the polluted environment, the alder progenies which had large leaves in both environments had more injured leaves (in the first of three grades on our scale) than small-leaved alders. Among small-leaved alders, the highest percentages of healthy leaves were found in alders raised in the open nursery. Thus, the mortality of alders with large leaves (which were raised in the tent) is small, whereas the damage to the leaves is greater than in the small-leaved alders. A similar relationship was found between the size of leaves and the growth increment of alders in the vicinity of the copper smelter. A smaller increment was observed in large-leaved alders raised in the tent than in small-leaved alders, and this was so regardless of where they were raised in the nursery period.

Content of copper, zinc and magnesium ions in the leaves

Copper

The content of the copper increased after two years of growth of the black alders in the zone of emissions to 1073μg Cu^{++}/g dw, which is about 200 times more than in the control material (5.2 μg/g dw). Individual progenies differed significantly from each other ($F = 2.3^{**}$) in their ability to accumulate this element, from 443μg to 1502μgCu^{++}/g dw. Differences between populations are also significant $F = 2.7^{*}$ (Tab. 3). We found more copper accumulation in control conditions in the population sample for speckled alder than in black alder (6.0 and 5.2 μg/g dw respectively), but half as much when the same trees were grown in the polluted area (504μg and 1073μg/g dw respectively). The progeny of the hybrid tree was similar to black alder in its ability to accumulate copper, having 1037μgCu^{++}/g dw.

Zinc

The control trees of the studied alder species accumulated this metal at a level of 73µg^{++} for black alder and 80µg in speckled alder. These values indicate a high average accumulation of this element, about four times higher than values normally considered as essential for the growth of other plant species (Kramer and Kozlowski 1979). In the zone of emission from the copper smelter, the accumulation of this element was on average three times higher than for the control trees of black alder and two times than the control tree for speckled alder (Tab.3). Hybrid alders accumulated more Zn than both parental species.

Magnesium and Nitrogen

This macroelement increased by 6% in the leaves of alders from the experimental area. This value is statistically significant (Tab. 3). There was less magnesium in the leaves of speckled alder than in black alder. As was the case for Zn and Cu, the hybrids had an intermediate concentration of Mg. In contrast to copper and zinc, differences between magnesium accumulated in leaves from polluted and control environments were not as significant.

The alders in the polluted environment produced nitrogen in quantities similar to the control trees. Also non-significant were any differences between populations. For this character there were significant differences only between families.

Table 3. Average content of Cu, Zn and Mg µg/g dw and % of N in the leaves of alders in the emission zone. Control results in brackets (after Mejnartowicz 1986) .

Black alder	**Cu**	**Zn**	**Mg**	**N**
Max. in family	1502 (17.0)	429 (144)	3733 (3340)	3.8 (3.6)
Min. in family	443 (1.0)	131 (38)	2213 (1840)	2.7 (2.2)
Grand mean	1073 (5.2)	246 (73)	2942 (2765)	3.1 (3.0)
T-test polluted / control	31**	17**	2.5*	NS
T-test between family	2.3**	NS	NS	2.4**
F- test between popul.	2.7*	NS	NS	NS
Hybrids	1037 (4.7)	259 (78)	2947 (2640)	3.2 (3.0)
Speckled alder	504 (6.0)	175 (80)	2467 (2540)	3.1 (2.7)

Transport and accumulation of copper, zinc and magnesium

It was not clear whether the metals entering plants through roots and leaves are transported to the roots after the end of the growing season or whether they also occur in the stem and shoots. Likewise, it is not known the extent to which they are transported to developing leaves at the beginning of the vegetative period. All mineral salts must move acropetally in the xylem. If heavy metals are bound to proteins, then as long as they are soluble they will be transported to developing leaves. In order to address this question, two-year-old shoots were cut from alders growing in the emission zone and placed in water in a phytotron together with control shoots from alders growing in a zone free of pollution.

Table 4. Content of Cu, Zn and Mg μg/g dw in shoots and leaves of alders grown in the emission zone. Shoots were detached before bud break and placed into distilled water. Control results in brackets.

Species		**Shoots**			**Leaves**	
	Cu	Zn	Mg	Cu	Zn	Mg
A. glutinosa	262 (10)	175 (110)	125 (75)	83 (7)	123 (70)	496 (120)
A. incana	225 (10)	148 (80)	116 (72)	76 (6)	176 (80)	512 (110)
Hybrid	230 (11)	248 (81)	176 (97)	92 (5)	172 (78)	560 (130)
Mean	239 (10)	190 (90)	139 (81)	84 (6)	157 (76)	522 (120)

The shoots were detached at the end of February. The developed leaves and shoots were analyzed for the content of Cu, Zn and Mg in an atomic absorption spectrophotometer. The results are presented in Table 4.

It was found that during winter dormancy in 1-year old and 2-year old shoots of alders they are deposited metals (Tab. 4). However their quantities are much smaller than in mature leaves at the end of the vegetative growth period (Tab. 3). On average in the winter shoots there was 239μg Cu^{++} compared to 1073μg Cu^{++}/g dw in the autumn leaves, which indicated that either a portion of the copper was removed together with leaf fall or that it was transported into the trunk or roots. When leaves started to develop on such shoots, some copper transport occurred, so that at the time they were fully expanded they already contained 84μg Cu/g dw, while the leaves of control shoots had only 6μg Cu/g dw.

Tree mortality and leaves damage

In the emission zone it was found that some progenies had substantial leaf injury, while others experienced little leaf injury. Toxic effects that resulted in damage of half the leaf blade and resulted in the death of the plants were clearly of a random nature. In both studied groups of alders (i.e. those raised in the tent and in the nursery) there were no significant differences between progenies in mortality rate. This trait, however, significantly differentiated populations of alders that had been raised in the plastic tent. The highest mortality was found in the population Babki (15%), whereas the other two populations had only 6.4 and 6.7% of tree mortality.

In the hybrid progeny 78% of the trees had undamaged leaves and the remaining 22% had injuries only in the 1st category. Injury in the hybrid progeny was less than for either black or speckled alder, where 66% of the trees had undamaged leaves, 27% were damaged to the 1st degree and 7% to the second degree. One progeny of black alder, which had trees with only undamaged leaves, also had a much lower content of copper in the leaves (867 μg Cu/g dw) and the highest content of zinc (429 μg/g dw). This finding may be related to the competitive absorption of Cu^{++} and Zn^{++}. This hypothesis was proposed by Bowen (1981) and supported by Harrison (1984).

Discussion

It is generally believed that seedlings growing well in the nursery also have a greater resistance to a polluted environment. Seedlings grown in a polyethylene tent were twice as a large as those grown in an open nursery. A plastic tent microclimate exerts some selection effects on alder progenies. Surprisingly, it appeared that some progenies that were among the tallest when grown in the open nursery were below average when grown under the tent. Some progenies are probably better able to utilize the tent microclimate than other progenies.

Even after two years of growth in the polluted environment, alder seedlings raised in the tent were still much taller (162cm) than those raised in the open nursery (132cm). The height increment of the nursery-grown seedlings in the polluted environment was greater (49cm) than for the seedlings from the tent (40cm – Table 1.). The reason for the better growth of alders from the nursery presumably lies in their better-developed root systems (Tab.2) and the greater susceptibility of the tent-grown alder's leaves to the pollutants. The leaves of the trees raised in the tent are much larger and accumulate more toxic elements than the leaves of alders raised in nursery.

Generally high level variation was found between progenies, and little differentiation between populations. Similar results have been found by Bousquet *et al.* (1988) for American alders: green alder [*Alnus crispa* (Pursh.)] and mature populations of speckled alder (*Alnus incana* ssp. *rugosa* DuRoi).

Conclusion

1. A plastic tent can exert some selection effects on alder seedlings raised in it.
2. Alders raised in an open nursery had greater height increments in a polluted area than those from the tent.

3. Large-leaved forms of alder are more susceptible to industrial emissions than small-leaved forms.
4. There are significant interprogeny and interpopulation differences in the ability to accumulate Cu^{2+} and no such differences in ability to accumulate Zn^{2+}.
5. Speckled alder grew better than black alder in the polluted environment.
6. Nitrogen did not differentiate significantly control trees from those growing in the polluted environment.

References

Bousquet, J., Cheliak, W.M. and Lalonde, M. 1988: Allozyme variation within and among mature populations of speckled alder (*Alnus rugosa*) and relationships with green alder (*Alnus crispa*). Am. J. Bot. 75 (11): 1678-1686

Bowen, J.E. 1981: Kinetics of active uptake of boron, zinc, copper and manganese in barley and sugarcane. J. Plant Nutrition 3: 215-233

Greweling, T. 1976: Biochemical analysis of plant tissue. Search Agriculture. Agronomy 6 (8): 1-35

Harrison, S.J., Lepp, N.W. and Phipps, D.A. 1984: Uptake of copper by excised roots. IV. Copper uptake from complexed source. Z. Phlanzenphysiologie. 113: 445-450

Hobjørg, A. 1975: Air pollution and vegetation III. Efects of air pollution on growth of *Alnus incana, Sorbus aucuparia* and *Ulmus glabra* grown in cotainers at the localities in southern Norway. Medlinger fra Norges Landbruchs høgskole. 54 (25): 22

Jamrich, V. 1968: Je stabilita chlorofylu faktorom odolnosti proti dymu? Zbornik Vedeckich Prac. X(1): 7-14

Kramer, P.J. and Kozlowski, T.T. 1979: Mineral nutrition and salt absorption. Pp. 334-372 In: Physiology of woody plants. Academic Press, New York San Francisco, London

Mejnartowicz, L. 1983: A genetic basis for the resistance of forest trees to antropopressure, with special study of the effect of some toxic gases. PAS Inst. of Dendrolgy Kórnik. 4-th Report for USA Dept. Agriculture. J-MOA-USDA-11:1-38

Mejnartowicz, L. 1986: Influence of emissions from a copper smelter on the content of copper, zinc and magnesium ions and nitrogen and protein in the leaves of alders. Arboretum Kórnickie 205-219

Mejnartowicz, L. 1999: Evidence for long-term heterosis phenomenon in the *Alnus incana* x *glutinosa* F1 hybrids. Sylvae Genetica 48: 100-103

Niklefeld, H. 1967: Plant sociological observation in the smoke-dameged area of an aluminium works. Cbl. Ges. Forstw. 84 (2) 6: 318-329

Perez, A., de-Anta, C.R. and Calvo-de-Anta, R. 1992: Soil pollution in copper sulphide mining areas in Galicia (N.W. Spain). Soil-Technology 3 (5):271-281

Peters, U. and Peitzmeier, E. 1989: Vergleichende Untersuchungen in Ruhrgebiet (Bochum und am Rand der Sudeifel. Verhandlungen-Geselschaft-für Ökologie. 18:131-138

Wenzel, K.F. 1968: Empfindlichkeit und Resistenzunterschiede der Pflanzengegenüber Luft verunreinigung. Forstarchiv 39: 189-194

Yee-Meiler, D. 1978: Der Einfluss von Kontinuerlichen, niedrigen SO_2–Begasungen auf den Phenolgehalt und die Phenoloxidase Activität in Blättern einiger Waldbaumarten. Europ. J. For. Path. 8 (1): 14-20

PART 2: GENETIC VARIATION UNDER DIVERSE ENVIRONMENTAL CONDITIONS

Small scale spatial genetic structure of six tropical tree species in French Guiana

B. DEGEN[1], H. CARON[2], E. BANDOU[1], C. DUTECH[3], P. BREYNE[4], M. VAN MONTAGU[4], L. MAGGIA[3] and A. KREMER[2]

[1] INRA Station de Recherches Forestières Guyane, Campus agronomique, 97387 - Kourou cedex, French Guiana; [2] INRA Station de Recherches Forestières, Laboratoire de Génétique des Arbres Forestiers, F-33611 Gazinet cedex, France; [3] CIRAD-Forêt, Campus agronomique, 97387 - Kourou cedex, French Guiana; [4] Department of Plant Genetics, Flanders Interuniversity Institute for Biotechnology, University of Gent, 9000 Gent, Belgium

Key words: AFLP, allozymes, autocorrelation, chloroplast-DNA, RAPD, tropical trees

Abstract: The small scale spatial genetic structure of six tropical tree species (*Carapa procera, Chrysophyllum sanguinolentum, Dicorynia guianensis, Eperua grandiflora, Virola michelii, Vouacapoua americana*) was studied for populations located within a sylvicultural trial in French Guiana. The analysed species have different biological and ecological characteristics with respect to spatial distribution, sexual system, pollen and seed dispersal agents, flowering phenology, and they differ in their environmental demands. The spatial position of trees and at least one data set on the genetic information for allozymes, cp-DNA marker, RAPDs and AFLPs served as basis for the analysis. As estimator for spatial genetic structure, Moran's index was computed for allozymes and cp-DNA markers, while for RAPDs and AFLPs mean Tanimoto distances between pairs of trees in different spatial distance classes were calculated. Each species showed significant spatial structure for at least one type of gene marker. The expansion of significant spatial genetic structures was relatively large and varied between 60m up to 800m. The strongest spatial autocorrelation was found for maternally inherited cp-DNA marker. Due to distinctly higher level of genetic variation, AFLPs and RAPDs were more sensitive to detect spatial structure than allozymes. Relations between the spatial genetic structures and the species' life characteristics are discussed.

Introduction

The spatial genetic structure of tree populations is influenced by several factors and processes. As shown by simulation studies, restricted gene flow via pollen and seeds, the spatial distribution pattern of trees and microhabitat selection can all contribute to spatial genetic structure (Doligez *et al.* 1998; Epperson 1990; Epperson and Li 1997; Sokal and Wartenberg 1983; Sokal *et al.* 1989). Information on the spatial genetic structure of tree populations helps to understand the driving forces for genetic dynamics (Epperson 1992) and can be

Müller-Starck, G. and Schubert, R. (eds.). Genetic Response of Forest Systems to Changing Environmental Conditions. Vol. 70 (For. Sci.), 75-92 © 2001 Kluwer Academic Publishers, Dordrecht, Boston, London

used for assessment of expected impact of forest exploitation and fragmentation (Young and Merriam 1994).

During the last ten years, an increasing number of experimental studies on the small scale spatial structure of tree populations has been published. Most of them were made in natural stands of temperate forest trees using allozymes (e.g. Geburek 1993; Leonardi *et al.* 1996; Schnabel *et al.* 1991; Young and Merriam 1994). In rare cases molecular genetic markers have been used for this purpose (Streiff *et al.* 1998). Beside some studies which did not find any spatial structure, the majority of studies reveal a weak positive autocorrelation only for the first distance classes up to 30 m and in rare cases up to 100 m.

Only few studies have been published on spatial genetic structure of tropical tree species (Boshier *et al.* 1995; Caron *et al.* 1998; Doligez and Joly 1997; Hamrick *et al.* 1993; Loiselle *et al.* 1995). As for most of the studies in temperate forests, these authors found either no structure at all or a spatial structure not greater than 80m.

The aim of the present study was to analyse the small scale spatial genetic structure of six tropical tree species (*Carapa procera, Chrysophyllum sanguinolentum, Dicorynia guianensis, Eperua grandiflora, Virola michelii, Vouacapoua americana*) in the sylvicultural trial Paracou in French Guiana. Due to low tree densities and relatively undisturbed history of the studied site, we expected stronger spatial structure for the analysed tropical tree species than have been found for temperate trees. The comparative application of different gene markers such as allozymes, cp-DNA markers, RAPDs and AFLPs, should reveal the impact of different modes of inheritance (biparental versus maternal) and different levels of genetic variation and gene action (codominant versus dominant) on the spatial pattern. We wanted to search for relations between the different life characteristics of the species, such as sexual system, pollen and seed dispersal agents and the spatial genetic pattern.

Material and Methods

Study site

All samples of the study were collected at the experimental trial Paracou (5°18'N, 52°53'W) near Sinnamary in French Guiana. The mean annual rainfall in this area is 3076 mm with peaks in April-June and December. There are two dry seasons: a short one in March and a longer one from August to November. The forest of Paracou is classified as lowland moist forest on ferralitic soil. The Experimental trial consists of 16 natural rainforest plots. Of 16 plots, 15 have a size of 6.25 ha and one covers an area of 25 ha. Since 1984, all trees with a d.b.h. < 10cm have been measured by CIRAD Forêt. The trees have been mapped and for many trees families, genera and species have been identified. Three sylvicultural treatments of different intensities were applied on different plots from 1986 to 1988. A part of the plots were left as untouched control. Currently, caused by the effective recruitment and strong micro environmental variation, adult densities in logged and unlogged plots do not differ greatly. Detailed description of the experimental trial and the different treatments can be found elsewhere (Forget *et al.* 1999; Schmitt and Bariteau 1988).

Species description

An overview of the characteristics of the six species studied is given in Table 1. All six species are canopy trees. With the exception of *Carapa procera*, which also occurs in West and Central Africa, all species are more or less geographically distributed in the Guyana plateau and the neighbouring territories of Brazil and Venezuela. To get an indicator of small scale spatial distribution of species, the index [R] of Clark and Evans (Ripley 1981) has been calculated for all 16 plots of the study site (Table 1). Values of index [R]<1 indicate a clumped and aggregated distribution, whereas random distribution corresponds to values of 1. Regular distribution is shown by values [R]>1. According to the mean [R] values, *Dicorynia guianensis*, *Vouacapoua americana* and *Eperua grandiflora* had a clumped spatial distribution. The diameter of the aggregates is relatively small for *Dicorynia guianensis* (50-100m) but relatively large for *Vouacapoua americana* and *Eperua grandiflora* (100-600m). The spatial distribution of the other three species is random. A detailed study on the spatial distribution of *Carapa procera* and *Vouacapoua americana* is given by Forget *et al.* (1999).

Table 1. Geographical distribution, mean and standard deviation of aggregation index [R] of Clark and Evans for spatial distribution within 16 plots of the experimental trial Paracou (R < 1 indicates clumped distribution, R = 1 indicates random distribution and R > 1 shows regular distribution of the tree species), mean and standard deviation of individuals per ha [N/ha] in 16 plots of Paracou, agents for pollen and seed dispersal, sexual system, flowering phenology, shade tolerance and water drainage demands of the six studied tree species

Scientific name	***Carapa procera***	***Chrysophyllum sanguinolentum***	***Dicorynia guianensis***	***Eperua grandiflora***	***Virola michelii***	***Vouacapoua americana***
Family	Meliaceae	Sapotaceae	Caesal-piniace	Caesal-piniace	Myristi-cacaceae	Caesal-piniace
Distri-bution	Guyana plateau, Amazonia, West and Central Africa	Guyana plateau, Amapa, South Venezuela	Guyana plateau, Amazonia	French Guiana, Surinam, Amapa	Guyana plateau, Para	French Guiana, Surinam, Para
R	1.02 ± 0.12	1.03 ± 0.19	0.73 ± 0.09	0.83 ± 0.28	1.06 ± 0.40	0.82 ± 0.18
N/ha	6.43 ± 1.56	3.17 ± 2.02	6.16 ± 2.60	6.80 ± 6.20	2.44 ± 1.39	8.38 ± 6.38
Pollinator	insects	insects, bats	insects	insects	insects	insects
Seed dispersal agents	gravity and wind, rodents < 30 m	monkeys > 100 m	gravity and wind, rodents < 50 m	gravity and wind < 30 m	birds, monkeys, kinkajou > 100 m	gravity and wind, rodents < 30 m
Sexual system	Bisexuality	Bisexuality	Bisexuality	Bisexuality	Dioecy	Bisexuality
Flowering Phenology	regular, spread	regular, spread	regular bianual, synchrony	irregular synchrony	regular spread	irregular synchrony
Shade tolerance	heliophilic	tolerant	hemitolerant	hemitolerant	heliophilic	tolerant
Water drainage	intermediate	dry	intermediate dry	intermediate dry	dry	intermediate

Sampling

The sampling strategy was based on former investigations that were made on the spatial distribution of the different species. It varied from rather systematic for aggregated species (so that trees of each patch would be represented) to random for species having a regular diffuse distribution. Trees were sampled in each plot proportional to their density. Because genetic inventories were done within different projects, not the same trees have been sampled for all gene makers. An overview of the sample size and analysed gene markers is given in Table 2.

Table 2. Sample size (N) for each type of gene marker of the six analysed species in the experimental trial Paracou, number of loci (L) and alleles (A) for allozymes, number of different chloroplast haplotypes (H) and number of different fragments (F) for RAPDs and AFLPs

	Carapa procera	*Chrysophyllum sanguinolentum*	*Dicorynia guianensis*	*Eperua grandiflora*	*Virola michelii*	*Vouacapoua americana*
Allozyme						
N		54	135	50	111	
L		4	4	3	2	
A		13	12	6	4	
Cp-haplotypes						
N			101		51	36
H			4		11	2
RAPD						
N	65	68	70	54	80	
F	46	48	61	49	33	
AFLP						
N		41		40		
F		124		236		

Genetic analysis

Allozymes

Allozymes were extracted from cotyledon leaves, fresh adults leaves and cambium tissue using an extraction buffer for tropical trees developed by Liengsiri *et al.* (1990). Procedures for enzyme separation by starch gel electrophoresis and staining were adapted from Zanetto *et al.* (1996). The best separation and detection of alleles was obtained by using large starch gels of 27 cm length with the Histidine-Citric Acid buffer system pH 7.3 described by Cardy *et al.* (1980). An overview of the analysed enzyme systems is given in Table 3.

DNA extraction

Due to the high density of trees in the tropical forest and the difficulties to reach the canopy to collect leaves, DNA was extracted from cambial tissue. Leaves were used only for a small number of individuals of *Dicorynia guianensis*. A thin piece of cambium (diameter: 2 cm) was stamped out and stored in a mixture of absolute ethanol and 2% CTAB in the ratio 70:30 vv/vv in small tubes. In the field the tubes were kept at ambient temperature and in the lab each sample was dried at 50°C overnight (E. Bandou, unpublished). Extraction methods were applied following a modified protocol of Doyle and Doyle (1987). The amount of extracted DNA was sufficient for PCR-RFLP, RAPD and AFLP analysis.

Table 3. Short name, E.C. code, structure and number of loci of the analysed allozyme systems; M = monomeric; D = dimeric; T = tetrameric

Species	Allozymes		Structure	No. of loci
	Name	**E. C. code**		
C. sanguinolentum	ADH	E.C.1.1.1.1	D	1
	MDH	E.C 1.1.1.37	D	2
	PGI	E.C.5.3.1.9	D	2
	PGM	E.C.2.7.5.1	M	1
D. guianensis	IDH	E.C.1.1.1.42	D	2
	PGI	E.C.5.3.1.9	D	2
	PGM	E.C.2.7.5.1	M	2
E. grandiflora	IDH	E.C.1.1.1.42	D	2
	MDH	E.C 1.1.1.37	D	2
	MR	EC 1.6.99.2	T	1
V. michelii	ADH	E.C.1.1.1.1	M	1
	MDH	E.C 1.1.1.37	D	2
V. americana	6PGDH	E.C.1.1.1.44	D	2
	PGI	E.C.5.3.1.9	D	4
	PGM	E.C.2.7.5.1	M	2

Cp-haplotypes

Molecular techniques derived from temperate species were adapted to allow a rapid screening of cp-DNA polymorphism. The method involves three successive steps: (1) amplification of cp-DNA fragments using universal primers, (2) digestion of the fragments with restriction enzymes, (3) electrophoresis to separate restriction fragments. For each species 16 universal primer pairs have been tested in combination with different restriction enzymes (Demesure *et al.* 1995; Dumolin-Lapègue *et al.* 1997). For *Dicorynia guianensis*, *Virola michelii* and *Vouacapoua americana*, a sufficient restriction fragment length polymorphism was detected within the Paracou samples. Using single tree progenies, inheritance of the cp-DNA polymorphism was tested for all species except *Vouacapoua americana*. In accordance with general results in angiosperms, the chloroplast polymorphism are mainly inherited maternally (Birky 1995). In the case of *Vouacapoua americana,* the high proportion of fixed populations indicated maternal inheritance.

RAPDs

PCR reactions (20 μl) contained 5-10 ng of genomic DNA; dATP, dCTP, dGTP, and dTTP (Pharmacia) each at 250 μM final concentration; 2.5 mM of $MgCl_2$, 2 μM of 10-mer primer; and 0.8 unit of *Taq* DNA polymerase (Gibco BRL). Between six and 15 Operon primers were used for each species. A thermal cycler (GeneAmp PCR system 9600 of Perkin Elmer) was programmed for one pre-amplification step of 1 min at 94 °C, followed by 20 cycles, with denaturation steps of 30 s at 94 °C, annealing steps of 50 s, progressively decreased by 1 s per cycle, at 37 °C, and extension steps of 1 min at 72 °C. Then, 30 cycles, with denaturation steps of 30 s at 93 °C, annealing steps of 30 s at 37 °C, and extension steps of 1 min at 72 °C. The final 15 extension steps were progressively extended by 5 s per cycle. The last cycle was followed by 2 min at 72 °C. After amplification, 5 μl of the samples was loaded and electrophoresed on 1.8% agarose gels using a 0.5X TBE running buffer.

AFLPs

AFLP analysis was performed according to Vos *et al.* (1995). Total genomic DNA was digested with the restriction enzymes *Eco*RI and *Mse*I. For each species, AFLP fingerprints were generated using six to eight different primer combinations with a total of six selective nucleotides (two at the *Eco*RI primer and four at the *Mse*I primer). All primers and adaptors were obtained from Genset (Paris, France). Only bands that could be reliably scored in all individuals were considered. The fingerprints were analysed for presence or absence of bands.

Data analysis

Spatial autocorrelation of allozymes and cp-haplotypes

Spatial autocorrelation was carried out on both allozyme data and cp-haplotypes using Moran's I statistic (Sokal and Oden 1978; Sokal and Wartenberg 1983). Moran's index [I] was computed for a given distance class [k]. Class k consists of all pairs of trees whose distance lies in the half-open interval $[(k-1)\cdot x, k\cdot x]$. For each allele or haplotype, respectively, with frequency higher than 5% in all samples, I_k was calculated as follows:

$$I_k = \frac{n\sum_{i=1}^{n}\sum_{j\neq i}^{n} w_{ij}(x_i-\bar{x})(x_j-\bar{x})}{\sum_{i=1}^{n}\sum_{j\neq i}^{n} w_{ij}\sum_{i=1}^{n}(x_i-\bar{x})^2}$$

where n is the total number of samples and $w_{ij} = 1$ if tree i and j both belong to the spatial interval k, otherwise $w_{ij} = 0$. For the allozyme data x_i equals 1 if tree i is homozygote for that allele and 0.5 if tree i is heterozygote for that allele and $x_i = 0$ if the tree has no copy of the allele. In case of cp-haplotypes x_i equals 1 if tree i represents the haplotype and $x_i = 0$ if the tree has a different haplotype. The value $\bar{x}$ corresponds to the mean value of x_i over all n trees (relative frequency of the allele or haplotype). Following Streiff *et al.* (1998) autocorrelation was calculated over all loci by summing the numerator and denominator of the first equation over the total number of genes. I_k is then calculated by dividing the sum of numerators by the sum of denominators. We computed Moran's I for different spatial distance classes. The distance classes were chosen so that each class includes at least 30 pairs of trees (D. Wartenberg cited in Doligez and Joly 1997). The expected values for the case of no autocorrelation are –1/(n-1). The plot of autocorrelation index against spatial distance classes is called a correlogram.

Spatial autocorrelation of RAPDs and AFLPs

The spatial genetic structure for RAPDs and AFLPs was calculated as mean genetic distances between all pairs of trees belonging to a given spatial distance class. In case of positive spatial autocorrelation, the genetic distances increase with spatial distances (Degen and Scholz 1998; Vendramin *et al.* 1999). First the RAPD and AFLP data of each individual were transformed into a matrix of 1 and 0. Value 1 indicated presence of a certain fragment and 0 coded absence of the fragment. The genetic dissimilarity [D_{ij}] between the binary matrix of two individuals i and j was computed using the Tanimoto distance (Deichsel and Trampisch 1985):

$$0 \leq D_{ij} = 1 - \frac{w_{ij}}{w_{ij} + y_i + y_j} \leq 1$$

where w_{ij} represents the number of common fragments of individual i and j, and the number of fragments which only exist in individual i or j is given by y_i and y_j, respectively. For random spatial structure the probability of two genotypes to build a pair is the same for all distance classes. It only depends on the frequencies of genotypes. Hence, the overall mean of genetic distances between all trees serves as a reference for random spatial structure. Distances below this reference indicate positive autocorrelation and values above negative autocorrelation. The plot of mean genetic distance against spatial distance classes is called a distogram.

Test

Permutation procedure using Monte-Carlo simulations was applied to test significant deviation from spatial random distribution of Moran's I and the mean Tanimoto distance D (Manly 1997). Each permutation consisted of a random redistribution of multilocus genotypes (allozymes), cp-haplotypes, RAPDs or AFLPs over the spatial co-ordinates of the sampled trees. For each of the spatial distance classes, observed values were compared with the distribution obtained after 1000 permutations. A 95 % confidence interval for the parameters was constructed as the interval from the 26th to the 975th ordered permutation estimates (Streiff *et al.* 1998).

Results

Carapa procera

The RAPD data of *Carapa procera* revealed a significant positive spatial autocorrelation up to 150 m (Table 4, Figure 1). The mean Tanimoto distances vary between spatial distance classes from 0.55 to 0.63 with an overall mean of 0.62.

Chrysophyllum sanguinolentum

For any distance class a significant Moran's index was found for the allozymes, but it should be noted that the value of the first distance class up to 75 m is very close to the boundary of the 95% confidence interval. At larger scales RAPDs and AFLPs had significant values up to 300m and 800m (Table 4, Figure 1). The differences of Tanimoto distance between spatial distance classes varied from 0.47 to 0.51 for the RAPDs and 0.216 to 0.227 for the AFLPs. This indicated a greater spatial heterogeneity for RAPDs and only minimal differences for AFLPs. At small scales RAPDs and AFLPs showed no significant autocorrelation.

Dicorynia guianensis

For allozymes, cp-haplotypes and RAPDs *Dicorynia guianensis* showed significant spatial autocorrelation (Table 4). For allozymes there is a spatial pattern up to 60 m (Figure 1), for cp-haplotypes on small scale extended to 150m (Figure 1) and also for large scale up to 600m (Table 4). The strongest autocorrelation was measured for the cp-haplotypes on small scale up to 100 m. Besides a significant positive autocorrelation there were also significant negative values between 250 and 350m. Those significant values indicated greater dissimilarity of genotypes than expected which can be simply explained by a symmetry of spatial genetic structure. The autocorrelation for cp-haplotypes on larger scales differed only between +0.12 and –0.08. The mean Tanimoto distances for RAPDs at different spatial distance classes altered from 0.34 to 0.38 (overall mean = 0.36).

Table 4. Spatial genetic autococorrelation of six tropical tree species for allozymes, cp-haplotypes, RAPDs and AFLPs. For allozymes and cp-haplotypes the Moran's Index and for RAPDs and AFLPs the mean Tanimoto distances were calculated as autocorrelation statistics. Only data sets with a sample size greater than 35 were used. The bold values indicate the expansion of significant positive autocorrelation.

	Carapa procera	*Chrysophyllum sanguinolentum*	*Dicorynia guianensis*	*Eperua grandiflora*	*Virola michelii*	*Vouacapoua americana*
Allozyme	______	**ns**	**60m**	**ns**	**ns**	______
Cp-haplotypes	______	______	**150m** **600m**	______	**150m**	**600m**
RAPD	**150m**	**300m**	**300m**	**ns**	**100m**	______
AFLP	______	**800m**	______	**500m**	______	______

Eperua grandiflora

The allozyme and RAPD data of *Eperua grandiflora* exhibited no significant spatial structure neither on small scales nor on larger scales (Table 4). For AFLPs significant spatial structure was detected at larger scales up to 500m (Figure 1). Here the mean genetic distances varied from 0.249 to 0.281 (overall mean

Virola michelii

The allozyme data of *Virola michelii* showed no significant autocorrelation, but cp-haplotypes and RAPDs structure was significant up to 150m and 100m (Table 4, Figure 1). The Moran's I of cp-haplotypes altered between +0.18 and –0.30 indicating a lower spatial heterogeneity than the cp-haplotypes of *Dicorynia guianensis*. Mean Tanimoto distances between RAPDs from individuals of different spatial distance class differed from 0.38 to 0.45 with an overall mean of 0.42.

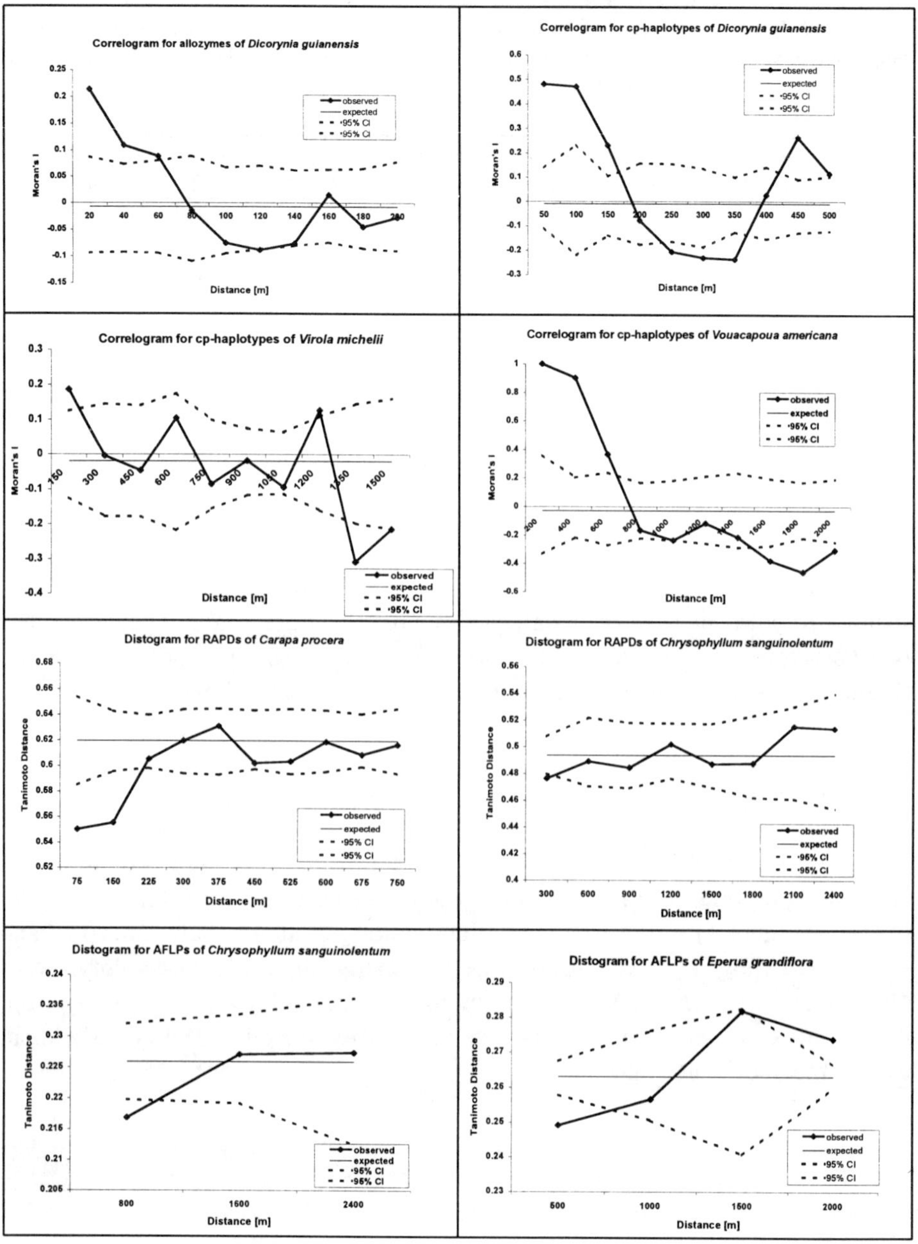

Figure 1. Correlograms and distograms for allozymes, cp-haplotypes, RAPDs and AFLPs of the analysed tree species

Vouacapoua americana

For *Vouacapoua americana* only data of cp-haplotypes could be analysed. The data exhibited significant spatial autocorrelation up to 600m (Table 4). The correlogram crossed the reference line at 800m (Figure 1). This can be used as an indicator for the diameter of spatial structure. The computed Moran's I varied between distance classes in broad range from +1 to –0.45 pointing a strong spatial structure.

Discussion

Relation between spatial structure and pollen and seed dispersal

According to theoretical studies we would expect significant spatial genetic structure in the first distance classes for species with restricted pollen and seed dispersal (Epperson and Li 1997; Sokal *et al.* 1989).
The seeds of *Carapa procera*, *Dicorynia guianensis*, *Eperua grandiflora* and *Vouacapoua americana* are primarily distributed by wind or gravity and secondarily dispersed by rodents. Usually seeds are distributed not more than 50m away from the mother tree (Forget 1992; Forget *et al.* 1999). Except for *Carapa procera*, the trees with d.b.h.>10 cm of those species had an aggregated spatial distribution in the experimental site Paracou (see Index R of Clark and Evans in Table 1). Those aggregates could also be caused by differences in microenvironment which favours different species at different patches (Forget *et al.* 1999).

Allozymes, cp-haplotypes and RAPDs of *Dicorynia guianensis* exhibited significant spatial autocorrelation in the first distance classes. This result could be explained by restricted seed dispersal. A previous study on spatial structure of allele frequencies in different pollen clouds of *Dicorynia guianensis* did not show any spatial structure in Paracou (Caron *et al.* 1998). Hence, restricted pollen flow as an explanation for the spatial structure of adult trees seems to be unlikely. Moreover the spatial genetic structure for cp-haplotypes cannot be produced by restricted pollen flow, since they are inherited maternally.

The results of *Eperua grandiflora* with significant spatial structure for AFLPs but not for allozymes and RAPDs are more difficult to interpret. Because the seeds are extremely heavy (mean 46 g) and mainly dispersed in an area not more than 20m away from the parent (Forget 1992), we would expect spatial genetic structure also for allozymes. Maybe there is a spatial structure but the sample size was not sufficient to detect it. Due to the relatively large aggregates of *Eperua grandiflora* (ca. 300-800m) in Paracou, the tree density differed strongly between different plots (see Table 1). In some plots, which are covered by such an aggregate, the density is relatively high (up to 22 trees per ha). It could be that despite the restricted seed dispersal, seed shadows in areas with high densities of individuals represent a mixture of different single tree progenies. In such cases only weak spatial structure is expected (Hamrick *et al.* 1993).

In Paracou, only two different cp-haplotypes were found for *Vouacapoua americana*. Different clumps of trees are fixed for one of them. This caused a very strong spatial structure which could be detected despite a relatively low sample size of 36 individuals. Since the cp-haplotypes are maternally inherited and therefore only distributed via seeds, there seems to be no seed exchange between the different aggregates (at least 200 m apart). A large scale study on cp-haplotype polymorphism in French Guiana exhibited fixation to one haplotype in many stands (Dutech *et al.* 2000). In this context, the spatial distribution of haplotypes in Paracou could be explained as combination of recent recolonisation from populations with different maternal origins and very low migration between populations.

In a study on spatial genetic structure of *Carapa procera* no spatial structure was found for allozymes in Paracou (Doligez and Joly 1997). In contrast, the present study detected significant spatial structure for RAPDs up to 150m. Differences in sample size cannot explain the opposite observations, since for allozymes more samples were analysed.

The seeds of *Chrysophyllum sanguinolentum* are distributed by monkeys, and the seeds of *Virola michelii* are dispersed by birds and monkeys. In comparison to the other studied species, they have a more distant seed dispersal (>100m). In both cases allozyme data did not show any significant spatial structure. However, *Virola michelii* exhibited spatial autocorrelation for cp-haplotypes in a first 150m distance class. RAPDs showed spatial structure up to 300m and AFLPs up to 800m for *Chrysophyllum sanguinolentum* and RAPDs revealed spatial structure up to 100m for *Virola michelii.* Ecological investigations on frugivory and seed dispersal of red howler monkeys in French Guiana showed that the monkeys regularly use a limited number of sleeping sites and exhibit a very long digestive time. Hence, concentrations of defecation including the seeds occur under sleeping trees (Julliot 1994). As a consequence of this behaviour, long distance seed dispersal by monkeys does not necessary imply a random distribution of seeds. They may produce aggregates of saplings which consist of single tree progenies.

Using allozymes, Hamrick *et al.* (1993) studied the influence of seed dispersal mechanisms on the genetic structure of four tropical tree species at Barro Colorado Island in Panama. They found for *Platypodium elegans*, a low-density, wind-dispersed species, the strongest spatial structure while *Swartzia simplex*, a high density and bird dispersed species, had only weak spatial structure. They also compared the spatial genetic structure at different life stages. For two species (*Platypodium elegans, Alseis blackiana*) the spatial structure disappeared in the older life stages. Loiselle *et al.* (1995) studied the spatial genetic structure of *Psychotria officinalis*, a tropical understory shrub in Costa Rica. This species is pollinated by insects, and seeds are primarily dispersed by birds, some species of which drop the seeds immediately while others transport them away from the parent plant. By use of allozymes they only found positive genetic correlations between individuals on a scale of 5m. For some tropical tree species, the effect of restricted seed dispersal is overshadowed by high mortality of seedlings in the neighbourhood of the parent tree. This phenomena is called "Janzen's spacing mechanism" according to the classical paper of Janzen (1970). It also have been studied for the bird-dispersed *Virola surinamensis* in Panama (Howe 1990). In this investigation, 54% of the seeds were scattered singly in the forest, while 46 % were dropped under the parent trees. Due to high densities of seed predators and pathogens, seed and seedling mortality under the parent trees exceeded 99.99%. Only seeds dispersed away from the parent tree had appreciable chances to survive. This could be one explanation why tree species with restricted seed dispersal do not exhibit spatial genetic autocorrelation in the first distance classes.

Spatial structure for different kinds of gene marker

Our study is one of the first applying different kinds of gene markers to study spatial genetic structure on the stand level. Another study has been done by Streiff *et al.* (1998). They used allozymes and microsatellites to study spatial genetic structure of *Quercus robur* and *Quercus petraea* in a French stand. Both co-dominant markers revealed similar spatial patterns. In our study, only allozymes are co-dominantly inherited, so a direct comparison is difficult. In the present investigation, allozymes showed significant values for two of four tested species, cp-haplotypes for all three tested species, RAPDs indicated significant spatial structure in four out of 5 tested species and the AFLPs had significant spatial structure for both tested species (Table 4). Simulation studies with a new version of the simulation model ECO-GENE (Degen *et al.* 1996) using data sets of Paracou indicated that samples for allozymes, in contrast to RAPDs and AFLPs, often failed to detect spatial structure (B. Degen in preparation). This is caused by the significantly lower variation of allozymes. Because they are inherited biparentally, they reflect the combined effect of pollen and seed dispersal. If seed dispersal is limited and pollen distribution is less restricted, then allozyme data probably show only weak spatial structure in the first distance classes. In contrast, maternally inherited markers, like chloroplast genes for many angiosperms, are useful markers to measure the impact of seed flow on spatial structures. However, chloroplast genomes only seldom exhibit within-population variation. Complete fixation can be quite rapidly obtained due to genetic drift and restricted seed flow among populations (Petit *et al.* 1997; Ferris *et al.* 1998). RAPDs are criticised because of reproducibility problems (Ford-Lloyd and Painting 1996). RAPDs are sensitive to small alterations in PCR conditions which can produce an unsystematical bias. Thus it may be that spatial structure remains undetected due to these errors, but there is no way that alternations in PCR could cause spatial structure as an artefact. In our study, RAPDs were found to be useful markers to detect spatial structure. For this purpose AFLPs may be the best kind of marker. They are highly reproducible, have a high multiplex ratio and represent a good sample of the whole genome (Breyne *et al.* 1999, Ford-Lloyd and Painting 1996). Despite the very low sample size, we could find spatial genetic structure for two species in Paracou.

Acknowledgement

This study was financially supported by two projects of the European Union (program INCO: TS3*CT94-0316, STD3 project) and two French grants: one from BRG (Bureau des Ressources Génétiques) and a grant from the French Ministry of Environment (SOFT-project 1995). We are thankful to Cirad-Forêt for giving us access to the experimental site Paracou and for support with the data. We wish to thank Catherine Julliot and Damien Bonal for helpful information on species ecology and Nathalie Colpaert for technical assistance in the AFLP analysis.

References

Birky, C. W. J. 1995: Uniparental inheritance of mitochondrial and chloroplast genes: mechanisms and evolution. Proceedings of National Academy of Sciences of the USA 92: 11331-11338

Boshier, D. H., Chase, M. R. and Bawa, K. S. 1995: Population genetics of *Cordia alliodora* (Boraginaceae), a neotropical tree. 3. Gene flow, neighborhood and population substructure. American Journal of Botany 82: 484-490

Breyne, P., Rombaut, D., Gysel, A. V. *et al.* 1999: AFLP analysis of genetic diversity within and between *Arabidopsis thaliana* ecotypes. Mol. Gen. Genet. 261: 627-634

Cardy, B. J., Stuber, C.and Goodman,M. M. 1980: Techniques for starch gel electrophoresis of enzymes from maize (Zea mays L.), pp. 31. North Carolina State University Raleigh, Raleigh

Caron, H., Dutech, C. and Bandou, E. 1998: Variations spatiotemporelles du régime de reproduction de *Dicorynia guianensis* Amshoff (Caesalpiniaceae) en forêt guyanaise. Genet. Sel. Evol. 30: 153-166

Degen, B., Gregorius, H-R. and Scholz, F. 1996: ECO-GENE, a model for simulation studies on the spatial and temporal dynamics of genetic structures of tree populations. Silvae Genetica 45: 323-329

Degen,B. and Scholz, F. 1998: Spatial genetic differentiation among populations of European beech (*Fagus sylvatica* L.) in Western Germany as identified by geostatistical analysis. Forest Genetics 5: 191-199

Deichsel, G. and Trampisch, H. J. 1985: Clusteranalyse und Diskriminanzanalyse. Gustav Fischer Verlag, Stuttgart

Demesure, B., Sodzi, N. and Petit, R. J. 1995: A set of universal primers for amplification of polymorphic non-coding regions of mitochondrial and chloroplast DNA in plants. Molecular Ecology 4: 129- 131

Doligez, A., Baril, C. and Joly, H. I. 1998: Fine-scale spatial genetic structure with nonuniform distribution of individuals. Genetics 148: 905-920

Doligez, A. and Joly, H. I. 1997: Genetic diversity and spatial structure within a natural stand of a tropical forest tree species, *Carapa procera* (Meliaceae), in French Guiana. Heredity 79: 72-82

Doyle, J. J. and Doyle, J. L. 1987: A rapid DNA isolation procedure for small quantities of fresh leaf tissue. Phythochemical Bulletin 19: 11-15

Dumolin-Lapègue, S., Pemonge, M-H. and Petit R. J. 1997: An enlarged set of consensus primers for the study of organelle DNA in plants. Molecular Ecology 6: 393-397

Dutech, C., Maggia, L. and Joly, H. I. 2000: Chloroplast diversity in *Vouacapoua americana* (Caesalpinniaceae), a neotropical forest tree. Molecular Ecology 9(9): 1427-1432

Epperson, B. K. 1990: Spatial autocorrelation of genotypes under directional selection. Genetics 124: 757-771

Epperson, B. K. 1992: Spatial structure of genetic variation within populations of forest trees. New Forests 6: 257-278

Epperson, B. K. and Li ,T. Q. 1997: Gene dispersal and spatial genetic structure. Evolution 51: 672-681

Ferris, C., King, R. A., Väinölä, R. and Hewitt, G. M. 1998: Chloroplast DNA recognizes three refugial sources of European oaks and suggests independent eastern and western immigrations to Finland. Heredity 80: 584-593

Ford-Lloyd, B. and Painting, K. 1996: Measuring Genetic Variation Using Molecular Markers. IPGRI, Rome

Forget, P. M. 1992: Regeneration Ecology of *Eperua grandiflora* (Caesalpiniaceae), a large-seeded tree in French Guiana. Biotropica 24: 146-156

Forget, P. M., Mercier, F. and Collinet, F. 1999: Spatial patterns of two rodent-dispersed rainforest trees, *Carapa procera* (Meliaceae) and *Vouacapoua americana* (Caesalpiniaceae) at Paracou, French Guiana. Journal of Tropical Ecology 15: 301-313

Geburek, T. 1993: Are genes randomly distributed over space in mature populations of sugar maple (Acer saccharum Marsh.)? Annals of Botany 71: 217-222

Hamrick, J. L., Murawski, D. A. and Nason, J. D. 1993: The influence of seed dispersal mechanisms on the genetic structure of tropical tree populations. Vegetatio 107: 281-297

Howe, H. F. 1990: Seed Dispersal By Birds and Mammals: Implications For Seedling Demography. Unesco, Paris and The Parthenon Publ. Group.

Janzen, D. H. 1970: Herbivores and the number of tree species in tropical forests. Am Nat 104: 501-528

Julliot, C. 1994: Frugivory and seed dispersal by red howler monkeys: Evolutionary aspect. Revue D Ecologie - La Terre et La Vie 49: 331-341

Leonardi, S., Raddi, S. and Borghetti, M. 1996: Spatial autocorrelation of allozyme traits in a Norway spruce (*Picea abies*) population. Can. J. For. Res. 26: 63-71

Liengsiri, C., Piewluang, C. and Boyle, T. J. B. 1990: Starch Gel Electrophoresis of Tropical Trees. ASEAN-Canada Forest Tree Seed Centre, Saraburi

Loiselle, R., Sork,V. and Nason, J. 1995: Spatial genetic structure of a tropical understory shrub, *Psychotria officinalis* (Rubiaceae). Am. J. Bot. 82: 1420-1425

Manly, B. F. J. 1997: Randomization, Bootstrap and Monte Carlo Methods in Biology. Chapman & Hall, London.

Petit, R. J., Pineau, E.and Démesure, B. *et al.* 1997: Chloroplast DNA footprints of postglacial recolonization by oaks. Proceedings of the National Academy of Sciences of the USA 94: 9996-10001

Ripley, B. D. 1981: Spatial Statistics. Wiley Series in Probability and Mathematical Statistics, New York.

Schmitt, L. and Bariteau, M. 1988: Gestion de l'ecosysteme forestier guyanais: Etude de la croissance et de la regeneration naturelle: Dispositif de Paracou. Bois et Foret des Tropiques 220: 3-24

Schnabel, A., Laushman, R. H. and Hamrick, J. L. 1991: Comparative genetic structure of two co-occurring tree species, *Maclura pomifera* (Moraceae) and *Gleditsia triacanthos* (Leguminosae). Heredity 67: 357-364

Sokal, R. R., Jacquez, G. M. and Wooten, M. C. 1989: Spatial autocorrelation analysis of migration and selection. Genetics 121: 845-855

Sokal, R. R. and Oden, N. L. 1978: Spatial autocorrelation in biology. 1. Methodology. Biol. J. Linnean Soc. 10: 199-228

Sokal, R. R. and Wartenberg. D. E. 1983: A test of spatial autocorrelation analysis using an isolation-by-distance model. Genetics 105: 219-237

Streiff, R., Labbe, T. and Bacilieri, R. *et al.* 1998: Within-population genetic structure in *Quercus robur* L. and *Quercus petraea* (Matt.) Liebl. assessed with isozymes and microsatellites. Molecular Ecology 7: 317-328

Vendramin, G. G., Degen, B. and Petit, R. J. *et al.* 1999: High level of variation at *Abies alba* chloroplast micrsatellite loci in Europe. Molecular Ecology 8: 1117-1126

Vos, P., Hogers, R.and Bleeker, M. *et al.* 1995: AFLP: a new technique for DNA fingerprinting. Nucleic Acids Res 23: 4407-4414

Young, A. G.and Merriam, H. G. 1994: Effects of forest fragmentation on the spatial genetic structure of *Acer saccharum* Marsh. (sugar maple) populations. Heredity 72: 201-208

Zanetto, A., Kremer, A., Müller-Starck, G. and Hattemer, H. H. 1996: Inheritance of isozymes in pedunculate oak (*Quercus robur* L.). Journal of Heredity 87: 364-370

Genetic variation in natural populations of *Araucaria angustifolia* (Bert.) O. Kuntze in Brazil

V. A. SOUSA[1,2] AND H. H. HATTEMER[2]

[1]*Centro Nacional de Pesquisa de Florestas (CNP). Empresa Brasileira de Pesquisa Agropecuária (EMBRAPA); CEP 83.411-000 – Colombo (PR); Brazil,* [2]*Institut für Forstgenetik und Forstpflanzenzüchtung, Georg-August- Universität Göttingen, D–37077 Göttingen, Germany*

Key words: *Araucaria angustifolia*, isoenzymes, genetic structure and variation

Abstract: The natural distribution of "pinheiro do Paraná" (*Araucaria angustifolia)* is restricted to the southern hemisphere, occurring mainly in Brazil, small areas in Argentina as well as Paraguay. The continuous distribution in Brazil is situated almost exclusively in the three southernmost Brazilian states: Paraná (PR), Santa Catarina (SC) and Rio Grande do Sul (RS). There are small populations in the northernmost part of the range where the forests can be found between 1200 and 1800 m above sea level. It is a species of economic importance for the southern region of Brazil, and for many years it has played an important role in generating export revenues. Due to exploitation, the primary forests were drastically reduced to less than 0.66 % of the original area. If the present rate of exploitation is maintained, the last remnants of the valuable primary forests containing *A. angustifolia* will soon disappear. Therefore, studies are necessary to elucidate the genetic basis of the remaining populations aiming at their genetic conservation. Seed from 130 trees from populations representing 3 geographic locations (Campos do Jordão – SP, Irati – PR and Caçador – SC) were analyzed using 8 enzyme gene loci (GOT–A, GOT–B, GOT–C, PGM–A, SKDH–A, SKDH–B, MDH–B and 6-PGDH–B) to determine patterns of genetic structure and variation. Populations were polymorphic at 49 % of the loci. They had an average of 1.58 alleles per locus and 2.10 alleles per polymorphic locus. Mean expected heterozygosity (H_e) was 0.142, and mean observed heterozygosity (H_a) was 0.141. The northern population was much more variable than the southern populations as measured by several parameters. The northern population was also strongly differentiated from the southern populations, which differed little from each other. A provisional evaluation of part of the data revealed a relationship between geographic distance among populations and their genetic distance.

Introduction

The genus *Araucaria* only occurs in the southern hemisphere, being represented by 16 species distributed over Australia, New Guinea and South America. In spite of its restricted occurrence, fossils are recorded in the northern hemisphere floras from the Mesozoic age (upper Triassic, Jurassic and Cretaceous) (Kendall, 1949; Hill,1995; 1949; Stockey, 1982). Thus, the family was more diverse and widespread during the Mesozoic than it is at present. According to

Müller-Starck, G. and Schubert, R. (eds.). Genetic Response of Forest Systems to Changing Environmental Conditions. Vol. 70 (For. Sci.), 93-103

Kershaw and McGlone (1995), in southern and central Brazil, *Araucaria angustifolia* revealed a substantial late Quaternary record. Pollen evidence indicates a substantial expansion of *Araucaria* forests into the lower latitudes, now occupied by savanna, during the late Pleistocene. This expansion suggests a temperature decrease of at least 7 °C as well as higher levels of humidity (de Oliveira and Colinvaux 1992 cited by Kershaw and Mcglone 1995). Also, Ledru (1993) recorded an expansion of *Araucaria* closer to its present range only within the very late Pleistocene and Early Holocene. Pollen assemblages from around 30,000 BP indicate warm and relatively wet conditions and a more open vegetation during the relatively wet glacial maximum. The subsequent *Araucaria* phase, dated to between 13,000 and 8,000 BP, is considered to indicate cool and wet conditions before increased temperatures resulted in the replacement of *Araucaria* forests by semideciduous forest (Kershaw and McGlone, 1995).

Araucaria angustifolia occurs mainly in Brazil, with additional small areas in the northeastern part of Argentina ("Provincia de Misiones" and the eastern part of Paraguay) (Hueck, 1952, Carvalho, 1994). In Brazil the species occurs in different regions from 19° 15′ S (Conselheiro Pena, MG) to 31° 30′ S (Canguçu-RS) and from 41° 30′ W to 54° 30′ W. Most of the natural range of *Araucaria* is located exclusively in the three southern–most Brazilian states: Paraná, Santa Catarina, and Rio Grande do Sul (Fähser, 1981). Hueck (1966) delimits its range as follows: the southern limit is formed by the Rio Grande do Sul and the eastern limit by the watershed on the Serra do Mar (São Paulo , Minas Gerais, 20° S, Rio de Janeiro, 22° S and Espírito Santo, 20° S, State). Its northern limit is in the valley of the Rio Doce (18° S) and its western limit is in the Argentinean province of Misiones 54° 15' W. Due to its larger distribution in Paraná State, *Araucaria angustifolia* is called "Paraná pine". In Brazil, the original area was almost 200,000 to 250,000 km^2 (Fähser, 1981) with irregular form, especially in Paraná State (40 % of the surface), Santa Catarina (31 %) and Rio Grande do Sul (25 %) and some small dispersed areas in the south of São Paulo State, Minas Gerais and Rio de Janeiro. The natural distribution range of *Araucaria angustifolia* was drastically reduced. "The Inventory of the Pine of the South of Brasil" (IBDF / FUPEF, 1978) estimated that an area of 565,418 ha *Araucaria* forests remained in the States of Paraná, Santa Catarina and Rio Grande do Sul. This value corresponds to about 3 % of the original area. Only 120,776 ha, or 0.66 %, of this total was primary forests. Studies are necessary to elucidate genetic structures of the remaining populations for purposes of at their genetic conservation. However, conservation action has not yet been taken nor has the rational basis for the conservation been available. Analysis of variation for several genetic markers such as RFLP, PCR or isoezymes can be carried out in order to compare the amounts of genetic variation within populations of most forest tree species. According to Bergmann and Hattemer (1998), isoenzyme gene markers are still the best indicators of genetic diversity within and differentiation between tree populations.

The aim of this study was to determine patterns of genetic structure and variation of *Araucaria angustifolia* in natural stands in Brazil. The results were to provide basic information for future activities in tree breeding and conservation.

Material and methods

Material

Seeds of 130 trees were collected from 13 stands representing 3 populations of *A. angustifolia* and stored at –30 °C until analysis. The three populations are located in protected areas, two national parks (Irati and Caçador) and one state park (Campos do Jordão). Details about the geographic location of the different populations and the numbers of trees analyzed for each population are shown in Tables 1 and 2, respectively.

Natural forest (preserved forest) was sampled for each population as well as isolated stands of different dimension (islands). In addition, seeds were also taken from a 37 year old plantation of the Irati population.

Table 1. Geographical position of three populations studied in Brazil

Populations	Altitude (m)	Latitude (S)	Longitude(W)	Area (ha)	State
Campos do Jordão	1,800	23° 45'	45° 30'	8,341.00	S. Paulo (SP)
Irati[1]	790	25° 10' 25° 35'	50° 25' 50° 55'	3,495.00	Paraná (PR)
Caçador[2]	1,100	26° 30' 27° 30'	51° 00' 51° 30'	710.44	S. Catarina (SC)

[1] 590 km away from Campos do Jordão
[2] 700 km away from Campos do Jordão and 150 km away from Irati

Table 2. Number of trees analyzed for each population

Population	Stands						Total
	Natural Forest	Forest Island 1	Forest Island 2	Forest Island 3	Forest Island 4	Forest Island 5	
Campos do Jordão	10	10*	-	-	-	-	20
Irati	10	10	10	10*	10**	-	50
Caçador	10	10	10	10	10	10*	60

* = Exploited Forest
** = Plantation

Electrophoresis

Embryo and megagametophyte were homogenized in 0.08–0.1 M Tris-HCl buffer with pH 7.5. Other compounds added were: 5 g saccharose, 150 mg DTT and 3 g PVP in 100 ml of buffer. Extracts from the megagametophyte and the embryo were positioned adjacent to each other in the gels. Horizontal starch gel electrophoresis (10.5 % starch concentration plus 2.5–3.5 % sucrose) was performed as described by Feret and Bergmann (1976), Conkle *et al.* (1982) and Liengsiri *et al.* (1990). The buffer system of Ashton pH 8.6 for GOT and PGM as well as Tris-Citro pH 7.3 were used as electrode and gel buffers for SKDH, MDH and 6-PGDH. The gene loci analyzed were: GOT–A, GOT–B, GOT–C, PGM–A, SKDH–A, SKDH–B, MDH–B and 6-PGDH–B. Details of the analysis of the inheritance mode of the zymograms, which was performed by testing the segregation of the different variants in the haploid megagametophytes of seeds from single trees, will be presented elsewhere.

Measures of variability

Data of seed trees, embryos and effective pollen were utilized in the statistical analysis. Since the species is dioecious, only the female trees could be genotyped. This was done by analyzing the macrogametophytes of 8 seeds per tree. These data were used for both inheritance analysis and for genotyping the seed trees. Genetic variation of seed trees and embryos/progenies was analyzed using the following measures: number of alleles per locus and (A/L); percentage of polymorphic loci (PPL); genetic diversity (ν) (Gregorius, 1987); gametic diversity (ν_{gam}) (Gregorius, 1978) and observed heterozygosity (H_o) or expected heterozygosity ($H_e = \delta_T$) (Gregorius, 1987). On the other hand, allelic frequencies and diversities (ν) were calculated for seed trees, embryos and effective pollen. The degree of differentiation among populations was analyzed with the parameters D_j and δ (Gregorius and Roberds, 1986) and the fixation index F_{ST} (Wright, 1965).

The computer programs of GSED version 1.1. (Gillet, 1998) and BIOSYS-2 (Swofford and Selander, 1997) were used to calculate the above measures.

Results and discussion

Genetic variation

Measures of genetic variation of the 13 investigated stands, based on seed and progenies, are presented in Table 3.

Table 3. Genetic variability measures of seed trees in the investigated populations (Campos do Jordão (C. Jord.), Irati and Caçador (Caç.)) of *Araucaria angustifolia*

Population	No. trees	Genetic Multiplicity		Allelic Diversity	Gametic Diversity	Heterozygosity	
	N	A/L	P	v	v_{gam}	H_a=H_o	H_e=δ_T
C. Jord. NF	35	1.875	0.750	1.370	15.042	0.279	0.274
C. Jord. NF-E	35	2.000	0.750	1.348	13.975	0.240	0.262
Irati NF	10	1.500	0.375	1.105	2.617	0.075	0.100
Irati NF-I1	10	1.250	0.250	1.095	2.415	0.093	0.089
Irati NF-I2	10	1.000	0.250	1.074	2.027	0.075	0.072
Irati NF-E	10	1.500	0.500	1.118	2.937	0.113	0.111
Irati Pl.	10	1.500	0.375	1.159	4.199	0.095	0.145
Caç. NF	10	1.750	0.625	1.159	3.929	0.175	0.145
Caç. NF-I.1	10	1.500	0.250	1.097	2.451	0.088	0.093
Caç. NF-I.2	10	1.875	0.750	1.228	5.872	0.200	0.195
Caç. NF-I.3	10	1.500	0.500	1.103	2.382	0.113	0.098
Caç. NF-I.4	10	1.625	0.500	1.118	2.700	0.125	0.111
Caç. NF-E	10	1.625	0.500	1.159	4.088	0.163	0.144
Total	180	1.580	0.490			0.141	0.142

Note : N = number of trees; A/L = number of alleles per locus; P = percentage of polymorphic loci (from 8 loci); NF = natural forest; NF-E = natural forest-exploited; NF-I1,2,3,4,5 = natural forest-island 1,2,3,4,5 and Pl. = plantation

On the average, the populations were polymorphic at 49% of the loci, they had 1.58 alleles per locus and 2.1 alleles per polymorphic locus. The mean expected heterozygosity (H_e) was 0.142 and the mean observed heterozygosity (H_a) was 0.141. Observed heterozygosity is not substantially different from expected heterozygosity. In general, the northern population (Campos do Jordão) was much more variable than the southern populations.

Genetic differentiation among stands and populations

The allelic differentiation δ for each locus and for the gene pool is presented in Table 4. Although most of the estimates of allelic structures of the seed trees are based on small samples, most of them prove to be significantly different. The exceptions occur at the gene locus GOT–C that possesses only little variation. The average differentiation measures of the seed trees are generally moderate but vary greatly among gene loci. The same is true for the embryos. With the exception of the gene locus SKDH–B, they are almost consistently less differentiated than the seed trees. This is perhaps due to the eightfold larger sample sizes or may simply reflect genetic change implied by the process of reproduction. The pollen contributions to the embryos of the 13 stands are just as clearly differentiated as the embryos themselves. This result is rather striking.

Table 4. Average allelic differentiation δ among the 13 population samples of *Araucaria angustifolia*, (Gregorius and Roberds, 1986) and fixation index F_{ST} (Wright, 1965).

Gene Loci	Seed trees		Embryos		Effective pollen	
	δ	F_{ST}	δ	F_{ST}	δ	F_{ST}
GOT-A	0.131	0.110	0.092	0.034	0.106	0.030
GOT-B	0.060	0.101	0.057	0.107	0.059	0.111
GOT-C	0.048	0.050	0.031	0.027	0.020	0.017
PGM-A	0.055	0.058	0.097	0.063	0.046	0.071
MDH-B	0.121	0.323	0.112	0.289	0.106	0.264
SKDH-A	0.051	0.130	0.038	0.102	0.036	0.098
SKDH-B	0.035	0.089	0.107	0.060	0.083	0.063
6PGDH-B	0.182	0.257	0.119	0.101	0.115	0.062
Gene pool	0.085	0.181	0.069	0.101	0.071	0.082

In the graph at the bottom of Figure 1, the seed trees and the progenies of the three populations were considered after pooling the material of the individual stands. The diagram shows clearly that the differentiation of the two southern populations is similar. There is not much difference between the seed trees and their progeny, as the two generations display similar differentiation (cf. the two graphs at the top of Figure 1). Nevertheless, the female trees were consistently slightly more differentiated than their progeny. The two snail diagrams at the top of Figure 1 depict the gene pool differentiation of the 13 stands. The two samples from the population of Campos do Jordão not only harbor more genetic variation. They are also more strongly differentiated from the others. Since exploitation in this case had no effect on differentiation, the two stands show extreme differentiation. However, the stands sampled in the other two populations show a very homogeneous pattern of differentiation.

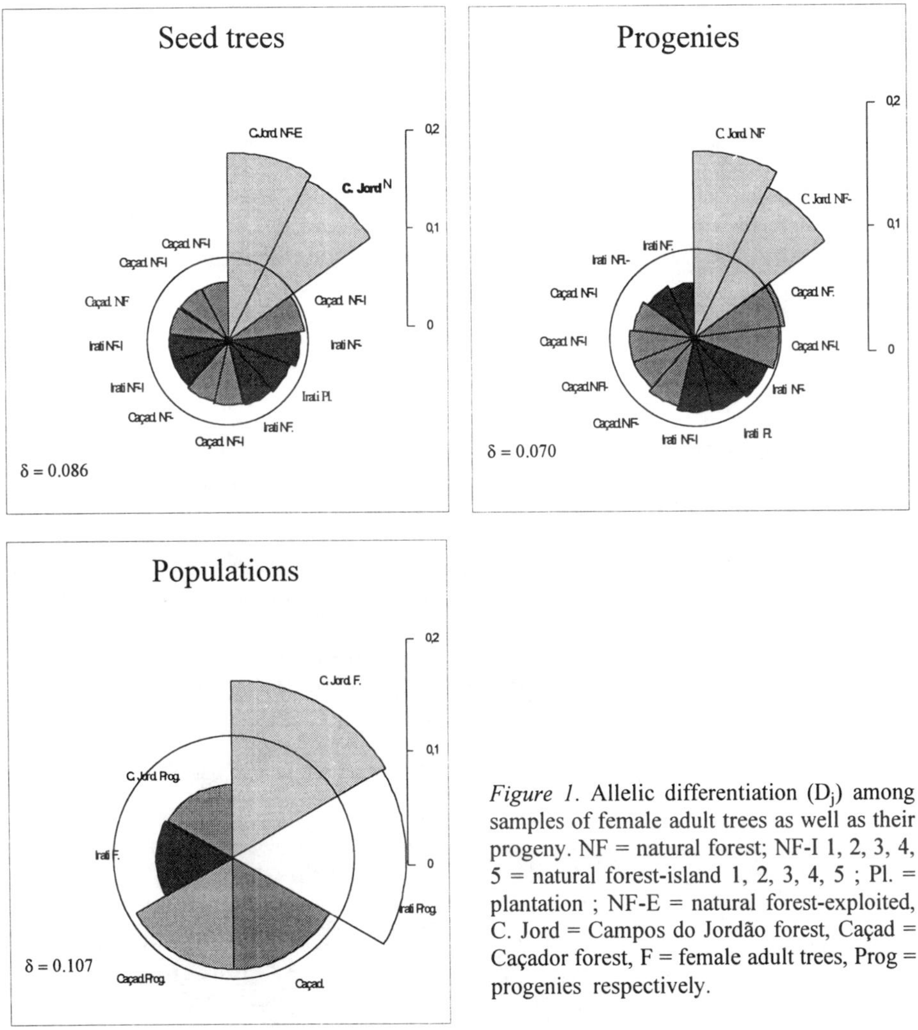

Figure 1. Allelic differentiation (D_j) among samples of female adult trees as well as their progeny. NF = natural forest; NF-I 1, 2, 3, 4, 5 = natural forest-island 1, 2, 3, 4, 5 ; Pl. = plantation ; NF-E = natural forest-exploited, C. Jord = Campos do Jordão forest, Caçad = Caçador forest, F = female adult trees, Prog = progenies respectively.

Genetic distances between populations

The most commonly applied indices of genetic distance where compared and dendrograms are presented in Figures 2 and 3.

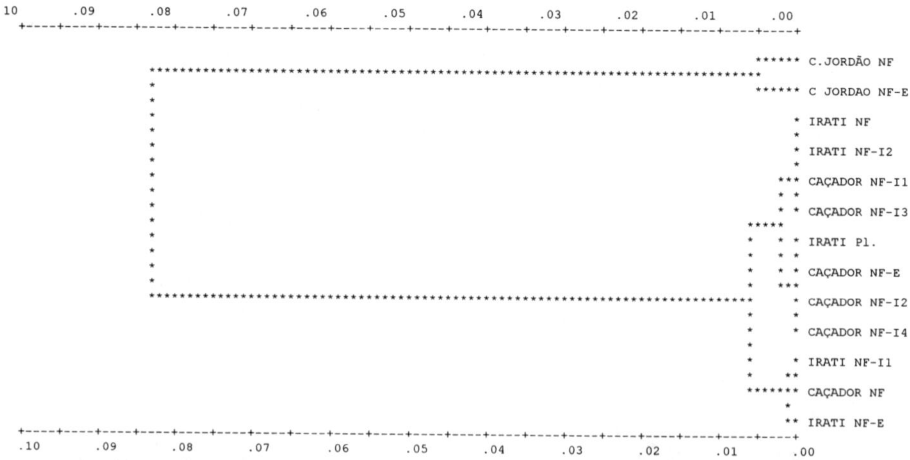

Figure 2. Dendrogram based on genetic distances, according to Nei (1972), among stands of *Araucaria angustifolia* (NF = natural forest, NF-E = natural forest-exploited, NF-I1, 2, 3, 4, 5 = natural forest- island 1, 2, 3, 4, 5 , respectively and Pl. = plantation).

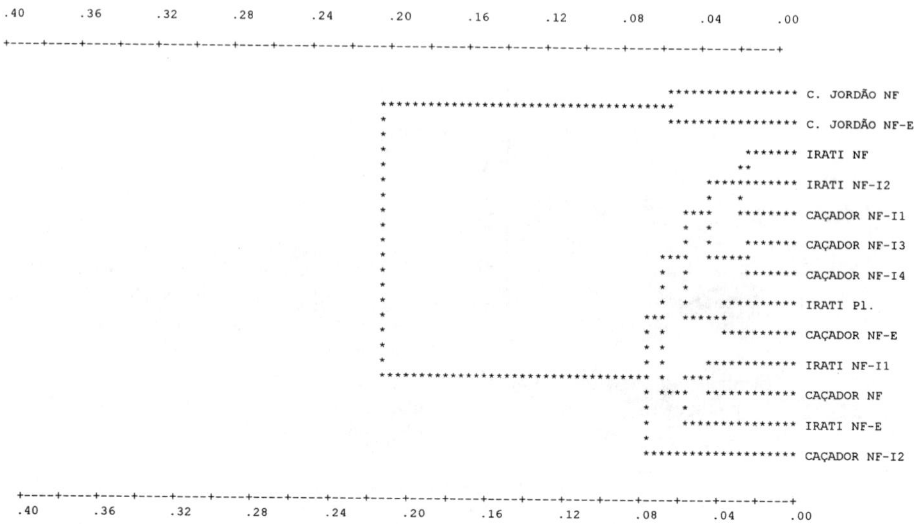

Figure 3. Dendrogram based on genetic distances d_o, according to Gregorius (1974) among stands of *Araucaria angustifolia*. The abbreviations are the same as those in Figure 2.

The northern population was strongly differentiated from the southern populations, if both genetic distances of Nei (1972) and Gregorius (1974) were used. The two southern populations did not differ much from each other (Irati and Caçador). Besides these agreements, the two dendrograms differ very much due to the employment of different measures of genetic distance. The reasons will be discussed elsewhere. Evaluation of the data revealed a relationship between geographic distance among populations and their genetic distance d_o. In Figure 4 this relationship is clearly to be seen. For some reason, a higher genetic variation was maintained in the scattered northern population during retraction and expansion of vegetation that occurred in the Pleistocene and Holocene. The southern populations found good conditions for southward expansion due to favorable climate and soils. According to geomorphological studies by Ledru (1991), Andrade–Lima (1982) and Ab'Saber (1982), at least two dry periods occurred in the past. They have disturbed the dynamics of vegetation and been able to provoke changes in the forest composition, thus explaining its endemism. One of these severe dry periods took place during the Pleistocene and another less severe one during the Holocene.

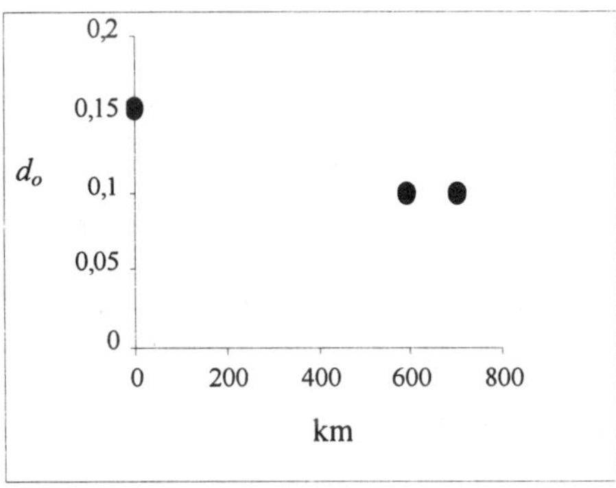

Figure 4. Geographic distance (km) and genetic distance (d_0) for the three populations

Even the exploited stand had higher values when compared to the natural southern stands. The southern stands did not differed much from each other regardless of forest types, i.e. natural, exploited or plantation. This result is similar to that for the progeny. The differentiation analysis yields slightly different results when the embryo genotypes of seeds are used. This is an indication of genetic change involved in reproduction.

References

Ab' Saber, A. N. 1982: The palaeoclimate and palaeocology of Brazilian Amazonia. pp. 41–59 in: Prance, G. T. (Ed.) Biological Diversification in the Tropics. Columbia Univ. Press, New York

Bigarella, J. J. and Andrade-Lima, D. 1982: Palaeoenvironmental changes in Brazil. pp. 27–40 in: Prance, G. T. (Ed.) Biological Diversification in the Tropics. Columbia Univ. Press, New York

Bergmann, F. and Hattemer, H. H. 1988: Isozymes in forest genetics research. pp. 227–238 in: Mandal, A. K. and Gibson, G. L (Eds.) Forest Genetics and Tree Breeding. CBS Publishers & Distributors, New Dehli

Carvalho, P. E. R. 1994: Espécies florestais brasileiras: recomendações silviculturais, potencialidades e uso da madeira. Empresa Brasileira de Pesquisa Agropecuária (EMBRAPA)/ Centro Nacional de Pesquisa de Florestas (CNPF). Brasília. 640 p.

Conkle, M. T. 1979: Isozyme variation and linkage of allozymes in six conifer species. pp 11–17 in: Conkle, M. T. (Eds.) Proceedings of the Symposium on Isozymes of North American Forest Trees and Forest Insects. Gen. Tech. Rep. PSW-48). USDA Forest Service. Berkeley, California

De Oliveira, P. E. and Colinvaux, P. A. 1992: Vegetational and climatic changes of the late Quaternary in southeastern Brazil. *8th International Palynological Congress Programme and Abstracts*, Aixen-Provence, p. 34

Fähser, L. 1981: Die Bewirtschaftung der letzten Brasil-Kiefer-Naturwälder, eine entwicklungspolitische Aufgabe. Forstarchiv 52: 22–26

Feret, P. P. and Bergmann, F. 1976: Gel electrophoresis of proteins and enzymes. pp 49–77 in: Miksche, J. P. (Ed.) Modern Methods in Forest Genetics. Springer, Berlin, Heidelberg

Gillet, E. M. 1998: GSED: Genetic Structure from Electrophoresis Data, Version 1.1. User's Manual. Institut für Forstgenetik und Forstpflanzenzüchtung, Universität Göttingen. Germany. http:// www.uni-forst.gwdg.de/forst/fg/index.htm

Gregorius, H. R 1974: Genetischer Abstand zwischen Populationen. I. Zur Konzeption der genetischen Abstandmessung. Silvae Genetica. 23: 22–27

Gregorius, H. R. 1978: The concept of genetic diversity and its formal relationship to heterozygosity and genetic distance. Math. Biosciences. 41: 253–271

Gregorius, H. R. 1987: The relationship between the concepts of genetic diversity and differentiation. Theor. Appl. Genet. 76: 947–951

Gregorius, H. R. and Roberds, J. H. 1986: Measurement of genetical differentiation among subpopulations. Theor. Appl. Genet. 71: 826–834

Hill, R. S. 1995: Conifer origin, evolution and diversification in the Southern hemisphere.pp: 10–29 in: Enright , N. J. and Hill, R. S. (Eds.) Ecology of the Southern Conifers. Melbourne University Press, Carlton

Hueck, K. 1952: Verbreitung und Standortansprüche der brasilianischen Araukarie (*Araucaria angustifolia*). Forststwiss. Centralblatt 70: 272–289

Hueck, K. 1966: Die Wälder Südamerikas. Stuttgart

Hueck, K. 1972: As florestas da América do Sul; ecologia, composição e importância econômica . Polígono. São Paulo. 466 p.

IBDF/FUPEF 1978: *Inventário Florestal do Pinheiro no Sul do Brasil.* Curitiba). 327 p

Kershaw, A. P. and McGlone, M. S 1995: The quaternary history of the Southern conifers. pp. 30–64 in: Enright, N. J. and Hill, R. S. (Eds.) Ecology of the Southern Conifers. Melbourne University Press, Melbourne

Kendall, A. 1949: Jurassic members of the Araucariaceae. Ann. Botany 13: 151–161

Ledru, M. P. 1993: Late quaternary environmental and climatic changes in central Brazil. Quaternary Research 39: 90–98

Liengsiri, C., Piewluang, C. and Boyle, T. J. B. 1990: Starch Gel Electrophoresis of Tropical Trees. A Manual. ASEAN–Canada Forest Tree Seed Centre and Petawawa National Forestry Institute. Muaklek, Saraburi, Thailand. pp. 51

Nei, M. 1972: Genetic distance between populations. Am. Nat. 106: 283–292

Stockey, R. A. 1982: The Araucariaceae: An evolutionary perspective. Rev. Palaeobotany & Palynology. 37: 133–154

Swofford, D. L. and Selander, R. B. 1997: BIOSYS-2 : A Computer Program for the Analysis of Allelic Variation in Genetics. User's Manual. Department of Genetics and Development, Univ. of Illinois at Urbana–Champaign, Urbana, Illinois. USA

Wright, S. 1965: The interpretation of population structure by F-statistics with special regard to systems of mating. Evolution 19: 395–420

Microsatellite DNA markers and their usefulness in poplars, and conservation of microsatellite DNA loci in Salicaceae*

O.P. RAJORA[1] AND M.H. RAHMAN[2]

[1] *Department of Biology, Life Sciences Centre, Dalhousie University, Halifax, Nova Scotia B3H 4J1, Canada;* [2]*Department of Renewable Resources, University of Alberta, Edmonton, Alberta T6G 2H1, Canada*

Key words: Microsatellite DNA markers, simple sequence repeats (SSR), *Populus*, *Salix,* clone fingerprinting, somaclonal variation, conservation of microsatellite DNA loci

Abstract: Highly informative genetic markers, such as microsatellites, would facilitate programs in genetics, breeding, biotechnology, and conservation of forest trees. Markers for 12 microsatellite DNA/simple sequence repeat (SSR) loci were developed and characterized in trembling aspen (*Populus tremuloides*). The compatibility and usefulness of the *P. tremuloides* SSR primers to resolve microsatellites in 24 other *Populus* species and three interspecific hybrids belonging to all six sections of the genus *Populus* and two *Salix* species were examined. The utility of the microsatellite DNA markers was also evaluated for clone and cultivar identification and for determining somaclonal variation. The $(TC/AG)_n$ microsatellites were the most abundant. The primers of one highly polymorphic dinucleotide SSR marker (*PTR7*) amplified two loci, and alleles could not be assigned to a specific locus. Fifty-four alleles were detected in 36 or 38 *P. tremuloides* individuals at the 10 polymorphic SSR loci showing single-locus patterns. The number of alleles at these10 SSR loci ranged from 2 to 11, with an average of 5.4 alleles per locus, and the observed heterozygosity ranged from 0.05 to 0.82, with a mean of 0.40 per locus. Although the highest polymorphism was observed for a dinucleotide SSR locus, the trinucleotide SSR loci also showed high polymorphism. Microsatellite DNA variants of four SSR loci followed a single-locus and those of *PTR7* a two-locus Mendelian inheritance patterns. With a few exceptions, all SSR primer pairs were successful in PCR amplification of genomic DNA and resolving microsatellites of comparable size in all 24 *Populus* species, three *Populus* hybrids, and two *Salix* species. Thus, all SSR loci characterized are conserved in the genus *Populus* as well as in *Salix* species. Each clone of six *Populus* species and each cultivar of *P.* x *canadensis* could be uniquely fingerprinted based on their genotypes at one to five SSR loci. Species-specific microsatellite DNA alleles for *P. deltoides*, *P. nigra*, and *P. balsamifera* were detected at one SSR locus. Microsatellite DNA somaclonal variation was detected in micropropagated *P. tremuloides* plants. The microsatellite DNA markers that we have developed could be used for clonal fingerprinting, tree forensics, species and hybrid identification, certification of controlled crosses, genome mapping, marker-assisted early selection, genetic diversity assessments, and conservation and sustainable forest management programs in poplars and also in willows.

---

*Contribution FGB0003 (RCGB0003) from the Forest Genetics and Biotechnology Group of Dalhousie University

Müller-Starck, G. and Schubert, R. (eds.). Genetic Response of Forest Systems to Changing Environmental Conditions. Vol. 70 (For. Sci.), 105-115 © 2001 Kluwer Academic Publishers, Dordrecht, Boston, London

Introduction

Highly informative genetic markers are required for assisting various programs in genetics, breeding, genome mapping, tree improvement, tree forensics, and conservation, restoration and sustainable management of forest genetic resources. Microsatellite DNA or "simple sequence repeats (SSR)" are highly informative genetic markers. Microsatellites are abundant polymorphic elements in plant nuclear genomes and consist of tandemly repeated, short DNA sequence motifs (Wang *et al.* 1994). These regions are interspersed throughout eukaryotic genomes, but generally embedded in unique DNA sequences. Polymorphism among individuals arises from changes in the number of repeats.

Microsatellite DNA markers are highly polymorphic, reproducible, and codominant in inheritance. These features make microsatellites as ideal markers for constructing high-resolution genetic maps and identifying loci controlling traits of interest (e.g., Hearne *et al.* 1992; Paglia *et al.* 1998), population and conservation genetics studies (Powell *et al.* 1995; Chase *et al.* 1996; Rajora *et al.* 2000), clonal identification (Dayanandan *et al.* 1998), certification of controlled crosses (Rajora *et al.* 2001), identification of species and hybrids (Rajora and Rahman 2001a), paternity determination (van de Ven and McNicol 1996), marker-assisted early selection (Weising *et al.* 1998), assessing genetic effects of forest management practices and developing strategies for conservation and sustainable management of forest genetic resources (Rajora *et al.* 2000).

Poplars (*Populus*) are fast growing, economically important multipurpose trees. Trembling aspen (*Populus tremuloides*) is a widespread tree species in North America, and commercially important for pulp, paper and oriented strand board industries in Canada. It is an ideal candidate for genetic improvement (Rajora 1991a), and active genetics and tree improvement programs for this species are in progress in North America. Highly informative genetic markers would facilitate these genetics and breeding programs.

Published information on microsatellites in forest trees is quite rudimentary. Due to the limited amount of sequence data available in databases, it is often necessary to construct genomic libraries and screen for specific SSRs. In addition to trembling aspen (Dayanandan *et al.* 1998; Rahman *et al.* 2000), microsatellite DNA markers have also been developed for *Populus nigra* (van der Schoot *et al.* 2000), a few pine (*Pinus*) (Smith and Devey 1994; Kostia *et al.* 1995; Echt *et al.* 1996; Fisher *et al.* 1998) and spruce (*Picea*) (van de Ven and McNicol 1996; Pfeiffer *et al.* 1997; Paglia *et al.* 1998; Rajora *et al.* 2001) species, and some other forest trees.

We have identified, isolated and characterized SSR loci/microsatellite DNA markers in *Populus tremuloides* by screening partial genomic libraries (Dayanandan *et al.* 1998; Rahman *et al.* 2000), estimated the genomic frequency of the identified SSRs and determined their informativeness and inheritance. The compatibility and usefulness of the *P. tremuloides* SSR primers to resolve microsatellites, and the conservation of microsatellite DNA loci in 24 other *Populus* species and three interspecific hybrids belonging to all six sections of the genus *Populus* and two *Salix* species were examined. Utility of the characterized SSRs in identification of clones and cultivars of six *Populus* species and an interspecific hybrid was assessed. The usefulness of the microsatellite DNA markers was also assessed for determining somaclonal variation in tissue-culture-micropropagated *P. tremuloides* plantlets. This paper summarizes the results of these studies.

Isolation, and characterization of microsatellite DNA markers

Fourteen microsatellite DNA loci were identified. Twelve microsatellite DNA loci could be resolved consistently (Table 1), which were characterized as described in Dayanandan *et al.* (1998) and Rahman *et al.* (2000). The di-nucleotide SSR, $(TC/AG)_n$, showed high abundance with five clones and $(TG/AC)_n$ showed low abundance with only one positive clone, suggesting that $(TC/AG)_n$ repeats are abundant in the genome of trembling aspen.

Seven SSR loci were of dinucleotide, four of trinucleotide, one of tetranucleotide and two of mononucleotide repeats (Dayanandan *et al.* 1998; Rahman *et al.* 2000; Table 1). Three clones with trinucleotide repeat sequences containing $(GGT/ACC)_n$, $(TGG/ACC)_n$, and $(GGA/CCT)_n$ were isolated using the dinucleotide probe $(TG)_{15}$. Although trinucleotide SSRs are considered to be relatively less abundant (Wang *et al.* 1994), detection of these three trinucleotide SSRs using a dinucleotide probe indicates that trinucleotide SSRs may be abundant in *P. tremuloides.*

Informativeness and inheritance of microsatellite markers

Informativeness of each microsatellite DNA locus was evaluated by determining allelic diversity and observed heterozygosity by genotyping 36 (*PTR1-PTR4*) or 38 (*PTR5-PTR15*) *Populus tremuloides* individuals. Inheritance of microsatellite DNA markers was examined for five loci (*PTR1*, *PTR2*, *PTR3*, *PTR4* and *PTR7)* in progenies of two controlled crosses.

Table 1. Repeat type, number (A) and size range of alleles, and observed heterozygosity (Ho) for 12 microsatellite DNA loci in *Populus tremuloides*

Locus	Repeat	A	Size (bp)	Ho
PTR1	$(GGT)_5N_{45}(AGG)_9$	5	254-272	0.53
PTR2	$(TGG)_8$	7	204-225	0.31
PTR3	$(TC)_{11}$	11	190-266	0.82
PTR4	$(TGG)_8$	6	200-230	0.19
PTR5	$(TG)_7$	3	250-256	0.61
PTR6	$(AT)_8$	6	186-240	0.53
PTR7	$(CT)_5$ AT $(CT)_6$	NA*	200-230	NA*
PTR8	$(A)_{11}(TC)_8$	4	132-140	0.37
PTR11	$(GT)_3G(GT)_2GGT$	2	138-142	0.05
PTR12	$(AAAG)_3$ $A_6n_7(AAAG)_2$	3	248-256	0.08
PTR14	$(TGG)_5$	7	130-204	0.50
PTR15	$(GA)_3AA(GA)_5$ $n_{10}(GA)_5$	1	220	0.00
Mean		**5.4**		**0.40**

*Resolved two highly polymorphic loci. Alleles could not be assigned to a specific locus and heterozygosity values could not be calculated.

With the exception of *PTR7*, all other primer pairs amplified products as expected for single-locus patterns (e.g., Fig. 1). The *PTR7* primer pair showed

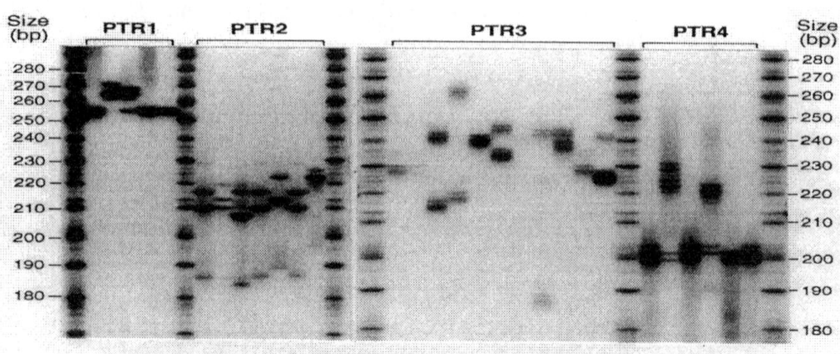

Figure 1. Allelic variation at the four SSR loci (*PTR1, PTR2, PTR3*, and *PTR4*) in selected individuals of *Populus tremuloides* (from Dayanandan *et al.* 1998).

amplification products of three to four sizes (Rahman *et al.* 2000) in the majority of individuals tested. Inheritance studies suggested that there are possibly two copies of *PTR7* microsatellite DNA sequences in the *P. tremuloides* genome. However, in other *Populus* species and interspecific hybrids, *PTR7* primers resolved microsatellite DNA variants typical of a single-locus. The microsatellite DNA variants of *PTR1-PTR4* each showed single-locus codominant Mendelian inheritance (Rahman *et al.* 2000).

A total of 54 alleles were identified at 10 polymorphic SSR loci in *P. tremuloides* individuals, with an average of 5.4 alleles per polymorphic locus (Table 1). The highest allelic diversity and heterozygosity were observed for the dinucleotide SSR locus *PTR3* and the lowest allelic diversity and heterozygosity for the locus *PTR11* (Table1). Although the *PTR3* locus was the most informative, the three other dinucleotide (*PTR5, PTR6*, and *PTR8*) and all four trinucleotide microsatellite DNA loci were highly informative (Table 1). Contrary to the general view that trinucleotide repeats are less polymorphic than dinucleotide repeats, trinucleotide repeats loci in *P. tremuloides* showed considerable polymorphism.

Conservation and usefulness of SSR loci in Salicaceae

Since development of microsatellite DNA markers is quite expensive and laborious in forest trees, identification of SSRs from a species that are informative and useful in related species or genera results in tremendous saving of time and resources. We tested the conservation of 10 microsatellite DNA loci (*PTR1-PTR8*, *PTR12*, and *PTR14*) and usefulness of the markers developed from *P. tremuloides* in 24 *Populus* species from all six sections, three interspecific hybrids, and two *Salix* species (Table 2; Rajora and Rahman 2001a). With the exception of *PTR2* and *PTR7* in *P. maxicana* and *PTR2* in *P. cathayana*, primers of all 10 SSR loci amplified genomic DNA and resolved microsatellites (mostly polymorphic) of comparable size in all 24 *Populus* species, three *Populus* hybrids and two *Salix* species tested (Table 2). This study suggests that the SSR loci identified from *P. tremuloides* are conserved in the genus *Populus* and most likely in the genus *Salix*. Since *Populus* and *Salix* are the two main genera of the family Salicaceae, it could be inferred that the *P. tremuloides* microsatellite DNA loci are well conserved in this family. The microsatellite DNA markers developed could potentially be used for various genetics, breeding, biotechnology, tree forensics, genome and QTL mapping studies, and conservation and sustainable forest management programs of the *Populus* species and hybrids and *Salix* species studied.

Table 2. Populus species and hybrids, and *Salix* species tested for usefulness of the microsatellite DNA markers and conservation of the SSR loci developed from *Populus tremuloides*

***Populus* species**		
Section ABASO	Section AIGEIROS	Section TACAMAHACA (cont.)
P. maxicana	*P. deltoides*	*P. trichocarpa*
	P. fremontii	*P. tristis*
Section TURANGA	*P. nigra*	*P. yunnanensis*
P. euphratica		
	Section TACAMAHACA	AIGEIROS x TACAMAHACA
Section LEUCOIDES	*P. angustifolia*	hybrids
P. ciliata	*P. balsamifera*	*P.* x *canadensis*
P. lasiocarpa	*P. cathayana*	(*P. deltoids* x *P. nigra*)
P. wilsonii	*P. koreana*	*P.* x *berolinensis*
	P. laurifolia	(*P. laurifolia* x *P. nigra)*
Section LEUCE	*P. maximowiczii*	
P. alba	*P. simonii*	***Salix* species**
P. grandidentata	*P. suaveolens*	*S. alba*
P. tremula	*P. szechuanica*	*S. fragilis*
P. sieboldii		
P. x *canescens*		

Clone and cultivar identification

Accurate identification of clones and cultivars is essential for effective breeding programs, varietal control, protection and registration, and day-to-day handling of planting and breeding stocks. Most *Populus* species are clonally propagated. Clonal identification based on traditional morphological and phenological criteria is difficult and subjective. Use of biochemical and molecular genetic markers provide powerful tools for clone and cultivar identification in plants. Allozyme and RAPD markers have been used to differentiate among clones and cultivars in poplars (Rajora 1988, 1989a, 1989b; Rajora and Zsuffa 1989; Castiglione *et al.* 1993; Lin *et al.* 1994). However, allozyme markers are limited by their polymorphism levels, and RAPD markers are limited by problems in reproducibility and their dominant nature. Highly polymorphic and codominant markers, such as microsatellites, provide excellent markers for clone and cultivar identification in poplars (Dayanandan *et al.* 1998; Rajora and Rahman 2000b; Rahman and Rajora 2001b) and other plants.

All of the clones of *P. nigra*, *P. deltoides*, *P. balsamifera, P. maximowiczii*, *P. trichocarpa*, and *P. grandidentata,* and cultivars of *Populus* x *canadensis* could be uniquely fingerprinted based on their singlelocus or multilocus microsatellite genotypes (Rajora and Rahman 2001b; Rahman and Rajora 2001b; Table 3; Fig. 2). Compared with allozyme and/or RAPD markers for the same *P. tremuloides, P. deltoides*, *P. nigra* and *P. maximowiczii* clones (Rajora 1988, 1989a, 1989b) and *P.* x *canadensis* cultivars (Rajora and Zsuffa 1989), microsatellite DNA markers were found to be more efficient and informative overall as well as on a per locus basis for clone and cultivar identification. For example, *P.* x *canadensis* cultivars 'Ostia' and "Canada Blanc' could not be differentiated by their genotypes at 31 allozyme (Rajora and Zsuffa 1989) and 248 RAPD (Rajora and Rahman 2001b) loci. However, these cultivars could be differentiated by their microsatellite genotypes at each of the four SSR loci (Rajora and Rahman 2001b).

Table 3. Usefulness of microsatellite DNA markers for clone and cultivar (CV) identification in *Populus** species and an interspecific hybrid*

Species/hybrid	No. clones/ CV studied	No. unique genotypes	No. SSR loci for clone/CV identification
P. tremuloides	38	38	5
P. grandidentata	5	5	1
P. deltoides	20	20	3
P. nigra	25	25	3
P. x *canadensis*	17	17	4
P. balsamifera	29	29	5
P. maximowiczii	11	11	2
P. trichocarpa	6	6	2

*Dayanandan *et al.* (1998); Rajora and Rahman (2001b); Rahman and Rajora (2001b)

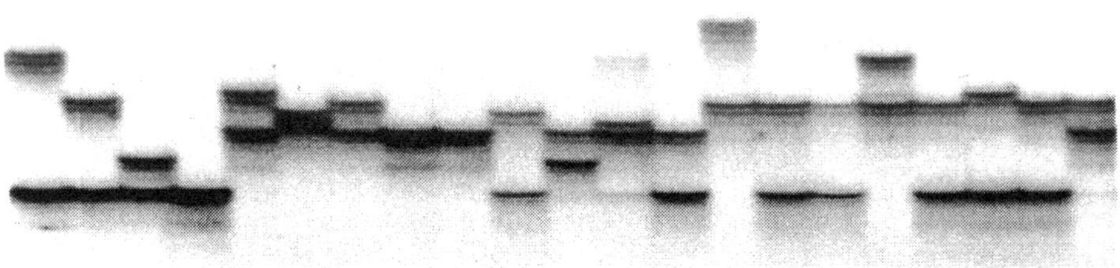

Figure 2. Microsatellite DNA fingerprints of 21 *Populus nigra* clones at the SSR locus *PTR7* (from Rahman and Rajora 2001b)

Populus species and interspecific hybrid identification

Microsatellite DNA markers could differentiate among *P. deltoides*, *P. nigra*, *P. balsamifera* and their interspecific hybrids. *Populus deltoides*, *P. nigra* and *P. balsamifera* had species-specific alleles at the SSR locus *PTR6*. *Populus deltoides* and *P. nigra* samples were completely and *P. balsamifera* samples almost fixed for a single species-specific allele each (Rajora and Rahman 2001a). The second rare allele observed in *P. balsamifera* was also species-specific.

Determination of somaclonal variation in poplars

It is essential to examine somaclonal variation in tissue-culture-micro-propagated plants for ascertaining their clonal fidelity as well as identifying somaclonal variants for potential use in plant improvement. Somaclonal variation in growth, morphological, physiological and disease resistance traits is well documented in poplars (Fry *et al.* 1997). In trembling aspen, somaclonal variation has been observed for isozymes (Noh and Minocha 1990) and disease resistance (Valentine *et al.* 1988) traits. However, there is no published information on the use of DNA markers for detecting somaclonal variation in tissue-culture-derived trembling aspen.

We have examined somaclonal variation in eight plantlets of one genotype and four plantlets of another genotype of trembling aspen by using microsatellite DNA markers of 10 loci (Rahman and Rajora 2001a). The plantlets were derived from micropropagation of trembling aspen trees through tissue cultures of vegetative buds (Rajora 1991b). No somaclonal variation was detected for eight SSR loci. However, *PTR5* identified one somaclonal variant among eight plantlets of tree (genotype) 1 and *PTR2* identified one somaclonal variant among four plantlets of tree (genotype) 2 (Fig. 3). It was not possible to ascertain the exact cause of somaclonal variation observed. However, it was inferred that one somaclonal variant among eight plantlets of tree 1 might have arisen by addition of a single chromosome or duplication of a chromosomal segment, whereas somaclonal variation in one plantlet of tree 2 may have been caused by insertion of microsatellite DNA sequences (Rahman and Rajora 2001a). Nevertheless, microsatellite DNA markers were found to be very valuable tools for detecting somaclonal variation in micropropagated plantlets of trembling aspen

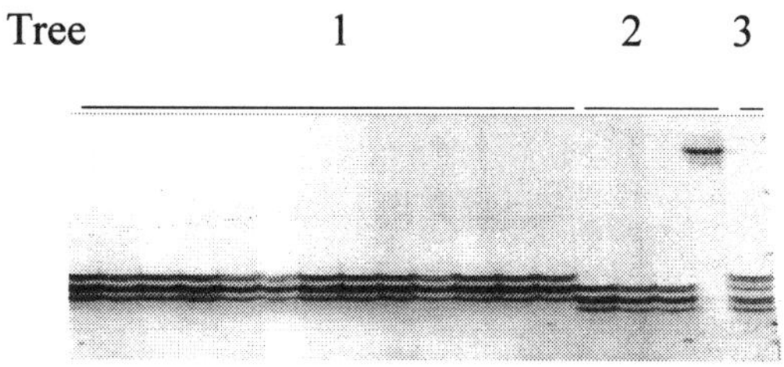

Figure 3. Somaclonal variation in one plantlet of tree 2 detected by the SSR locus *PTR2* (from Rahman and Rajora 2001a)

Conclusions

We have isolated and characterized highly informative microsatellite DNA loci and markers in *P. tremuloides* that are useful in the genus *Populus* as well as in *Salix* species. These microsatellite DNA loci were found to be conserved in Salicaceae. Almost all of the microsatellite DNA markers developed for *P. tremuloides* can be used for various genetics and breeding studies in all the 24 *Populus* species and two *Salix* species examined. The microsatellite DNA markers developed can be used in *Populus* for DNA fingerprinting and identification of clones, cultivars, species, hybrids and somaclonal variants, as demonstrated in this study. These markers can also be used for tree forensics, certification of controlled crosses, genome mapping, marker-assisted early selection, genetic diversity assessment, and conservation and sustainable management of poplar genetic resources.

Acknowledgements

This research was funded by a Natural Sciences and Engineering Research Council of Canada Individual Research Grant (OGP 0170651) to O.P. Rajora. We thank Dr. Wout Boerjan of University of Gent, Belgium and Dr. Malathi Lakshmikumaran of the Tata Energy Research Institute, India for supplying DNA samples of many *Populus* and *Salix* species, and Dr. Alex Mosseler of the Canadian Forest Service for reviewing this paper.

References

Castiglione, S., Wang, G., Damiani, G., Bandi, C., Bisoffi, S. and Sala, F. 1993: RAPD fingerprints for identification and for taxonomic studies of elite poplar (*Populus* spp.) clones. Theor. Appl. Genet. 87: 54-59

Chase, M.R., Kesseli, R.V. and Bawa, K.S. 1996: Microsatellite markers for population and conservation genetics of tropical trees. Am. J. Bot. 83: 51-57

Dayanandan, S., Rajora, O.P. and Bawa, K.S. 1998: Isolation and characterization of microsatellites in trembling aspen (*Populus tremuloides*). Theor. Appl. Genet. 96: 950-956

Echt, C.S., May-Marquardt, P., Hseih, M. and Zahorchak, R. 1996: Characterization of microsatellite markers in eastern white pine. Genome 39:1102-1108

Fisher, P.J., Richardson, T.E. and Gardner, R.C. 1998: Characteristics of single- and multi-copy microsatellites from *Pinus radiata*. Theor. Appl. Genet. 96: 969-979

Fry, D.J., Douglas, G.C. and Saieed, N.T. 1997: Somaclonal variation in *Populus*: An evaluation. pp 33-43, in Klopfenstein, N.B., Chun, Y.W., Kim, M.-S., and Ahuja, M.R. (Eds.), Micropropagation, genetic engineering, and molecular biology of *Populus*. Gen. Tech. Rep. RM-GTR-297. USDA, For. Serv., Rocky Mountain Research Station, Fort Collins, CO

Hearne, C., Ghosh, S. and Todd, J. 1992: Microsatellites for linkage analysis of genetic traits. Trends in Genetics 8: 288-294

Kostia, S., Sirkaa-Liisa, V., Vakkari, P. and Pulkkinen, P. 1995: Microsatellite sequences in a conifer, *Pinus sylvestris*. Genome 38: 1244 - 1248

Lin, D., Hubbes, M. and Zsuffa, L. 1994: Differentiation of poplar and willow clones using RAPD fingerprints. Tree Physiol. 14: 1097-1105

Noh, E.-W. and Minocha, S.C. 1990: Pigment and isozyme variation in aspen shoots regenerated from callus culture. Plant Cell, Tissue and Organ Culture 23: 39-44

Paglia, G.P., Olivieri, A.M. and Morgante, M. 1998: Towards second-generation STS (sequence-tagged sites) linkage maps in conifers: a genetic map of Norway spruce (*Picea abies* K.). Mol. Gen. Genet. 258: 466-478

Pfeiffer, A., Olivieri, A.M. and Morgante, M. 1997: Identification and characterization of microsatellites in Norway spruce (*Picea abies* K). Genome 40: 411-419

Powell, W., Morgante, M., McDevitt, R., Vendramin, G.G. and Rafalski, J.A. 1995: Polymorphic simple sequence repeat regions in chloroplast genomes: Application to population genetics of pines. Proc. Natl. Acad. Sci. USA 92: 7759-7763

Rahman M.H., Dayanandan, S. and Rajora, O.P. 2000: Microsatellite DNA markers in *Populus tremuloides*. Genome 43: 293-297

Rahman, M.H. and Rajora, O.P. 2001a: Microsatellite DNA somaclonal variation in tissue culture micropropagated trembling aspen. Plant Cell Reports 20 (in press); published online 20 July 2001, DOI 10.1007/s002990100365

Rahman, M.H. and Rajora, O.P. 2001b: Microsatellite DNA fingerprinting and identification of clones of six poplar species. Can. J. For. Res. (to be submitted)

Rajora, O.P. 1988: Allozymes as aids for identification and differentiation of some *Populus maximowiczii* Henry clonal varieties. Biochem. Syst. Ecol. 16: 635-640

Rajora, O.P. 1989a: Genetic structure and identification of *Populus deltoides* Marsh. clones based on allozymes. Genome 32: 440-448

Rajora, O.P. 1989b: Characterization of 43 *Populus nigra* L. clones representing selections, cultivars and botanical varieties based on their multilocus allozyme genotypes. Euphytica 43: 197-206

Rajora, O.P. and Zsuffa, L. 1989: Multilocus genetic structure, characterization and relationships of *Populus* x *canadensis* cultivars. Genome 32: 99-108

Rajora, O.P. 1991a: A long-term program plan for genetic improvement of aspen and other poplar hardwoods in Alberta. Alberta Forest Service, and BioGenetica Inc., Edmonton, Alberta, Canada

Rajora, O.P. 1991b: Manual of tissue culture micropropagation techniques for trembling aspen (*Populus tremuloides*). Alberta Forest Service, and BioGenetica Inc., Edmonton, Alberta, Canada

Rajora, O.P. and Rahman, M.H. 2001a: Conservation of microsatellite DNA loci in Salicaceae. Genome/Heredity (to be submitted)

Rajora, O.P. and Rahman, M.H. 2001b: Microsatellite DNA and RAPD fingerprinting and genetic relationships of hybrid poplar (*Populus* x *canadensis*) cultivars. Theor. Appl. Genet. (to be submitted)

Rajora, O.P., Rahman, M.H., Buchert, G.P. and Dancik, B.P. 2000: Microsatellite DNA analysis of genetic effects of harvesting in old-growth eastern white pine (*Pinus strobus*) in Ontario. Molecular Ecology 39: 339-348.

Rajora, O.P., Rahman, M.H., Dayanandan, S. and Mosseler, A. 2001: Isolation, characterization, inheritance and linkage of microsatellite DNA markers in white spruce (*Picea glauca*) and their usefulness in other spruce species. Mol. Gen. Genet. 264: 871-882

Smith, D.N. and Devey, M.E. 1994: Occurrence and inheritance of microsatellites in *Pinus radiata*. Genome 37: 977-983

Valentine, F., Baker, S., Belanger, R., Manion, P. and Griffin, D. 1988: Screening for resistance to *Hypoxylon mammatum* in *Populus tremuloides* callus and micropropagated plantlets. In: Ahuja, M.R., (Ed.) Somatic cell genetics of woody plants. Kluwer Acad. Publ., Dordrecht

van de Ven, W.T.G. and McNicol, R.J. 1996: Microsatellites as DNA markers in Sitka spruce. Theor. Appl. Genet. 93: 613-617

van der Schoot, J., Pospiskova, M., Vosman, B. and Smulders, M.J.M. 2000: Development and characterization of microsatellite markers in black poplar (*Populus nigra* L.). Theor. Appl. Genet. 101: 317-322

Wang, Z., Weber, J.L., Zhong, G. and Tanksley, S.D. 1994): Survey of plant short tandem repeats. Theor. Appl. Genet. 88: 1-6

Weising, K., Winter, P., Huttel, B. and Kahl, G. 1998: Microsatellite markers for molecular breeding. J. Crop Production 1: 113-143

PCR-RFLP analysis of introns of nuclear genes in *Populus* and *Prunus*

B. HEINZE

Institute of Forest Genetics, Federal Forest Research Centre, A-1140 Vienna, Austria

Key words: PCR, nuclear genes, intron, polymorphism, *Prunus*, *Populus*

Abstract: Many PCR primer sequences are available in the literature that amplify genes, introns and spacers from plant chloroplast and mitochondrial DNA, which are reasonably conserved over evolutionary distances. Similar attempts with nuclear DNA are hampered by greater diversity in exon sequences, the presence of gene families and pseudogenes, and often a lack of suitable sequences in the public data banks. However, within a species or a genus, using published sequences to design primers that amplify presumably polymorphic introns is more straightforward. It is often possible to find primers that amplify gene fragments from a number of angiosperm taxa, although specific PCR approaches are needed in many situations. These include using polymerases with different characteristics, and specific temperature profiles and optimization strategies. Primer sequences based on sequence data from even a broad range of genera may still fail to amplify the desired fragment from specific species, genera, or families - a finding in accordance with DNA sequence evolution theory. Amplified fragments can be analysed by restriction digestion followed by agarose electrophoresis as a simple but often effective way of detecting polymorphisms. Examples are given from such approaches including peroxidases in *Populus* sp., specific and broad-range primers for phenylalanine ammonia lyase (PAL) in both *Populus* and *Prunus*, the hypervariable S-RNAases in *Prunus avium*, and cinnamoyl alcohol dehydrogenase (CAD) in a number of angiosperm taxa. In *Populus*, using these primers allowed us to detect introgression of genes from hybrid clones into the native *Populus nigra* genepool, while in *Prunus avium*, we expect to gain insight into the peculiar mating system of this species.

Introduction

Why analyse introns?

The analysis of polymorphic spacers and introns of the organelle DNA of plants using PCR approaches has revealed a wealth of information for evolution, geographic variation, and population genetics studies (Taberlet *et al.* 1998). Nuclear genes also contain conserved regions of DNA or protein sequence, albeit less phylogenetically stable ones. The lesser extent of sequence conservation is due to the more rapid evolution of biparentally

Müller-Starck, G. and Schubert, R. (eds.). Genetic Response of Forest Systems to Changing Environmental Conditions. Vol. 70, (For. Sci.) 117-127 © 2001 Kluwer Academic Publishers, Dordrecht, Boston, London

inherited genetic components. Nervertheless, certain protein sequences (sometimes called domains) are often conserved throughout evolution. As there is less selection pressure for intron DNA sequences, these stretches probably harbour more polymorphism than for instance isozymes, which are at present widely used in the fields of research mentioned above. Under this assumption, introns would constitute fairly neutral genetic markers. Sequence variation in the intron should be detectable with a simple method like restriction digestion-gel electrophoresis. Investigating possible conserved sequences in exons as primer sites for PCR amplification of introns is supported by the growing number of sequences present in the public data banks.

Possible problems in this approach include: (1) the greater diversity of exons at the DNA level (compared to protein sequences, and to organellar DNA); (2) the presence of gene duplications in many species (The European Union Arabidopsis Genome Sequencing Consortium and The Cold Spring Harbor, Washington University in St Louis and PE Biosystems Arabidopsis Sequencing Consortium 1999; Lin *et al.* 1999; Meyerowitz 1999; Perry and Furnier 1996); (3) gene families and pseudogenes that might act as multiple PCR priming sites; and (4) the lack of specific sequence data from a wide array of genera and families, which hampers delineating truely conserved regions of the sequence. Additionally, the public sequence data banks are strongly biassed in favour of model organisms such as *Arabidopsis thaliana*, tobacco (*Nicotiana tabaccum*) and rice (*Oryza sativa*).

In the course of several unrelated research projects, I have tried to design and use PCR primers for the amplification of introns in nuclear genes in woody angiosperms. This report summarizes results and experience gained, and discusses the future potential of such an approach.

Populus

For this 'model' genus, many specific DNA sequences have been accumulated in the sequence data banks. Suitable plant material, including offspring from controlled pollinations, is available from breeding research stations. Clones are available that are hybrids between distant species, thus harbouring much genetic variation. The approach taken here was to design specific primers for certain genes for the genus or at least for certain sections of the genus.

PAL (phenyl alanine ammonia lyase) is a well-studied enzyme, due to its involvement in wood genesis and pathogen response. Sequences present in GenBank (e.g, POPPALGA) indicate a conserved intron position. In *Populus*, 1 or 2 loci encoding this enzyme are present (Subramaniam *et al.* 1993). Peroxidases are numerous in plants. Sequences of anionic peroxidases were present in GenBank from *P. nigra* and *P. trichocarpa*, e.g. POPP01, POPP02, PTXP1PER, PTXP2PER, etc.

Prunus

S-RNAases are involved in self-incompatibilty (pollen rejection) in this genus. Putative mRNA sequences are available (Tao *et al.* 1999), and they contain conserved and hypervariable domains; the presence of an intron is possible. PAL sequences are also present in GenBank.

Conserved primers for many species ?

Sequence data from many gene families in GenBank can be aligned to design conserved primers. This approach was taken by Strand *et al.* (1997) who succeeded in amplifying parts of ADH, AAT, CAM, CHI, CHS, and other genes. The authors often obtained multiple bands useful for further cloning and sequence analysis, but no straightforward genetic interpretation (designation of alleles) was possible. In this paper, similar attempts are described for the cinnamoyl alcohol dehydrogenase (CAD) and phenylalanine ammonia lyase (PAL), designing primers that flank identified introns.

Materials and Methods

The homologous DNA and protein sequences considered in this paper were obtained from GenBank and analysed with a matrix-comparison approach implemented by Antoniw (1995). His suite of computer programs compares windows of DNA sequence stretches of two sequences against each other. The window sizes are adjustable. Regions of higher homology are indicated graphically, and possible priming sites suggested. Based on data obtained with Antoniw's programs, sequences were further aligned manually or with the help of the PC/Gene 6.85 - NALIGN or OMIGA 1.1.3 software packages (both from Oxford Molecular, Great Britain), to identify conserved intron positions. Primer sequences were selected manually or using PCRPLAN in PC/Gene 6.85.

Plant DNA was prepared from leaf, bud and seed tissues of *Populus* sp., *Prunus* sp., *Fagus* sp., and *Quercus* sp. This was done by homogenizing the material in 1.5 mL vials with a motor-driven pestle, and consequently applying standard phenol-chloroform extraction methods (Croy 1993) and/or commercial kits (NucleoSpin Blood Kit, Machery and Nagel, Germany; Qiagen Blood Kit, Qiagen, Germany; Phytopure Kit, Amersham, Great Britain).

Polymerase chain reactions were set up, unless otherwise indicated, using 1-20 ng DNA in 1X reaction buffer (supplied with the enzyme), 2 mM $MgCl_2$, primers at 200 nM each, dNTPs at 200 μM each, and 0.3-0.5 units of polymerase (DynaZyme, Finnzymes Oy, Finland; Platinum Taq, Life Technologies, USA; or AmpliTaq Gold, Perkin-Elmer, USA) per 10 μL reaction. A generic PCR program used for most experiments (exceptions are indicated in the following sections) consited of a pre-amplification phase of 10 cycles, alternating between 94°C and 70°C (1 min. each), and a proper cycling phase of 94°-50 sec, 55°C-1 min, and 70°C-2 min, repeated 32 times. Reaction products were analysed in 1.5-2% agarose gels. Restriction digests employed 5 μL of the whole PCR reactions and 3μL of a 2x restriction buffer/enzyme mix (approx. 0.1 unit enzyme; restriction buffers as supplied along with the respective enzymes). Restricted fragments were resolved on 2-3% agarose gels.

The primers PPAL-P1 (CAA GGC GGT GCT CTT CAG AAG G) and PPAL-M1 (GCT GAG TGA GGC AGT GTG TGG C) span the only intron in the *Populus* PAL gene. A stepdown PCR program (Hecker and Roux 1995; 10 cycles of 92.5°C - 45 sec, 72°C - 1 min were followed by 35 cycles of 92.5°C - 30 sec, 65°C - 1 min, 72°C - 1 min) was used. Primers POPX-P5 (GTC CAA ATG TGA GCA GCA TTA TCC) and POPX-M25 (ATC AGT CTG AAG CAG GCC TCG) span a conserved intron in *Populus* peroxidase genes. PCR amplification (using AmpliTaq Gold, Perkin-Elmer, USA, and the 'stepdown' programming mentioned above for PPAL) was used.

The PCR primers proposed by Tao *et al.* (1999) for amplification of the S-RNAase locus in *Prunus avium*, and those of Strand *et al.* (1997) for amplifying parts of ADH, AAT, CAM, CHI, CHS, and other genes, in many species, were also tested.

Specific primers on the basis of a *Prunus* GenBank PAL mRNA sequence were designed (PruPal-P: CAA GGG GCT GCC CTT CAG AAG G, and PruPal-M: GCC TGG TGA GGC AAA GTG TGG C).

For amplification of PAL gene fragments from many angiosperm species, several primers were designed after a matrix comparison (Antoniw 1995) and alignments. Primers U-PAL P0 (ATG GTD GCT GAG TWY AGG AA), U-PAL M3 (ACR TCT TGR TTR TGY TGC TC), U-PAL M (CCR AAN AYY CCR GCR TTM ARR AAY CT) and U-PAL M2 (GCC TTG TTY CTY GAM ACA TC) were tested.

CAD conserved primers were designed to amplify the four identified introns. Among them were CAD-af (CYC CTT ACA CYT AYA SYC TCA GGT), CAD-br (ACC TAC CCA GGR ACC ATW GG), CAD-dr1 (GCW ATC TTS ACM CCC ATG TG), CAD cr2 (GYA RGA CCA RAT YTT CTT GTT GCA), CAD-bf (GTC HMA YTA YCC WAT GGT YCC TGG), and finally,

CAD-cf (TWV YDG AGR TKG GWW SMG AKG TG). In this nomenclature, a-d denote the exons, and f and r forward and reverse primers, respectively.

Results

Specific genes in Populus (PAL and PPOX) and Prunus (S-RNAases and PAL)

In *Populus*, one fragment of approximately 1000 basepairs (bp) was obtained after PCR for PAL. Restriction digests using enzyme *Hae* III resulted in variable bands at approximately 350 bp, where hybrids between *P. nigra*, *P. deltoides* and some balsam poplars (*Populus* section *Tacamahaca*: *P. maximowiczii*, *P. trichocarpa*) had an extra band. This is most probably a heteroduplex DNA molecule forming in the later PCR cycles (Heinze 1997). A second locus was detectable in some *P. deltoides* clones. The marker proved useful for introgression studies in Austria (Heinze 1998a). In poplars, introgression occurrs either between species with overlapping geographical ranges, or between introduced exotic species or hybrids, and the native species. The latter situation is encountered in Central Europe, were *P. nigra* is native and *P. deltoides x P. nigra* hybrids have been introduced. The extra band specific to heterozygotes could be detected in some naturally established seedlings morphologically similar to *P. nigra*. In combination with other markers (Heinze 1997, 1998b), this assay thus allowed detecting alleles from introduced hybrids in naturally regenerating *P. nigra*.

Again in *Populus*, the primers POPX-P5 and POPX-M25 resulted in several fragments between 1100 and 1250 bp, but only with the polymerase enzyme AmpliTaq Gold. PCR results were inconsistent with other polymerase enzymes. Restriction digests with *Msp* I or *Hinf* I revealed many variable and invariant bands. Certain variable bands showed almost complete linkage (e.g., *Msp* I 650 bp + *Hinf* I 400 bp). Some of the fragments were fixed in certain species, and some were variable within species. The rather complicated banding patterns could be resolved by using progeny arrays and indicated the presence of several loci and alleles. These markers proved useful for classification of the Austrian clone collection of cultivated hybrids (Heinze 1998a).

In *Prunus avium* collected in Austria, reaction products for the S-RNAases involved in the incompatibility system were extremely variable in band sizes after PCR, and exhibited almost individual fingerprint-like patterns after restriction. While it seemed there was somewhat reduced variability in cultivated sweet cherry, it was also difficult to interpret the results in terms of alleles. Because of the incompatibility system, only non-self pollen can successfully fertilize egg cells, and each individual should be heterozygous at this locus. However, this incompatibility is incomplete. Arrays of open pollinated progeny contained some individuals with only one apparent allele. These may either be explained by the presence of null alleles (i.e, variation at primer binding sites that hamper proper amplification of all allelic combinations), by the hypothesis that not all alleles differ in banding patterns from each other, or by incompleteness of incompatibility in these progenies, i.e. some self-fertilization.

PAL in *Prunus avium* appears to be an opposite example, of low variation: the specific primers PruPal-P and PruPal-M resulted in invariable bands. An apparent second locus was detected in *Prunus cerasus*, which is a tetraploid species.

Conserved primers for many species

The conserved primers of Strand *et al.* (1997), in our hands, revealed relatively low complexity banding patterns in *Prunus*. No variation was apparent. At this point, however, it should be stressed that PCR optimization strategies are often crucial for the successful amplification. A hot start with either antibodies against the Taq Polymerase enzyme (as present in Life Technologies' Platinum Taq), or later addition of the polymerase (after initial denaturation), the use of different enzyme brands, and 'touchdown' or 'stepdown' programming (Don *et al.* 1991; Hecker and Roux 1996) are often necessary. Re-amplification (a second round of PCR using part of the first reaction) may be helpful.

Primers U-PAL P0 and U-PAL M3 were the most successful among the newly designed 'universal' PAL primers (Fig. 1), but primers U-PAL M and U-PAL M2 also gave satisfactory results in combination with U-PAL P0. In general, a somewhat weak amplification was obtained in the the species tested, with low complexity in *Prunus* and *Populus,* and no amplification in *Quercus* and *Fagus*. This may be due to mutations at the priming sites in the *Fagaceae* family, highlighting the need for a wide range of sequences for the initial alignments. The *Populus* and *Prunus* fragments, however, showed similar restriction patterns as those obtained with the specific primers (PPAL and Pru-PAL, respectively). The only differences were due to the primer positions along the sequence.

Among the successfully tested CAD primer pairs were CAD-af + CAD-br, CAD-af + CAD-dr1, CAD-af + CAD cr2, CAD-bf + CAD-dr1, and finally, CAD-cf + CAD-dr1. CAD introns in *Prunus, Populus,* and *Quercus* with some primer pairs gave invariant bands, while especially the longer fragments sometimes were variable. With CAD-af and CAD-cr2 (Fig. 2) in *Prunus avium* it seems posssible to distinguish alleles. Sequencing or high-resolution electrophoresis may reveal further polymorphisms.

Position	1 30
Arabidopsis L PAL 2 exon 1	ATGAAAGGAAGTCATTTAGATGAAGTGAAG
Arabidopsis L PAL exon 1	ATGAAAGGTAGCCATTTGGATGAAGTGAAG
Capsicum PAL1 exon 1	ATGACCGGAAGCCATCTTGATGAAGCGAAG
Prunus avium PAL exon 1	CTAAAAGGGAGCCACTTGGATGAGGTGAAG
Daucus carota PAL exon 1	TTGACAGGAAGCCATTTGGATGAAGTCAAG
Nicotiana tabacum PAL A exon 1	TTGAAAGGAAGCCATTTAGATGAAGTGAAG
Populus PAL A exon 1	ATGAAAGGAAGCCATTTGGACGAAGTGAAA
Vitis PAL exon 1	TTGAAGGGAAGCCATTTGGATGAAGTGAAG
Primer U-PAL-P0	**ATGGTDGCTGAGTWYAGGAA**
Position	31 60
Arabidopsis L PAL 2 exon 1	AAG**ATGGT**C**G**AA**GAGT**A**T**C**G**T**A**GACCAGTC
Arabidopsis L PAL exon 1	AGA**ATGGTTGCTGAGTTTAGGAA**GCCAGTT
Capsicum PAL1 exon 1	AAA**ATGGTGG**AG**GAGTT**C**AGGAA**TCCTGTT
Prunus avium PAL exon 1	CGC**ATGGTGGCTGAGT**AC**AGGAA**GCCGGTG
Daucus carota PAL exon 1	AGG**ATGGTGGCTGAGTTTAGGAA**GCCTATG
Nicotiana tabacum PAL A exon 1	AAA**ATGGTGGCTGAGTTTAGGAA**ACCAGTA
Populus PAL A exon 1	CGT**ATGGT**A**GCTGAGT**A**TAGGAA**GCCTGTC
Vitis PAL exon 1	CGC**ATGGTGGC**G**GAGT**ACC**GGAA**GCCGGTG
Position	61 90
Arabidopsis L PAL 2 exon 1	GTGAATCTTGGCGGAGAAACACTGACGATC
Arabidopsis L PAL exon 1	GTGAATCTTGGTGGTGAGACTCTGACCATT
Capsicum PAL1 exon 1	GTCAGCTTAGCAGGTGCCAACCTGAAGATC
Prunus avium PAL exon 1	GTGAAGCTCGGTGGAGAGAGCCTGACCATT
Daucus carota PAL exon 1	GTGCAGCTTGGCGGGGAGACACTCACCGTC
Nicotiana tabacum PAL A exon 1	GTGAAACTTGGAGGTGAGACTTTGACAGTG
Populus PAL A exon 1	GTGAATCTTGCTGGTCAGAACCTGATTATC
Vitis PAL exon 1	GTTCGGCTCGGTGGTGAGACGCTTACCATA
Position	91 120
Arabidopsis L PAL 2 exon 1	GGACAAGTTGCTGCCATCTCCACC------
Arabidopsis L PAL exon 1	GGACAAGTGGCTGCGATCTCAACT------
Capsicum PAL1 exon 1	TCTCAGGTAGCTGCAGTGGCTGCCGCTGCG
Prunus avium PAL exon 1	TCCCAAGTGGCGGCCATAGCCACT------
Daucus carota PAL exon 1	TCTCAGGTGGCTGCTATTGCTG--------
Nicotiana tabacum PAL A exon 1	GCTCAAGTTGCGGCTATTGCTGCA------
Populus PAL A exon 1	GCTCAGGTTGCATCTATTGCTGGC------
Vitis PAL exon 1	TCGCAGGTGGCGGCAATCGCTGGG------

Figure 1. Partial alignment of exon 1 sequences from several PAL genes present in GenBank. The sequence chosen for primer U-PAL-P0 is indicated (standard IUPAC codes are used for degenerate positions). Conserved residues in this region are printed in bold.

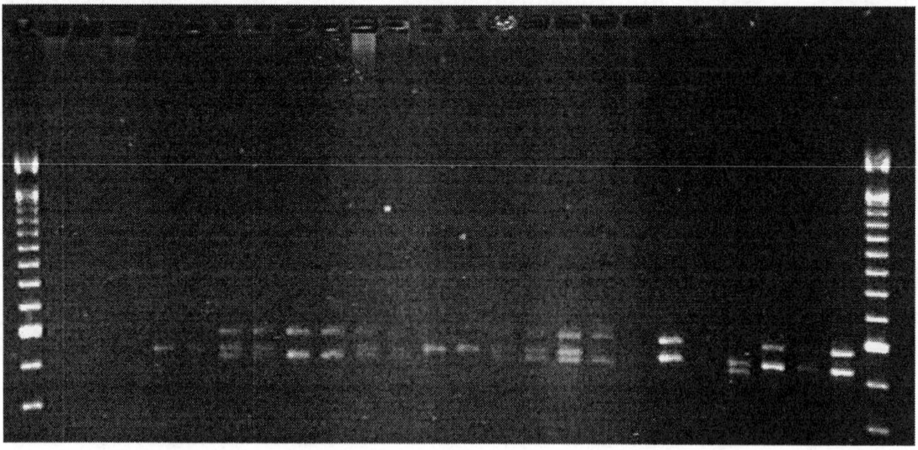

Figure 2. Analysis of parts of the CAD gene in *Prunus avium*. Primers CAD-af and CAD-cr2 amplified the first intron, the second exon, and the second intron in a number of unrelated *P. avium* clones. PCR products were restriction digested using *Hinf* I and *Rsa* I and resolved in 2% agarose. Outer lanes are size markers (100 bp ladder, lower bright band is 600 bp).

Discussion

The results indicate that in principle, a PCR/gel electrophoresis approach is possible for detecting of alleles for a number of genes. In the ideal situation, target genes should have highly conserved exon sequences interrupted by highly variable introns. Such situations are unlikely, but as shown above, the approach is still worth pursuing. Enzymes often contain fairly conserved domains that often correspond to exons. One important point to consider, however, is the rather unconstrained mutability of the third codon position in DNA. To take account of this, primers will often contain degenerate positions, at the cost of a higher probability of false priming and the generation of PCR artefacts. Such effects can be circumvented, to a degree, by establishing special PCR conditions – hot start techniques, touchdown annealing, and re-amplification. In turn, this makes the routine analysis of larger samples more complicated and shoud therefore be kept to the minimum. A protocol that can be applied to many primer sets in many species would be preferred, as this would minimize handling, and consequently, labour and cost. The 'stepdown' and 'touchdown' strategies adopted here (Hecker and Roux 1996; Don *et al.* 1991) are good starting points for further optimization.

It is important to keep phylogenetic relationships between the species and sequences in mind. In a phylogenetic tree, any of the branches may mutate and thereby destroy possible priming sites. Therefore, primer sites that are suitable for a wide range of taxa may by chance turn out as useless in a particular species or genus. Such a situation was encountered in PAL, where the two members of the *Fagaceae*, *Quercus* and *Fagus*, did not respond to PCR primers that yielded favourable results in genera as distant as *Populus* (*Salicaceae*) and *Prunus* (*Rosaceae*). Phylogenetic relationships also exist between members of a gene family within species. The aligment in Figure 1 contains two *Arabidopsis* PAL sequences, one of which (PAL 2) seems to move away from the consensus. Gene duplication was recently discovered as a major feature in *Arabidopsis* (and possibly more generally, plant?) genome evolution (The European Union Arabidopsis Genome Sequencing Consortium and The Cold Spring Harbor, Washington University in St Louis and PE Biosystems Arabidopsis Sequencing Consortium 1999; Lin *et al.* 1999; Meyerowitz 1999). In the presence of gene families, it would be advantageous to identify priming sites specific to single members. For instance, amplification of only a limited number of peroxidase sequences in *Populus* required the use of a special enzme, AmpliTaq Gold (Perkin-Elmer). Apparently this enzyme, which is delivered as a liposome complex, amplifies only from perfect primer-target matches. In this way, DNA is only amplified from a small number of loci. The resulting fragment patterns are less complex and can be easily interpreted genetically.

In general, however, once primers are available, the PCR - restriction digestion - agarose gel electrophoresis approach outlined in this paper is not very demanding in a technical sense. This feature is an advantage compared to microsatellite or AFLP analysis. It will ease the interchange of methods (only primer sequences and PCR condititons need to be exchanged, which can be done over e-mail) and results (patterns are simple and easy to communicate to others). Hopefully this will make future genetic studies comparable to a higher degree than previously realized.

Conclusions and Outlook

PCR primers could be designed for genetic analysis for specific species or genera, but also for widespread use in angiosperms. They nevertherless have to be tested and optimized by trial and error. Success with such a primer designing strategy requires a thorough understanding of the underlying molecular genetics and phylogeny, and a proficiency in PCR methods. Once 'fairly universal' primers and methods have been established, it may be possible to substantially supplement, if not replace, isozyme analysis in angiosperm genetics. As developing and testing such primers is a major effort, it requires a good deal of collaboration within the scientific community, and free exchange of sequence and primer data through data banks or the scientifc literature. With common efforts, a set of suitable primers for angiosperms may one day be available to all.

Acknowledgement

I am grateful to Elmar Kickingereder for laboratory work.

References

Antoniw, J. 1995: A new method for designing PCR primers specific for groups of sequences and its application to plant viruses. Mol. Biotechnol. 4: 111-119

Croy, R.R.D. (Ed.) 1993: Plant Molecular Biology Labfax. BIOS Scientific Publishers Ltd., Oxford

Don, R.H., Cox, P.Y.T., Wainwright, B.J., Baker, K., Mattick, J.S. 1991: Touchdown PCR to circumvent spurious priming during gene amplification. Nucleic Acids Res. 19: 4008

European Union Arabidopsis Genome Sequencing Consortium, The Cold Spring Harbor, Washington University in St Louis and PE Biosystems Arabidopsis Sequencing Consortium 1999: Sequence and analysis of chromosome 4 of the plant Arabidopsis thaliana. Nature 402: 769-777

Hecker, K.H. and Roux, K.H. 1996: High and low annealing temperatures increase both specifity and yield in touchdown and stepdown PCR. Biotechniques 20: 478-485

Heinze, B. 1998a: Molekulargenetische Unterscheidung und Identifizierung von Schwarzpappeln und Hybridpappelklonen (Molecular genetic differentiation and identification of black poplars and hybrid poplar clones). FBVA-Berichte (Schriftenreihe der Forstlichen Bundesversuchsanstalt Wien/Vienna) 105 (44 pp.)

Heinze, B. 1998b: PCR-based chloroplast DNA assays for the identification of native Populus nigra and introduced poplar hybrids in Europe. Forest Genetics 5: 31-38

Heinze, B. 1997: A PCR marker for a Populus deltoides allele and its use in studying introgression with native European Populus nigra. Belgian J. Bot. 129: 123-130

Lin, X., Kaul, S., Rounsley, S., Shea, T.P., Benito, M.-I., Town, C.D., Fujii, C.Y., Mason, T., Bowman, C.L., Barnstead, M., Feldblyum, T.V., Buell, C.R., Ketchum, K.A., Lee, J., Ronning, C.M., Koo, H.L., Moffat, K.S., Cronin, L.A., Shen, M., Pai, G., Van Aken, S., Umayam, L., Tallon, L.J., Gill, J.E., Adams, M.D., Carrera, A.J., Creasy, T.H., Goodman, H.M., Somerville, C.R., Copenhaver, G.P., Preuss, D., Nierman, W.C., White, O., Eisen, J.A., Salzberg, S.L., Fraser, C.M. and Venter, J.C. 1999: Sequence and analysis of chromosome 2 of the plant Arabidopsis thaliana. Nature 402: 761-768

Meyerowitz, E.M. 1999: Today we have naming of parts. Nature 402: 731-732

Perry, D.J. and Furnier, G.R. 1996: Pinus banksiana has at least seven expressed alcohol dehydrogenase genes in two linked groups. Proc. Natl. Acad. Sci. USA 93: 13020-13023

Strand, A.E., Leebensmack, J. and Milligan, B.G. 1997: Nuclear DNA-based markers for plant evolutionary biology. Mol. Ecol. 6: 113-118

Subramaniam, R., Reinold, S., Molitor, E.K. and Douglas, C.J. 1993: Structure, inheritance, and expression of hybrid poplar (Populus trichocarpa x Populus deltoides) phenylalanine ammonialyase genes. Plant Physiol. 102: 71-83

Taberlet, P., Fumagalli, L., Wust-Saucy, A.G. and Cosson, J.F. 1998: Comparative phylogeography and postglacial colonization routes in Europe. Molecular Ecology 7: 453-464

Tao, R., Yamane, H., Sugiura, A., Murayama, H., Sassa, H., and Mori, H. 1999: Molecular typing of S-alleles through identification, characterization and cDNA cloning for S-RNases in sweet cherry. J. Am. Soc. Hort. Sci. 124: 224-233

Genetic types in white oak populations north of the Alps and in the Danube valley

U.M. CSAIKL[1] AND A.O. KÖNIG[2]

[1]*Österreichisches Forschungszentrum Seibersdorf (ARCS), A - 2444 Seibersdorf, Austria;*
[2]*Bundesforschungsanstalt für Forst- und Holzwirtschaft (BFH), Institut für Forstgenetik und Forstpflanzenzüchtung , D - 22927 Grosshansdorf, Germany*

Key words: white oaks, *Quercus robur*, *Q. petraea*, haplotypes, genetic variation

Abstract: 55 populations of white oak were investigated in South Germany and parts of Austria in order to detect restriction fragment length polymorphisms (RFLPs) in the chloroplast DNA (cpDNA) of oak leaves. Four primer-enzyme systems were used. The variation in the observed fragments was scored resulting in a ten digit code, which allowed the determination of 20 haplotypes. They are in accordance with results of earlier investigations and are associated with the glacial refugia and the postglacial re-migration history. Several haplotypes show a regional distribution pattern which can be of use for the characterisation of forest reproductive material.

Introduction

During the last glacial period plant species had to withdraw to southern Europe to areas with climatic conditions that allowed a survival of the species. When climatic conditions improved a re-colonisation process started. Many plant species died either during the glacial or postglacial period, because they could not withdraw to suited refugia or could not re-migrate to areas with favourable climatic conditions after the warming process. The European mountain ranges like the Pyrenees and the Alps extend in West-East direction and thus formed a serious barrier for the re-migration.

The re-colonisation process has been intensively studied by analysing pollen profiles in the layers of bogs and moors and postglacial pollen distribution maps have been established, allowing conclusions with regard to re-migration (Firbas 1949). Due to the lack of bogs it is not always possible to delimit exactly the refugia of all species. Probable refugial areas can sometimes be retraced from the re-migration routes. Oaks are assumed to have survived in refugia on the Iberian peninsula, the Apennine peninsula and on the Balkans (Huntley and Birks 1983). Newer investigations propose also a refuge west of the Black Sea (Brewer *et al.* 2002).

Müller-Starck, G. and Schubert, R. (eds.). Genetic Response of Forest Systems to Changing Environmental Conditions. Vol. 70, (For. Sci.), 129-138 © *2001 Kluwer Academic Publishers, Dordrecht, Boston, London*

Independent from paleobotanic research, investigations have been carried out on the genetic variation in oak species throughout their distribution range in Europe (Petit *et al.* 1993, Dumolin-Lapègue *et al.* 1997b). They have revealed that the distribution of genetic types, as assessed by polymorphisms of chloroplast DNA fragments, show a geographic pattern, and that this pattern has its origin in re-colonisation routes.

As oaks are species requiring moderate temperatures, it is evident that re-migration into the present distribution area must have taken place mainly in plains and hilly regions and especially along big river valleys. The Danube valley and the area north of the Alps might have played a role in the colonisation of Middle Europe. In the course of a project, called "oakgenetics" (see Acknowledgement), supported by the Commission of the European Communities, the distribution of genetic types has been studied in the laboratories of twelve partner institutes by a standardised method (see Dumolin-Lapègue *et al.* 1997b). The results have been compiled and will be published in a special issue of Forest Ecology and Management. In the present paper a part of the samples from the region north of the Alps and in the Danube valley were re-analysed with the help of a refined method developed in the Austrian Research Center Seibersdorf (ARCS). Being a crossroad of different re-migration routes, it was assumed that in this region a high number of variants of genetic types could be detected. To clarify the picture, additionally some populations at the borders of the described region were included which do not belong to the Danube valley in a strict sense.

Material and Methods

During the "oakgenetics"-project populations were selected following a 50 km grid collection system. Additionally samples from provenance experiments in the field or in the nursery were included. In regions where the occurrence of oak is sparse, like in the Alpine part of the Inn valley, no stands could be selected but single trees had to be sampled. With the assistance of local foresters stands or trees were selected which were assumed not being established from alien material. Old trees were preferred to young trees. The minimum distance between trees should be 50 m in order to minimise the sampling of related trees. As an average five trees (three to seven) per location were sampled. From each tree green leaves or twigs with leaves were picked and kept separately in labelled plastic bags in order to keep them fresh during transport to the laboratory. There they were processed immediately or shock-frozen and stored at -20° or -80°C. For the present paper 33 populations from South Germany (mainly Bavaria) and 22 from Tyrol and Upper Austria with altogether 260 trees were analysed.

The chloroplast chromosome consists of a circular molecule, which has been well investigated for many plant species (Clegg *et al.* 1994). There are rather large similarities in the gene content and structure of the chloroplast chromosome between the so far investigated species. Despite the conservative rate of evolution of chloroplast DNA (cpDNA), molecular analyses reveal complex patterns of mutational change (Clegg *et al.* 1994). Insertions of cpDNA pieces lead to longer and deletions to shorter sections of given DNA regions. Additionally point mutations may have occurred, which can also be detected if a restriction enzyme recognition site is involved. Highly conserved regions are frequently flanked by variable non coding DNA regions. This fact can be used to detect and describe genetic polymorphisms. According to Wagner (1992), cpDNA variation in forest trees has received greater attention than nuclear DNA polymorphisms.

The basic laboratory procedures used are described by Taberlet *et al.* (1991), Demesure *et al.* (1995) and Dumolin-Lapègue *et al.* (1997a). Here only a summary of the methods is presented. Discs of deep frozen leaves are homogenised and total DNA is extracted. For further analysis specific regions of cpDNA are amplified in order to get the necessary amount of DNA for further analysis. This is done during the polymerase chain reaction (PCR) with the help of so-called primers, which are short stretches of nucleotides matching the DNA sequences in the region of interest. In this process the DNA is subjected to numerous cycles of denaturing, attachment of primers and subsequent elongation of the DNA strand producing a multitude of exact copies. In the present project four regions of the oak cpDNA were analysed with the help of four primer pairs. In this paper they will be designed only by abbreviations. For the region T-F the two universal 20 base pair primers developed by Taberlet *et al.* (1991) were used and for the regions D-T, A-S and C-D those developed by Demesure *et al.* (1995). The length of the studied regions varies between approximately 1200 and 3700 bp. Subsequently the amplified DNA-products are digested by restriction endonucleases. This means that each of the amplified DNA segments is cut into pieces, so-called fragments, of different length, according to the location of the recognition sequence for the respective enzymes. In summary for the present investigations the following primer-enzyme-combinations were applied: T-F *Alu* I, D-T *Taq* I, A-S *Hinf* I and C-D *Taq* I.

In the next step the fragments are subjected to a polyacrylamid-gelelectrophoresis, in which long fragments migrate slower in an electric field than the shorter ones resulting in bands at different positions. At the ARCS the electrophoresis was modified by using longer gels and a shark tooth comb. This resulted in a higher resolution as compared to the method used in the "oakgenetics"-project. Certain fragments show variation which is called a restriction fragment length polymorphism (RFLP). Usually the polymorphisms are scored with a 1 for the longest fragment on the top of the observed gel region in ascending order for the following ones. A point mutation, which leads to a new restriction site, will split an amplified DNA strand into two fragments. This is scored by a 9, or if an other point mutation is occurring by an 8. For the determination of the haplotypes three fragments of each of the D-T, C-D and A-S regions, and one fragment of the T-F region were taken into consideration. If the information on all scores is combined, a ten digit code is obtained, which enables the determination of different genetic types. In the case of cpDNA they are called haplotypes, because the chloroplast genes are usually inherited uniparentally; in oaks they contain only the maternal information (Dumolin *et al.* 1995).

Results

The results are compiled in table 1. In the left hand column the numbers of the eight haplotypes are listed which have been found with the method of Dumolin-Lapègue *et al.* (1997b). Additionally the letters of the lineages to which they belong were added. 181 trees belong to lineage A (presumed refugium Balkans), six ones to lineage B (Iberian peninsula) and 73 ones to lineage C (presumed refugium Apennine). The columns in the middle of table 1 show the 10 digit code for each haplotype established, using a method with higher resolution. Each different code presents an individual haplotype. This resulted for example in the detection of eight new types instead of the former haplotype 1 and four new ones for haplotype 7. Within the region investigated in total 20 different haplotypes could be distinguished.

To get an impression on the spatial distribution of the haplotypes they were mapped together with the border lines of South Germany and Austria (figure 1). The black and white presentation does unfortunately not allow to assign a distinctive pattern to each of the 20 haplotypes as it was possible in the oral presentation with the help of colour transparencies. So only the more frequent haplotypes are shown by distinct patterns, whereas for the rare ones the circles were left open. The map shows at least for two haplotypes a regional cluster: one is occurring in Lower Bavaria (new haplotype 91, horizontally hatched

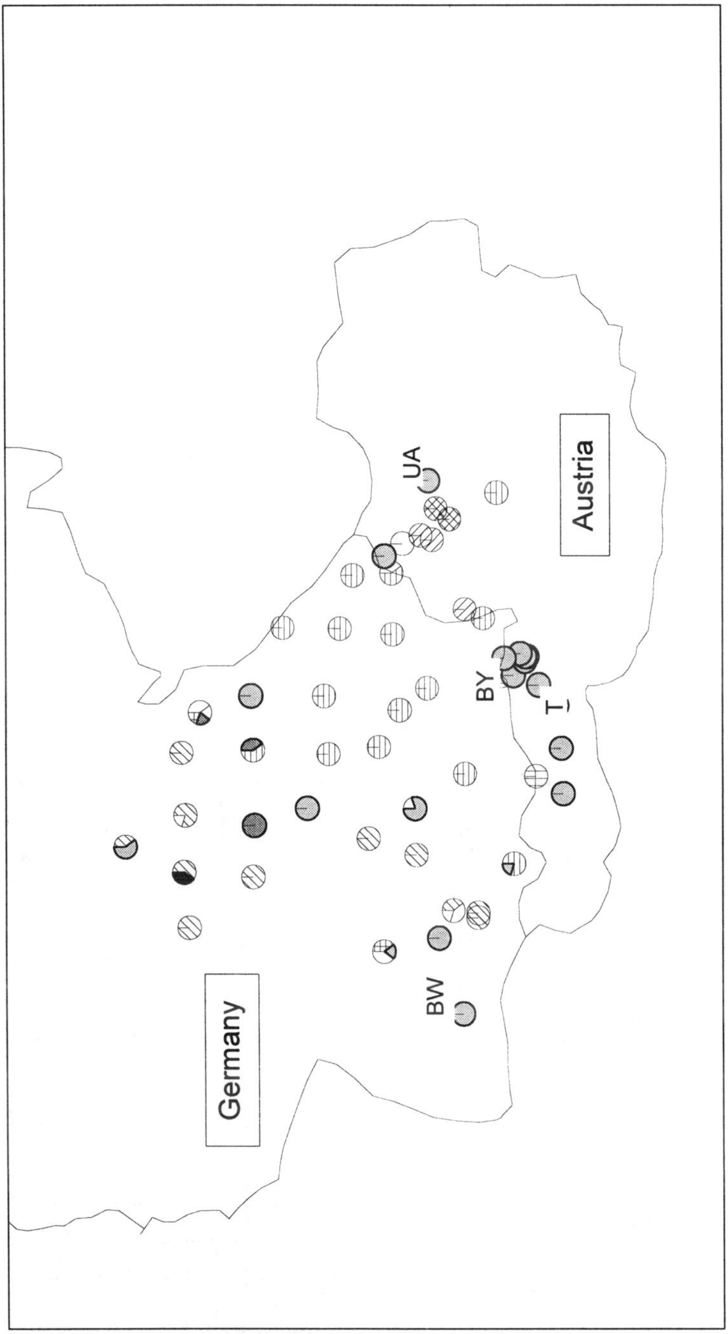

Figure 1. Distribution of haplotypes in South Germany (BW and BY), Tyrol (T) and Upper Austria (UA). A circle with a uniform pattern shows that only one haplotype was found at the respective location, whereas a division of a circle into sectors of different patterns indicates the detection of more than one haplotype.

Table 1. Haplotypes according to Dumolin Lapègue *et al.* (1997) and haplotypes detected with a refined method. The 10 digit code reflects the variation observed in the banding pattern of individual fragments and shows which variants separate the haplotypes

Standard haplotypes				New 10 digit code											Results		
Cons. no.	Hap. no. in EC-project	Total no. of trees	Lineages *		D-T			C-D			A-S		T-F	New haplotype no.	No. of trees in D	No. of trees in A	Total
1	1		C	2	1	1	6	2	3	3	2	2	2	14	2	0	2
2	1		C	9	2	2	2	2	3	3	2	2	2	43	0	3	3
3	1		C	9	1	2	5	8	4	1	2	9	2	51	2	0	2
4	1		C	9	1	2	6	2	3	3	2	2	2	56	46	0	46
5	1		C	9	1	2	6	2	3	3	2	2	3	57	1	0	1
6	1		C	9	1	2	2	2	3	3	2	2	2	69	0	11	11
7	1		C	9	1	2	2	2	3	3	2	2	3	70	0	1	1
8	1	67	C	2	1	5	6	2	3	3	2	2	2	87	1	0	1
9	2		C	9	1	2	5	1	2	1	2	9	2	62	0	3	3
10	2	6	C	9	1	2	5	9	2	1	2	9	2	65	0	3	3
11	4	8	A	2	1	1	6	1	2	1	3	2	3	16	8	0	8
12	5	2	A	2	1	2	6	1	2	1	3	2	3	81	2	0	2
13	7		A	9	1	2	6	2	3	1	3	2	2	58	3	0	3
14	7		A	2	1	5	5	1	2	1	3	2	3	90	0	15	15
15	7		A	2	1	5	6	1	2	1	3	2	3	91	65	6	71
16	7	171	A	2	1	5	3	1	2	1	3	2	3	95	29	53	82
17	10		B	4	2	3	6	1	2	1	2	2	3	36	2	0	2
18	10	4	B	9	1	2	6	2	3	1	2	2	2	55	2	0	2
19	11	1	B	4	2	3	6	1	2	1	2	2	1	35	1	0	1
20	12	1	B	4	2	4	6	1	2	1	2	2	3	21	1	0	1

*) = Lineages with presumed origin on the Balkans (A), the Iberian peninsula (B), and the Apennine (C), respectively

circle and one in the Inn valley in Tyrol (new haplotype 95, grey filled circles). In Frankonia (north-western part of the map) types from the Apennine refugium are prevailing (new haplotype 56, diagonally hatched) in mixture with types originating from the south-eastern European refugia.

Discussion

The method of Dumolin-Lapègue *et al.* (1997b) that had been applied in the research project "oakgenetics" in some twelve laboratories has led to remarkable results in relation to the postglacial colonisation of oak in Europe (special issue of "Forest Ecology and Management", Elsevier, in press). Even in the small geographic sector presented in this study it could be seen that the re-migration from the Iberian refugia (haplotypes 10, 11 and 12) stopped in South Germany in the border region of Baden-Württemberg (BW) and Bavaria (BY). The ancestors of these populations had expanded through France and penetrated probably through the Porte de Bourgogne until they arrived in Southwest Germany. This thesis would be in accordance with conclusions of Küster (1998) with regard to the re-migration of forest species. Farther east in Bavaria and in Austria these types were not detected.

Haplotype 1 (lineage C, Apennine, according to Dumolin-Lapègue *et al.* 1997b) entered into Germany in the South-West from where it expanded mainly to the North. On the other hand haplotype 2 (new haplotypes 62 and 65 in this study) belonging to the same lineage, occurs mainly in Upper Austria. Its re-migration route is quite different to that of haplotype 1. Haplotype 2 probably passed the Alps somewhere in the eastern part and expanded then to the north and west. In the west its distribution stopped in the German-Austrian border region. In the rest of Germany it was found only occasionally in one stand in the northwest of the country. (König *et al.* 1998). It can therefore be assumed that it had been established there with alien seed or plant material.

It is remarkable that also the new haplotypes show a geographic structure. Since there are no topographic barriers in the investigated region which could have hindered West-East-migration or *vice versa,* no hypotheses can be set up and tested. But it is evident from table 1 and demonstrated additionally in figure 2 that there are differences in the frequency of occurrence of several haplotypes among Germany and Austria. The well pronounced clusters support the thesis that the analysed stands of that areas might be autochthonous. Chloroplast genes of oaks are inherited maternally and are therefore only distributed in the acorns, and not by pollen like genes of the nuclear chromosomes (Dumolin *et al.* 1995). The situation of stands consisting of one single haplotype is favoured by the

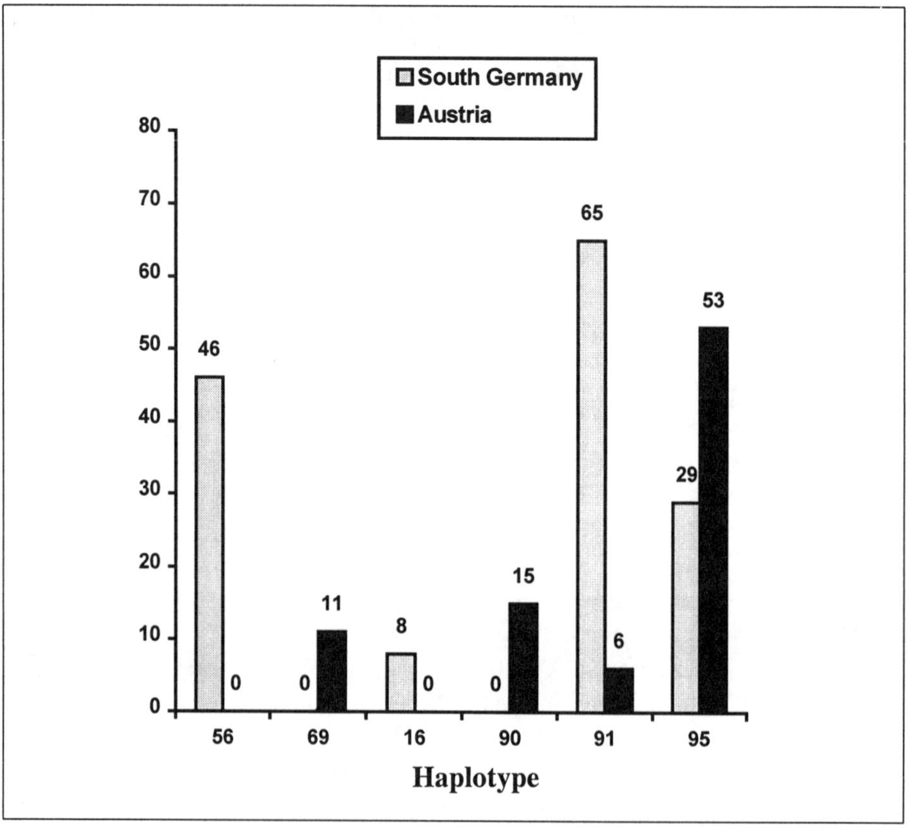

Figure 2. Frequencies of haplotypes (more than five trees investigated) among South Germany and Austria

founder effect. That haplotype which arrives first, produces the acorns for the next generation. Acorns are heavy and the majority of them germinate in the surrounding of the mother tree, thus establishing a population of the same type. The situation is of course different in areas where different lineages meet or where man has interfered and introduced seed from regions where different haplotypes are prevailing. The first situation might occur in the border region of Baden-Württemberg and Bavaria where the Iberian lineage comes in touch with the Balkan and the Apennine lineages. The second case might be true for the analysed stand in the forest district of Kemnath (north-eastern circle in figure 1), where four different haplotypes were detected. An additional example comes from the Inn valley, where a non matching haplotype has been found in the park of an old castle.

Conclusions

The results are a further example that molecular methods can be successfully applied for the detection and evaluation of genetic variation in forest trees. With regard to subtyping they show further the rapid development of these methods.

The chloroplast DNA has proved to be of special value for the detection of genetic variation in oak. The results of the European project "oakgenetics" are not only of interest for elucidating postglacial re-migration routes, but can also be used for the characterisation of regions and areas. The knowledge of the geographic distribution pattern supports certain conclusions on the origin (autochthony) of stands. In some cases the method reveals human influence with regard to stand establishment.

With the method of subtyping it was possible to detect a finer geographic pattern. A further development of the method, by taking additional fragments into consideration, will probably allow a clear characterisation of oak seed stands at least for specific subregions. As a consequence this would enable a more effective identity control of forest reproductive material moving in trade. In the case of doubts on the origin or provenance it could be analysed if the haplotype of the seed lot and the stand match. If they do not, some mix up or manipulation cannot be excluded.

With regard to these aspects it is therefore recommended to apply the described method to the characterisation of registered oak seed stands and to put efforts into its further development.

Acknowledgement

The full title of the mentioned "oakgenetics" project is: Synthetic maps of gene diversity and provenance performance for utilization of oak resources in Europe". It was supported by the Commission of the EC (FAIR1-CT95-0297) and co-ordinated by A. Kremer (INRA, France).

The technical assistance of N. Ziegler in the collection of the material in Bavaria is greatly acknowledged. The material from Baden-Württemberg was provided by A. Franke, FVA Freiburg. In Austria numerous foresters contributed to the collection and transfer of the material to the ARCS. The laboratory analyses were carried out by K. Hohl, K. Groppe and N. Ziegler, and are as well greatly acknowledged.

References

Brewer, S., Cheddadi, R. and Beaulieu, J. L. 2002: The migration of deciduous *Quercus* throughout Europe since the last glacial period (special issue of "Forest Ecology and Management" Elsevier, in press)

Clegg, M. T., Brandon, S. G., Learn, G. H. and Morton B. 1994: Rates and patterns of chloroplast DNA evolution. Proc. Natl. Acad. Sci. USA, Vol. 91: 6795-6801

Demesure, B., Sodzi, N. and Petit, R. J. 1995: A set of universal primers for amplification of polymorphic non-coding regions of mitochondrial and chloroplast DNA in plants. Molecular Ecology 4: 129-131

Dumolin, S., Demesure B. and Petit, R. J. 1995: Inheritance of chloroplast and mitochondrial genomes in pedunculate oak investigated with an efficient PCR method. Theoretical and Applied Genetics 91: 1253-1256

Dumolin-Lapègue, S., Pemonge, M.-H. and Petit, R. J. 1997a: An enlarged set of consensus primers for the study of organelle DNA in plants. Molecular Ecology 6: 393-397

Dumolin-Lapègue, S., Demesure, B., Fineschi, S., Le Corre, V. and Petit, R. J. 1997b: Phylogeographic structure of white oaks throughout the European continent. Genetics 146: 1475-1487

Firbas, F. 1949: Spät- und nacheiszeitliche Waldgeschichte Mitteleuropas nördlich der Alpen. Verlag von Gustav Fischer in Jena

Huntley, B. and Birks, H. J. B. 1983: An atlas of past and present pollen maps for Europe, 0-13000 years ago. Cambridge University Press, Cambridge

König, A. O., Groppe, K. and Ziegenhagen, B. 1998: First results of chloroplast-DNA investigations in German populations of *Quercus robur* and *Q. petraea*. In: Diversity and Adaptation in Oak Species. Proc. of the second meeting of WP 2.08.05, Genetics of *Quercus*, of the IUFRO. University Park (State College), Pennsylvania, U.S.A., Oct 12-17, 1997. Ed. K. C. Steiner. The Pennsylvania State University, University Park, PA, 102-108

Küster, H. 1998: Geschichte des Waldes. C. H. Beck, München

Petit, R. J., Kremer, A. and Wagner, D. B. 1993: Geographic structure of chloroplast DNA polymorphisms in European oaks. Theoretical and Applied Genetics 87: 122-128

Taberlet, P., Gielly, L., Pautou, G. and Bouvet, J. 1991: Universal primers for amplification of three non-coding regions of chloroplast DNA. Plant Molecular Biology 17: 1105-1109

Wagner, D. B. 1992: Nuclear, chloroplast, and mitochondrial DNA polymorphisms as biochemical markers in population genetic analyses of forest trees. New Forests 6. Kluwer Academic Publishers: 373-390

Highly polymorphic uniparentally inherited DNA markers for spatial genetic analysis of silver fir (*Abies alba* Mill.) populations

B. ZIEGENHAGEN[1], B. DEGEN[2,1], R.J. PETIT[3], M. ANZIDEI[4], A. MADAGHIELE[4], F. SCHOLZ[1] AND G.G. VENDRAMIN[4]

[1]Federal Research Centre for Forestry and Forest Products, Institute for Forest Genetics and Forest Tree Breeding, D-22927 Grosshansdorf, Germany; [2]INRA, Station de Recherches Forestières, Campus agronomique, Kourou cedex, French Guyana; [3]INRA, Laboratoire de Génétique et Amélioration des Arbres Forestières, 3361 Gazinet cedex, France; [4]Consiglio Nazionale delle Ricerche, Istituto Miglioramento Genetico delle Piante Forestali, I-50135 Firenze, Italy

Key words: conifers, silver fir, biodiversity, conservation, chloroplast microsatellites, spatial genetic diversity, paternity

Abstract: The present study is a summary on the development and inheritance analysis and application of chloroplast microsatellite markers in spatial genetic analysis of the conifer species silver fir (*Abies alba* Mill.). It aims at discussing and concluding the indicative value of these uniparentally paternally inherited markers in biodiversity and conservation studies at two different spatial scales. When applied among natural populations throughout the natural range the highly polymorphic markers harboured considerable sensitivity towards isolation and/or genetic drift. A pilot study at the stand level revealed the potential of the marker for paternity analysis and reconstruction of the mating system in an endangered relic population. Perspectives are given for a combined usage of DNA markers with contrasting modes of inheritance.

Introduction

Biodiversity studies and resulting efforts of conservation have moved into the focus of international research activities. At the within-species level, understanding and maintaining of biological diversity needs knowledge on the intraspecific genetic diversity (UN Convention on Biological Diversity, Rio 1992). In European forest tree species, the intraspecific genetic diversity is supposed to reveal distinct spatial patterns due to the glacial and post-glacial history of the populations (e.g., Cones and Kadereit 1998, Taberlet *et al.* 1998, for general considerations, see Slatkin 1987). Furthermore, anthropogenic influences such as seed transfer or silvicultural practices are supposed to have interfered with the natural processes of distribution of the intraspecific genetic

Müller-Starck, G. and Schubert, R. (eds.). Genetic Response of Forest Systems to Changing Environmental Conditions. Vol.70, (For. Sci.), 139-149 © 2001 Kluwer Academic Publishers, Dordrecht, Boston, London

variation (e.g., Degen and Scholz 1998, Bucci and Vendramin 2000). Dealing with conservation issues consequently means to measure genetic diversity as well as to derive indications for its spatio-temporal dynamics (Namkoong *et al.* 1996). For experimental implementation, a stepwise procedure is proposed starting with analysing the spatial distribution of intraspecific genetic diversity by means of genetic markers. The usage of different genetic markers will allow to reconstruct different processes that may have led to the present spatial genetic variation, e.g. processes like migration, mating, isolation, genetic drift or selection. Particularly, novel DNA markers harbouring contrasting modes of inheritance, different mutation rates, and originating from non-coding *vs.* coding regions of nuclear or organelle genomes are supposed to enable the reconstruction of different population genetic processes, separately (Gillet 1999a). The obtained data may be composed into a more completed picture reflecting the different processes that have become effective. In a last step, these data may provide the basis for deriving the above mentioned indications and for parameterising computer simulation models. The latter are expected to either test hypotheses or to predict the spatio-temporal dynamics of intraspecific genetic diversity under simulated conditions.

The present study summarises and draws conclusions from mainly three publications by the authors that have been dealing with the development, characterisation and application of a uniparentally inherited DNA marker in spatial genetic investigations of silver fir (*Abies alba* Mill.). For this species, a strong human impact by long-distance seed transfer may be excluded as silver fir populations are supposed to be natural in most parts of their natural range. Two chloroplast microsatellite (= cp-microsatellite) loci had been characterised and confirmed to be paternally inherited in fir (*Abies* spp., Vendramin and Ziegenhagen 1997). The application of the marker in 17 selected populations throughout the natural range of silver fir (*Abies alba* Mill.) exhibited a high degree of polymorphism (Vendramin *et al.* 1999a). Within a relic silver fir population, the variation at the same cp-microsatellite loci had been used for a pilot paternity analysis (Ziegenhagen *et al.* 1998).

In this paper, the potential of a DNA marker which is uniparentally inherited and supposed to be neutral is summarised and discussed for its indicative value of reconstructing isolation and genetic drift as well as for its potential of reconstructing uniparental gene flow including paternity analysis. Furthermore, ongoing experiments are described and discussed regarding the analysis of the DNA sequences that are underlying the observed size variation of the loci. Prospects are given for a combined usage of DNA markers with contrasting modes of inheritance in biodiversity studies of silver fir.

Material and Methods

Detailed protocols on DNA extraction, specific PCR amplification of the cp-microsatellite loci under study, cloning and sequencing of the PCR products, sizing of the PCR products as well as details on genetic data analysis and geographical visualisation are given in the relevant papers (Vendramin and Ziegenhagen 1997, Ziegenhagen *et al.* 1998, Vendramin *et al.* 1999a).

Isolation, characterisation and inheritance analysis of cp-microsatellites in fir

For isolation of cp-microsatellite loci in fir, twenty pairs of universal primers derived from the completely sequenced *Pinus thunbergii* chloroplast genome (Wakasugi *et al.* 1994) and known to amplify cp-microsatellite loci in *Pinus* species (Powell *et al.* 1995, Vendramin *et al.* 1996) had been tested on individuals of different *Abies* species involved as parental individuals in intra- and interspecific controlled crosses. Eleven out of the twenty pairs were found to generate PCR products, but only two of them (Pt 30204 and Pt 71936) exhibited length polymorphism within and among *Abies* species (Vendramin and Ziegenhagen 1997). Sequence analysis of the obtained PCR products with length polymorphism confirmed the presence of microsatellites at both loci. Inheritance analysis revealed the predominant transmission of the paternal variant from the parents to a total of 75 F1 offspring, representing both, embryos and seedlings. In the megagametophytes of those seeds from which the embryos had been analysed from, the predominant occurrence of the maternal variant was scored (Vendramin and Ziegenhagen 1997, also see Ziegenhagen *et al.* 1996).

Assessment of the geographical distribution of haplotypic genetic diversity

Seventeen populations scattered throughout the natural range of silver fir (*Abies alba* Mill.) were analysed for size variation at the two cp-microsatellite loci Pt 30204 and Pt 71936 (Vendramin *et al.* 1999a). Location, sample size and ontogenetic stage of the samples are given in Figure 1 (for more details see Vendramin *et al.* 1999a). The analysis of a total of 714 individuals resulted in the detection of 18 different size variants at locus Pt 30204 and of 8 different size variants at locus Pt 71936 combining to 90 different two-locus-haplotypes (Vendramin *et al.* 1999a). Estimations of the two-locus haplotypic diversity were obtained by measuring allelic richness $r_{(n)}$ after rarefaction to a uniform sample size of n, as described in El Mousadik and Petit (1996). G_{ST} calculations were performed as proposed for haploid loci by Pons and Petit (1995). For details of data analysis, see Vendramin *et al.* (1999a).

Pilot paternity analysis in a relic silver fir population

Study site was a mixed stand of beech, Norway spruce and relic silver fir individuals in a valley of the Saxonian Ore Mountains (Polenztal, Forstamt Lohmen). Adult silver fir trees occurred as solitary individuals and in a few groups, some of them with surrounding naturally regenerated young trees. Needle samples were available and investigated from two relatively isolated adult silver fir trees (A and B) of about 30 m distant from each other and from all 24 young trees present in their immediate neighbourhood (8 in the understory of A, and 16 in the understory of B, also see schematic drawing in Fig. 2). The nearest neighbouring group of other adult firs in the stand is located at a distance of more than 100 m, the nearest solitary adult individual at about 50 m apart from the analysed trees. Altogether 26 individuals were analysed at the polymorphic microsatellite loci, Pt 30204 and Pt 71936.

Results, Discussion and Conclusions

Uniparental inheritance vs. heteroplasmy

Before using DNA polymorphisms as genetic markers in population genetic investigations they need to be validated by means of inheritance analysis (Gillet 1999b). Our polymorphic cp-microsatellite loci were confirmed to be predominantly uniparentally paternally inherited. This finding of paternal transmission of plastid DNA is in line with previous results in silver fir (Ziegenhagen *et al.* 1995) and with what has been found in most other conifer species investigated so far (Wagner 1992). Altogether these results give genetic evidence for cytological findings on the paternal transmission of the plastid organelle during fertilisation, zygotic and proembryo development in another member of the Pinaceae, douglas fir (Owens and Morris 1990 and 1991). In our cp-microsatellite analysis the occurrence of 1% heteroplasmic embryos was observed. This is similar to previous findings of 1% heteroplasmy as detected by cp-microsatellite analysis in the conifer species *Pinus radiata* (Cato and Richardson 1996). Both findings may genetically confirm the rare event of 'maternal leakage' postulated in the cytological studies by Owens and Morris (1990 and 1991), although in our studies we cannot exclude the possibility of contamination of embryonic tissue by endosperm tissue during the preparation procedure.

Thus for the usage of the marker in population genetics, in general, the following conclusions are drawn with respect to the plant material. When taking seeds as plant material, heteroplasmic embryos that are interfering with the reconstruction of pure paternal lineages can easily be rejected from analysis. This may be more difficult for adult trees where chimera cannot be ruled out that may develop from heteroplasmic embryos in the due course of tissue segregating for the one or the other haplotype (e.g., Govindaraju and Wagner 1988). Such chimera may be hard to identify when just a few needles are analysed. Therefore, when using needle material as a basis for paternity analysis, respectively for analysis of paternal lineages false positive results may be expected, though with low probability.

In contrast, diversity studies e.g., measuring `allelic richness´ are not affected by the phenomenon of maternal leakage.

Geographical distribution of haplotypic genetic diversity

The validated cp-microsatellite markers were applied in spatial genetic analysis by scoring and analysing the detected size variants. Throughout the natural range an extremely large number of two-locus-haplotypes were detected due to the comparably great number of size variants at the composed microsatellite locus Pt 30204, which seems to be a particularity of silver fir (Vendramin *et al.* 1999b). Figure 1 demonstrates the geographical distribution of the haplotypic diversity estimated in terms of allelic richness and based on the two-locus-haplotype variation. This 'absolute' measure is suggested for conservation biology studies (Petit *et al.* 1998) and in our study revealed a significant rank correlation among the two loci. Furthermore, there are great differences among the populations analysed concerning the parameter 'allelic richness'. For elucidating the indicative value of the marker we focus on the comparison of just two populations, one in the Southern Pyrenees (Fig. 1, No. 1) and the other one in Croatia (Fig. 1, No. 9) characterised by the same sampling strategy and similar sample sizes. The two natural populations are supposed to be located in areas of glacial refugia in the Pyrenees and Balkan peninsula (Konnert and Bergmann 1995). However, they are opposite in haplotypic diversity spanning a broad range from $r_{(n)}$ = 2.8 (Pyrenees) to $r_{(n)}$ = 8.3 (Croatia). Following the hypothesis of an accumulation of genetic variants in glacial refugia we may interpret a long term drift-migration-equilibrium of the Balkan population and in contrast processes of isolation and/or genetic drift due to bottlenecks in the history of the Pyrenean population. Isolation of the Pyrenees was also concluded from isozyme gene analysis (Konnert and Bergmann 1995).

Even though the results need to be strengthened by an improved sampling strategy throughout the natural range of silver fir, the potential of the marker for biogeography has become evident: It is highly sensitive towards stochastic population genetic or demographic processes that may cause losses of genetic variation. This is expected highest when isolation and drift both have become effective. The indicative value of cp-microsatellites is based on the combination of three features: they are a) selectively neutral, b) highly polymorphic, and c) uniparentally inherited. The latter reduces the reproductively effective population size to half the size of that when biparentally inherited markers are used (Birky *et al.* 1989) and is a basic reason for its sensitivity towards genetic drift/isolation (Morgante *et al.* 1997, Vendramin *et al.* 1998).

Figure 1. (extracted from Vendramin et al. 1999a) Distribution of haplotypic genetic diversity in

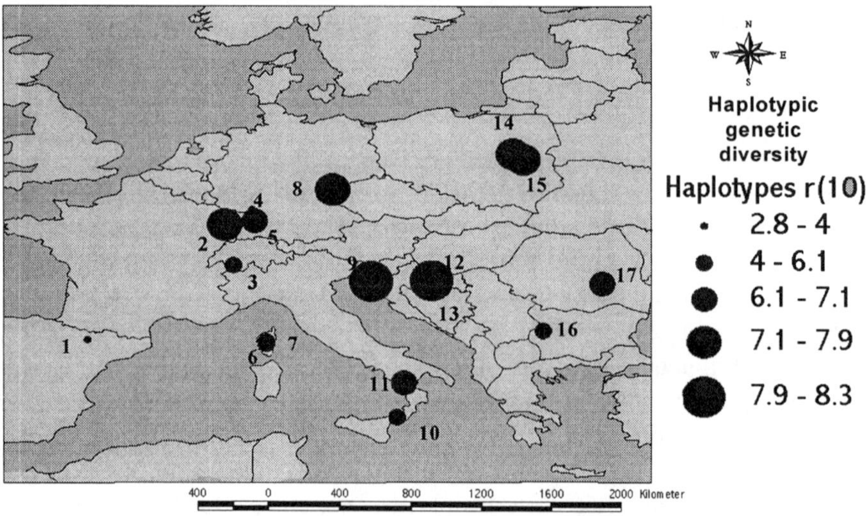

terms of allelic richness $r_{(n)}$. 1 - Pyrenees (N = 110, embryos from bulked seed collection), 2 - Vosges Mts. (N = 27, adult trees, sampled at random), 3 - Wallis (N = 28, provenance trial), 4 - Palatinate Mts. (N = 30, adult trees, sampled at random), 5 - Black Forest (N = 22, adult trees sampled at random), 6 - Corsica Marmano (N = 29, provenance trial), 7 - Corsica Rospa Sorba (N = 98, provenance trial), 8 - Ore Mts. (N = 33, adult and young trees at random), 9 - Croatia Delnice (N = 95, seedlings from bulked seed collection), 10 - Calabria F. Nardello (N = 10, provenance trial), 11 - Calabria Sotto Piano (N = 30, provenance trial), 12 - Croatia Nasice (N= 10, seedlings from bulked seed collection), 13 - Croatia Pozega (N = 24, seedlings from bulked seed collection), 14 - Poland Mins-Mazowiecki (N = 29, seedlings from bulked seed collection), 15 - Poland Lukow-Jata (N = 109, seedlings from bulked seed collection), 16 - Bulgaria (N = 10, adult trees sampled at random), 17 - Carpathian Mts., Romania (N = 20, adult trees, sampled at random).

Uniparental gene flow at the stand level

As there was evidence for an appropriate level of haplotypic diversity in the Ore Mountains (Fig. 1, No. 8), the marker was tested for paternity analysis in a relic Ore Mountain silver fir stand. In such relic stands, silver fir is supposed to suffer from inbreeding deficiency (Llamas Gómez and Braun 1994), self compatibility resulting into vital selfs cannot be excluded in the worst case (Moulalis 1986). The results of our analysis at the two highly polymorphic cp-microsatellite loci are visualised in Figure 2. Thereafter, at least 6 out of 24 young trees unambiguously have fathers different from A and B, furthermore the contribution of at least three different fathers is determined. This result was equally obtained by using either of the two microsatellite loci. For an unambiguous exclusion of adult trees A or B as fathers of the young trees carrying the haplotype of A or B, or an acceptance of them being selfs, further analyses are needed that exceed this pilot study.

For the potential of the marker for paternity analysis the following conclusions are drawn: In a small-sized relic population being part of a hypothetically existing metapopulation with an appropriate degree of polymorphism our cp-microsatellites are supposed to be useful markers to analyse spatial distances of pollen trapping by isolated trees or groups of trees. In cases where polymorphism is too low for unambiguous determination of selfs, an additional attempt should be made by applying highly polymorphic codominant nuclear microsatellites. However, these have still to be developed in fir. Once they are available, a combined application of chloroplast and nuclear microsatellites would at the same time be suited to rule out chimeric individuals due to maternal leakage of plastid DNA. Furthermore this opens up an occasion to analyse maternal gene flow as well and thus to learn about symmetry/asymmetry of paternal and maternal gene flow. To some extent this is already possible using our chloroplast microsatellites by analysing the maternal size variants in the megagametophytes of outspread seeds. At least this system is ready to get used for reconstructing within-stand F1 maternal gene flow. Of course, a highly polymorphic maternally inherited marker would be the marker of choice for maternity analysis and maternal gene flow. Recently, mitochondrial DNA microsatellites have been detected and characterised in pine species. So far, however, they exhibited only inter- and not intraspecific variation (Soranzo *et al.* 1999).

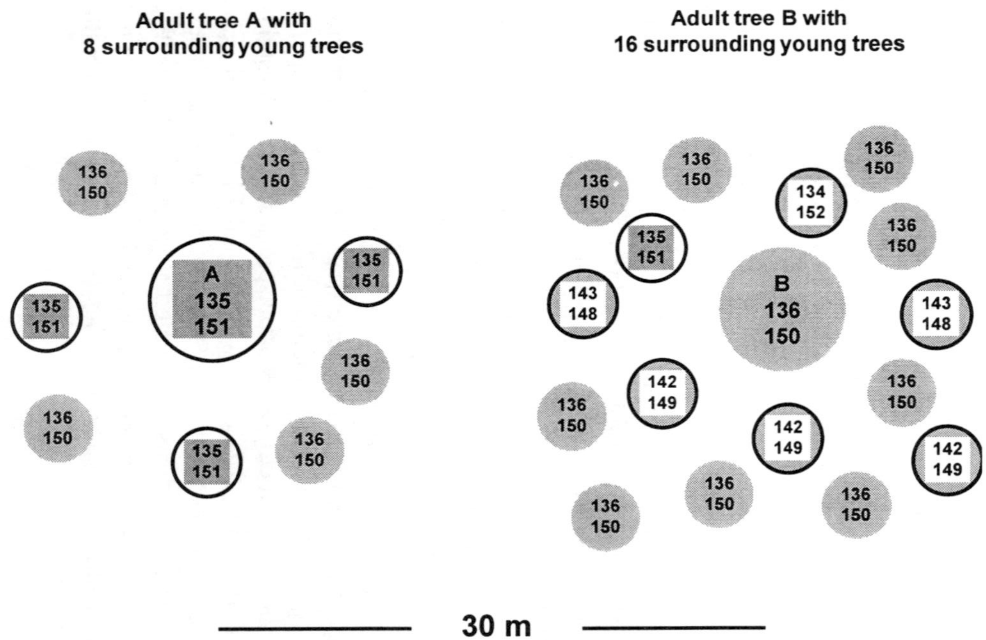

Figure 2. (extracted from Ziegenhagen *et al.* 1998): Schematic drawing of positions of adult trees A and B with surrounding young trees. Results of cp-microsatellite analysis are given as size variants (length in bp): upper value = size variant at locus Pt 30204; lower value = size variant at locus Pt 71936

Perspectives

The present paper summarises the usefulness of chloroplast microsatellite markers in spatial genetics of silver fir. Thereafter, they reveal a considerable potential for identifying populations that have undergone past processes of isolation and/or genetic drift. Also, they are useful in studies of the mating system in relic, respectively fragmented populations.

Ongoing experiments are dealing with sequence analysis of the 'mutational hot' locus Pt 30204. Actually, this locus is a complex one consisting of two different mononucleotide repeat stretches that were found to contribute to the size variation. Additionally, point mutations and small insertion/deletions were scored in the non-repetitive regions of the locus. Indeed, preliminary results reveal that variants of the same size are not necessarily variants of identical sequences. Thus sequence analysis will allow to detect homoplasy and in consequence offers an approach towards interpreting spatial genetic patterns in terms of spatial phylogeny.

To further strengthen investigations in the field of phylogeography covering the natural range, maternally inherited mitochondrial DNA markers are being developed in silver fir. These are supposed to more strongly differentiate among populations probably raising G_{ST} above 0.13, a value that was calculated from analysing our 17 populations using cp-microsatellites. In Norway spruce, estimates of genetic differentiation were about five fold higher when using mitochondrial DNA markers instead of cpDNA markers (Sperisen and Vendramin, personal communication). In the long run, also including more populations from the hypothised refugia and remigration routes, the glacial/postglacial history of silver fir may be better understood.

Acknowledgement

We would like to express our gratitude again to all who have provided plant material: Andrej Kormuták (Slovak Republic) Peter Rotach (Switzerland), Enrique Ritter (Spain), Dimitar Peev (Bulgaria), Ion Barbu (Romania), Joso Gracan (Croatia), Hubert Braun and Lorenzo Llamas Gómez (Germany), G.J. Wilhelm (Germany) and Walter Eder (Germany). Excellent technical assistance by Vivian Kuhlenkamp is highly appreciated. The studies were financially supported by the EU (Biotechnology Programme BIO4 CT96 0706).

References

Birky jr., C.W., Fuerst, P. and Maruyama, T. 1989: Organelle gene diversity under migration, mutation, and drift: equilibrium expectations, approach to equilibrium, effects of heteroplasmatic cells, and comparison to nuclear genes. Genetics 121: 613-627

Bucci, G. and Vendramin, G.G. 2000: Delineation of genetic zones in the European Norway spruce natural range: preliminary evidences. Molecular Ecology 9(7): 923-935

Cato, S.A. and Richardson, T.E. 1996: Inter- and intraspecific polymorphism at chloroplast SSR loci and the inheritance of plastids in *Pinus radiata* D. Don. Theoretical and Applied Genetics 93: 587-592

Cones, H.P. and Kadereit, J.W. 1998: The effect of Quarternery climatic changes on plant distribution and evolution. Trends in Plant Science 3: 432-438

Degen, B. and Scholz, F. 1998: Spatial genetic differentiation among populations of European beech *(Fagus sylvatica* L.) in Western Germany as identified by geostatistical analysis. Forest Genetics 5(3): 191-199

El Mousadik, A. and Petit R.J. 1996: High level of genetic differentiation for allelic richness among populations of the argan tree [*Argania spinosa* (L.) Skeels] endemic to Morocco. Theoretical and Applied Genetics 92: 832-839

Gillet, E.M. (Ed.) 1999a: Which DNA marker for which purpose? Final Compendium of the Research Project Development, optimisation and validation of molecular tools for assessment of biodiversity in forest trees in the European Union DGXII Biotechnology FW IV Research Programme Molecular Tools for Biodiversity.
URL http://webdoc.sub.gwdg.de/ebook/y/1999/whichmarker/index.htm

Gillet, E.M. 1999b: DNA markers - Concepts and characteristics. In: - Which DNA marker for which purpose? - Final Compendium of the Research Project Development, optimisation and validation of molecular tools for assessment of biodiversity in forest trees in the European Union DGXII Biotechnology FW IV Research Programme Molecular Tools for Biodiversity.
URL http://webdoc.sub.gwdg.de/ebook/y/1999/whichmarker/index.htm

Govindaraju, D.R. and Wagner, D.B. 1988: Chloroplast DNA variation within individual trees of *Pinus banksiana* - *Pinus contorta* sympatric regions. Can. J. For. Res. 18: 1347-1350

Konnert, M. and Bergmann, F. 1995: The geographical distribution of genetic variation of silver fir (*Abies alba, Pinaceaea*) in relation to its migration history. Plant Syst. Evol. 195: 19-30

Llamas Gómez, L. and Braun, H. 1994: Part A: Untersuchungen über ökologisch-genetische Anpassungsvorgänge bei der Tanne (*Abies alba* Mill.) in unterschiedlich immissionsbelasteten Regionen unter besonderer Berücksichtigung des Erzgebirges. pp. 1-64 in: Schriftenreihe der Sächsischen Landesanstalt für Forsten (Sächsische Landesanstalt für Forsten, Ed.) - Genetik und Waldbau der Weißtanne. Graupa

Morgante, M., Felice, N. and Vendramin, G.G. 1997: Analysis of hypervariable chloroplast microsatellites in *Pinus halepensis* reveals a dramatic genetic bottleneck. pp. 407-412 in: Karp, A., Isaac, P.G. and Ingram, D.S. (Eds.) - Molecular Tools for Screening Biodiversity - Plants and Animals. Chapman and Hall, London

Moulalis, D. 1986: Selbstinkompatibilität und Inzucht bei der Weißtanne. Forstw. Cbl. 105: 487-494

Namkoong, G., Boyle, T., Gregorius, H.-R., Joly, H., Savolainen, O., Wickneswari, R. and Young, A. 1996: Testing criteria and indicators for assessing the sustainability of forest management: Genetic criteria and indicators. 13 pp. In: Workshop of the International Forest Genetics Research Associates, Prabhu, R. (Organisation) - CIFOR project: Assessing the sustainability of forest management: Testing criteria and indicators. West Java, Indonesia, 8-12 April, 1996

Owens, J.N., Morris and S.J. 1990: Cytological basis for cytoplasmic inheritance in *Pseudotsuga menziesii*. I. Pollen tube and archegonial development. Amer. J. Bot. 77(49): 433-445

Owens, J.N. and Morris, S.J. 1991: Cytological basis for cytoplasmic inheritance in *Pseudotsuga menziesii*. II. Fertilization and proembryo development. Amer. J. Bot. 78(11): 1515-1527

Petit, R.J., El Mousadik, A. and Pons, O. 1998: Identifying populations for conservation on the basis of genetic markers. Conservation Biology 12: 844-855

Pons, O. and Petit, R.J. 1995: Estimation, variance and optimal sampling of gene diversity. I. Haploid locus. Theoretical and Applied Genetics 90: 462-470

Powell, W., Morgante, M., McDevitt, R., Vendramin, G.G. and Rafalski, J.A. 1995: Polymorphic simple sequence repeat regions in chloroplast genomes: Applications to the population genetics of pines. Proc. Natl. Acad. Sci. USA 99: 7759-7763

Slatkin, M. 1987: Gene flow and geographic structure of natural populations. Science 236: 787-792

Soranzo, N., Provan, J. and Powell, W. 1999: An example of microsatellite length variation in the mitochondrial genome of conifers. Genome 42: 158-161

Taberlet, P., Fumagalli, L., Wust-Saucy, A.-G. and Cossons, J.-F. 1998: Comparative phylogeography and postglacial colonization routes in Europe. Molecular Ecology 7: 453-464

Vendramin, G.G., Lelli, L., Rossi, P. and Morgante, M. 1996: A set of primers for the amplification of 20 chloroplast microsatellites in Pinaceae. Molecular Ecology 5: 111-114

Vendramin, G.G. and Ziegenhagen, B. 1997: Characterization and inheritance of polymorphic plastid microsatellites in *Abies*. Genome 40: 857-864

Vendramin, G.G., Anzidei, M., Madaghiele, A. and Bucci G. 1998: Distribution of genetic diversity in *Pinus pinaster* Ait. as revealed by chloroplast microsatellites. Theoretical and Applied Genetics 97: 456-463

Vendramin, G.G., Degen, B., Petit, R.J., Anzidei, M., Madaghiele, A. and Ziegenhagen, B. 1999a: High level of variation at *Abies alba* chloroplast microsatellite loci in Europe. Molecular Ecology 8: 1117-1126

Vendramin. G.G., Anzidei, M., Echt, C.S., Madaghiele, A., Sperisen C. and Ziegenhagen, B. 1999b: Chloroplast microsatellite approach for the study of diversity in conifers. pp 83-90 in: Douglas G.C. (Ed.) - Strategies for Improvement of Forest Tree Species. Coford, Dublin

Wagner, D.B. 1992: Nuclear, chloroplast, and mitochondrial DNA polymorphisms as biochemical markers in population genetic analyses. New Forests 6: 373-390

Wakasugi, T., Tsudzuki, J., Ito, S., Shibata, M. and Sugiura, M. 1994: A physical map and clone bank of the black pine (*Pinus thunbergii*) chloroplast genome. Plant Mol. Biol. Reptr. 12: 227-241

Ziegenhagen, B., Kormuták, A., Schauerte, M. and Scholz, F 1995: Restriction site polymorphism in chloroplast DNA of silver fir (*Abies alba* Mill.). Forest Genetics 2(2): 99-107

Ziegenhagen, B., Schauerte, M., Kormuták, A. and Scholz, F 1996: Plastid DNA polymorphism of megagametophytes and pollen in two *Abies* species. Silvae Genetica 45: 355-358

Ziegenhagen, B., Scholz, F., Madaghiele, A. and Vendramin, G.G. 1998: Chloroplast microsatellites as markers for paternity analysis in *Abies alba*. Can. J. For. Res. 28: 317-321

Levels of genetic differentiation in *Pinus halepensis* Mill. in Spain using quantitative traits, isozymes, RAPDs and cp-microsatellites.

R. ALIA[1], A. GÓMEZ[1], M. D. AGÚNDEZ[1], M. A. BUENO[1] and E. NOTIVOL[2]

[1]*Forest Research Centre, INIA,. 28040 Madrid, Spain;* [2]*Unit of Forest Resources. SIA, Diputacion General de Aragon, 50080 Zaragoza, Spain*

Key words: genetic differentiation, Fst, phenotypic variation

Abstract: The pattern of geographic variation and the relationship among genetic variation in quantitative and molecular traits have been investigated in 15 populations of Aleppo pine covering the range of its distribution in Spain. The F_{ST} was estimated by means of 5 polymorphic isozyme loci, and, in a subset of 6 populations, by 24 RAPDs and 9 Cp-microsatellites loci. The phenotypic traits measured were the following: germination parameters obtained by the Gompertz function, height growth at age of 4 years in two provenance-progeny test and height growth phenology parameters in one of the two sites. The average F_{ST} value estimated using isozymes analysis (0.11) was compared with that obtained from phenotypic traits. Results showed a weak selection acting in some of the traits (phenology) and a strong selection in germination and height growth. Results obtained when analysing 15 populations were comparable with those obtained by RAPDs and Cp-microsatellites in a smaller subset of populations. The importance of selection in the pattern of variation of the species was discussed.

Introduction

Mediterranean forests, and in particular those of Aleppo pine (*Pinus halepensis* Mill.), have been influenced by human activity over a long period of time. This human activity, combined with both selective and non-selective genetic processes, has produced a complex pattern of variation in this species. In addition, large differences exist among populations in terms of ecological conditions that, together with historical records, have been used to differentiate regions of provenance of the native stands (Gil *et al.* 1996). Fragmentation of the native area of Aleppo pine has resulted in a scattered distribution of the species, with isolated populations in central and southern Spain.

Several studies dealing with different molecular and biochemical markers (Schiller *et al.* 1986; Schiller and Grunwald 1987; Baradat *et al.* 1995; Vendramin *et al.* 1996), and the role of adaptation to drought and frost (Bariteau 1992; Weinstein 1989) are available. In the Iberian peninsula,

Müller-Starck, G. and Schubert, R. (eds.). Genetic Response of Forest Systems to Changing Environmental Conditions. Vol. 70, (For. Sci.), 151-160

significant genetic variation has been detected by using isozymes, RAPDs and Cp-microsatellites (Agúndez *et al.* 1997; Gómez 1998). This genetic variation is geographically structured showing a clinal pattern that suggests a spread of Aleppo pine since the last glaciation from native refugia located in the Iberian peninsula (Agúndez *et al.* 1999).

Despite investigation having been carried out into a wide range of aspects of this species, there are no studies concerned with variation in both molecular markers and quantitative traits. The comparison of levels of diversity and differentiation between molecular markers and phenotypic traits provides an important tool for the understanding of population structure and evolutionary processes (Beaumont and Bruford 1999). The role of selection detected by means of comparing molecular (i.e. neutral or nearly neutral markers) and quantitative differentiation have been shown in different species (Prout and Barker 1993; Spitze 1993; Yang *et al.* 1996; Podolski and Holtsford 1995; Kremer *et al.* 1997). Generally, there is no correlation between the two types of data, suggesting selection to be highly important in most of the species analysed. In the case of Aleppo pine, the use of different kind of molecular markers could give complementary information about the processes involved. While nuclear markers reveal contemporary structure, organelle DNA markers reflect more precisely the historical structure of gene diversity (Avise, 1994).

The aim of this study was to analyse the role of selective processes in the differentiation of phenotypic traits in Aleppo pine. This objective can be achieved by comparing differentiation measures between phenotypic traits and the neutral expectation provided by different kinds of molecular markers (isozymes, RAPDs and Cp-microsatellites).

Material and Methods

Populations analysed.

The populations analysed correspond to 15 native stands of Aleppo pine covering its natural range in Spain (Figure 1). Single tree seed collection was carried out from 25 trees in each stand, whilst the identification of individual seedlots was maintained for 5 trees in each population. A structuration of stands and half-sib families within stands are available for the different studies.

Different sets of populations were used depending on the type of study (isozymes, RAPDs, Cp-microsatellites and different traits of phenotypic variation). A summary of the populations analysed by type of study is presented in Table 1.

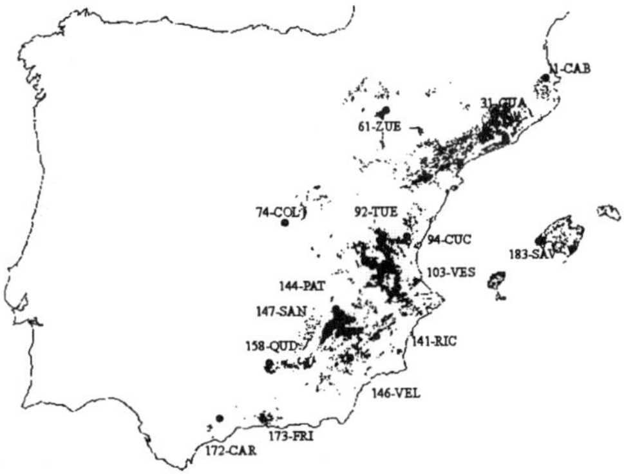

Figure 1. Native stands of Aleppo pine (shadowed), location of the populations analysed, and situation of the experimental provenance-progeny tests.

Table 1. Populations analysed by type of study.

	Code of Population														
Study	11	31	61	74	92	94	103	141	144	146	147	158	172	173	183
(1)	*	*	*	*	*	*	*	*	*	*	*	*	*	*	*
(2)	*		*				*	*					*		*
(3)	*	*	*	*	*	*	*	*	*	*	*	*	*	*	*
(4)	*	*	*	*		*	*	*	*		*		*		*

(1): isozymes, **(2)**: RAPDs, Cp-SSR, **(3)**: height at 4th year, germination, **(4)**: growth phenology

Isozyme analysis

Starch gel electrophoresis has been used to analyse the 15 populations, following the methods described by Agúndez *et al.* (1997, 1999). Five enzymatic systems encoded by five polymorphic loci were analysed in fifty to sixty seeds. The loci scored were: phosphoglucomutase (*Pgm*; EC 2.7.5.1), hexokinase (*Hk*; EC 2.7.1.1), aconitase (*Aco*; EC 4.2.1.3), 6-phosphogluconate dehydrogenase (*6Pgd*; EC 1.1.1.44) and malic dehydrogenase (*Mdh*; EC 1.1.1.37). The coefficient of genetic differentiation in the structured populations (F_{ST}) was computed for each loci, and 1.000 bootstrap samples were obtained to compute the 95% confidence intervals (Weir 1996). The composite multilocus coefficient of differentiation (CF_{ST}) was computed using a multivariate ANOVA as described by Kremer *et al.* (1997).

RAPDs and Cp-microsatellites analysis

DNA markers were analysed in a set of six populations. A total number of 24 RAPD loci, which tested Mendelian inheritance, were analysed in 60 samples from each population (Gómez 1998). Genetic differentiation (Gst) was calculated following Nei (1973). Nine chloroplast microsatellite markers were analysed in 24 samples from each population following Vendramin *et al.* (1996). Two genetic differentiation parameters were calculated: F_{ST} (AMOVA, Excoffier *et al.* 1992) and R_{ST} (Slatkin 1995).

Phenotypic traits

Growth variables were measured in two provenance-progeny tests, including 30 provenances and 5 families. The trials follow a RCB design with 7 replicates and two-trees plots, and are located in Madrid (Site 1) and Zaragoza (Site 2), Central Spain (Figure 1). The growth traits measured in the different sets of populations (Table 1) are: total height at 4 years for the two sites and height growth phenology at this age for site 1. A Gompertz model was adjusted for each tree:

$$ht = a \times e^{-e^{(b - c \times t)}}$$

where a is the total growth in the 4th year, b is related to the time of maximum growth rate and c is a shape parameter of the curve.

Germination parameters were obtained from 100 seed samples, following a RCB design with 4 replicates. As with the growth parameters case, the germination values were adjusted to a Gompertz function. In this case, a is the germination, b is related to the time of maximum germination rate and c is a shape parameter.

The NLIN procedure of SAS (SAS 1996) was used to adjust the Gompertz functions. The variance components in each experiment were estimated using the REML algorithm. The standard error of the estimation was computed using the delta method (Kendall and Stuart 1976). Methods to calculate Q_{ST} from variance components are extensively described in Wright (1951), Weir and Cockerham (1984) and Kremer *et al.* (1997). For the growth traits, the narrow-sense heritability was computed directly from a nested ANOVA model that includes the population and family within population effects. For the germination parameters, a heritability value of 0.7 was assumed.

Results

Table 2 shows the values of the coefficients of genetic differentiation from the three types of molecular markers analysed. The levels of genetic differentiation were varied slightly depending on the type of marker. The highest values were found using RAPDs and Cp-microsatellites, which were of the same magnitude as that of the composite measure of differentiation obtained by isozymes when all five loci were considered.

Table 2. Summary of measures of genetic differentiation for isozyme loci (15 populations), and RAPDs and Cp-microsatellites (6 populations).

Marker	Locus	Nb. Pop	F_{ST}	95% confidence interval	CF_{ST}
Isozymes	*Mdh*	15	0.1337	0.069, 0.174	
	6Pgdh	15	0.0250	-0.011, 0.052	
	Hk	15	0.1451	0.080, 0.185	
	Pgm	15	0.1164	0.017, 0.139	
	Aco	15	0.0924	0.002, 0.171	
Average		*15*	*0.1120*	*0.025, 0.160*	*0.1689*
RAPDs		*6*	*0.164*[1]		
Cp-SSRs		*6*	*0.189*[2] *0.164*[3]		

[1] G_{ST}, Nei (1973); [2] AMOVA (Excofier *et al.* 1992); [3] R_{ST} (Slatkin 1995)

The variance components for the phenotypic traits and the quantitative differentiation estimations are included in Table 3. All the traits analysed have a significant effect on the population, except the *b* and *c* parameters of the height growth curve. The effect on families within populations was studied only in growth variables showing a significant effect in all the traits, with the exception of the *c*-shape parameter.

Table 3. Variance components and Q_{ST} values of the phenotypic traits in Aleppo pine populations.

Trial	Trait	V_P	$V_{F(P)}$	V_E	Q_{ST}	SD	>0
Total height at age 4							
Site 1	HT4	395.8797	85.0801	341.201	0.131	0.090	Y
Site 2	HT4	602.0070	247.834	1134.99	0.127	0.123	Y
Growth phenology							
Site 1	*A*	100.3332	162.514	771.289	0.128	0.110	Y
Site 1	*B*	4.131597	19.6070	106.658	0.000	-	N
Site 1	*C*	0	0	0	-	-	-
Germination parameters							
	A	209.314		64.509	0.698	0.445	Y
	B	0.82650		0.8610	0.406	0.345	Y
	C	0.00023		5.2E-05	0.757	0.643	Y

The comparison of the coefficient of genetic differentiation among phenotypic traits and genetic markers is presented in Figure 2. A weak selection acting in some of the traits (phenology) and a strong selection acting in germination and height growth were detected.

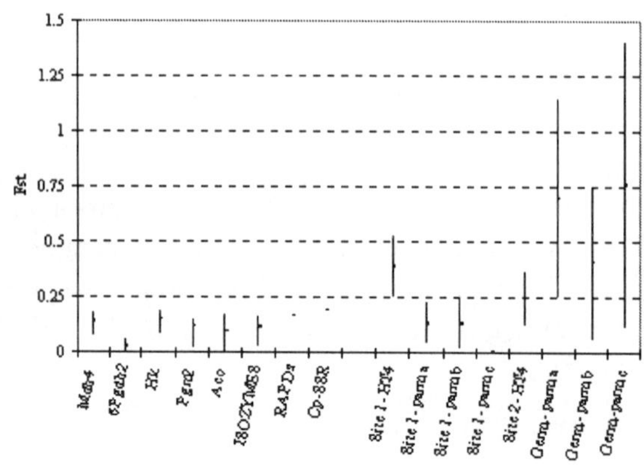

Figure 2. Values of the coefficient of genetic differentiation of phenotypic traits, isozymes, RAPDs and Cp-microsatellites.

Discussion

It is striking to note that the coefficient of genetic differentiation of nuclear loci has a similar value to that obtained with cytoplasmic genes. Because these two types of genes have different levels of ploidy and modes of inheritance they should be differently affected by evolutionary processes. In this context, the similarity between values could be interpreted as being due to a similar pattern of differentiation between contemporary and historical structures of gene variation. This fact could be explained by the postglacial spread of the species from the refugia in southern and northern Spain described by taking into account pollen and fossil records (Gil *et al.* 1996). However, the levels of subdivision for chloroplast genes in conifers where the organelle is paternally inherited and a high gene flow via pollen is expected are not clear, particularly under stepping-stone or isolation by distance models (Petit *et al.* 1993).

Because the isozyme loci analysed fit the hypothesis of neutrality according to the Ewens-Waterson test (Agúndez *et al.* 1999), the overall isozyme Fst estimate can be employed as a null hypothesis (not selection acting) to test the role of selection in the divergence of phenotypic traits among populations. The quantitative Fst computed for the germination parameters and total height at 4 years are clearly higher than those of the isozyme loci. These differences indicate that selective processes are involved in the phenotypic differentiation of the species. However, these processes are less important than previously reported in other forest species such as *Quercus* (Kremer *et al.* 1997) or *Pinus* (Yang *et al.* 1996; Eriksson 1998), probably due to the young age of evaluation of the quantitative traits in this study.

However, some peculiarities of the species seem to be of some importance. Aleppo pine is usually found in extreme environmental conditions where no other forest species can live. The main factor affecting the survival of the species in the range where the populations were sampled is drought tolerance. In such conditions, height growth is highly dependant on water availability as shown in field trials where, depending on climatic conditions, up to 4-5 cycles of growth have been described (Pardos *et al.*, 1995; Prada, 1999). The presented results suggest that height growth does not represent a strong selective factor affecting survival of the species, and therefore, height growth curves differ significantly in total growth, but not in shape or timing of maximum growth rate.

Germination traits have been studied in this species, in connection with drought stress. Calamassi *et al.* (1980, 1984), Falusi *et al.* (1983), and Schiller and Waisel (1989) found that the species can germinate even at very low water potentials or temperature, and that high variation exists among populations. Because these two factors condition the seedlings survival, they are of great importance for the natural regeneration of the species. The high values of differentiation detected in Aleppo pine populations in terms of germination and time of maximum germination, reflect the importance of germination traits both in the adaptation of the species to poor sites and in the dispersal of the seeds after forest fires.

In conclusion, the comparison among phenotypic traits and molecular markers indicates the existence of significant selective factors in Aleppo pine populations in Spain. However, it is necessary to conduct a more detailed study into pollen mediated gene flow and the levels of diversity of the different traits and loci to gain a clear picture of the mechanisms involved in the processes of differentiation of the species in the Western range of the Mediterranean area.

Acknowledgements

The study was funded by the European Commission, Brussels, project FAIR CT95-0097 "Mediterranean *Pinus* and *Cedrus* for sustainable afforestation of marginal lands", and by the Cooperation project DGCONA –INIA CC95-0097 "Mejora genética de coníferas".

References

Agundez, D., Degen, B., von Wuelish, G. and Alia, R. 1997: Genetic variation of Aleppo pine (*Pinus halepensis* Mill.) in Spain. Forest Genetics. 4(4): 201-209

Agundez, D., Degen, B., von Wuelish, G. and Alia, R. 1999: Multilocus analysis of *Pinus halepensis* Mill. from Spain: genetic diversity and clinal variation. Silvae Genetica

Avise, J.C. 1994: Molecular markers, natural history and evolution. Chapman and Hall, New York

Baradat, Ph., Michelozzi, M., Tognetti, R. and Khouja, M.L. 1995: Geographical variation in the terpene composition of Pinus halépensis Mill. In: Baradat Ph., Adams, W.T., Muller-Stark. G.(Eds). Populations genetics and genetic conservation of forest trees.pp: 141-158

Bariteau, M. 1992: Variabiliité géographicque et adaptation aux contraintes du milieu méditerranéen des pins de la section halepensis: résultats (provisoires) d'un essai en plantations comparatives en France. Ann Sci For 49: 261-276

Beaumont, M.A. and Bruford, M.W. 1999: Microsatellites in conservation genetics. In: Microsatellites (Goldstein, D.B. and Schlötterer, C. Eds.). Oxford University Press pp 165-182

Calamassi, R., Falusi, M. and Tocci, A. 1984: Effect de la température de germination et de la stratification sur la germination des semences de Pinus halepensis Mill. Silvae Genetica. 33(4/5):133-139

Eriksson, G. 1998: Evolutionary forces influencing variation among populations of *Pinus sylvestris*. Silva Fennica. 32(2): 173-184

Excoffier, L., Smouse, P.E. and Quattro, J.M. 1992: Analysis of molecular variance inferred from metric distances among DNA haplotypes: application to human mitochondrial DNA restriction data. Genetics. 136: 346-359

Falusi, M., Calamassi, R. and Tocci, A. 1983: Sensitivity of seed germination and seedling root growth to moisture stress in four provenances of Pinus halepensis Mill. Silvae genetica 32(1/2):4-9

Gil, L., Diaz-Fernandez, P., Jimenez, M. P., Roldan, M., Alia, R., Agundez, D., de Miguel, J., Martin, S. and de Tuero, M. 1996: Regiones de procedencia de *Pinus halepensis* Mill. ICONA. Madrid

Gomez, A. 1998: Variabilidad genética de *Pinus halepensis* en España mediante el uso de marcadores de ADN: RAPDs y Cp-microsatélites. Tesis Doctoral. UPM. Madri.

Kendall, M. and Stuart, A. 1976: The advanced theory of statistics. Vol 1. Distribution theory. Ed. Macmillan. New York

Kremer, A., Zanetto, A. and Ducousso, A. 1997: Multilocus and multitrait measures of differentiation for gene markers and phenotypic traits. Genetics. 145: 1229-1241

Nei, M. 1973: Analysis of gene diversity in subdivided populations. Proc.Natl.Acad.Sci.USA. 70: 457-462

Pardos, M., Climent, J., Gil, L., Pardos, J.A. 1995: Fenologia de Pinus halepensis en un huerto semillero clonal. IV Congresso Luso-Espanhol de Fisiologia Vegetal. Estoril, 3-6- de octubre de 1995. SPFV-SEFV Nacional

Petit, R. J., Kremer, A. and Wagner, D.B. 1993: Finite island model for organelle and nuclear genes in plants. Heredity 71: 630-641

Podolski, R. H. and Holtsford, T. P. 1995: Population structure of morphological traits in *Clarkia dudleyana*. I. Comparison of Fst between allozymes and morphological traits. Genetics. 140: 733-744

Prada, A. 1999: Mejora genética de Pinus halepensis Mill en la Comunidad Valenciana. PhD Thesis. UPM. Madrid (not published)

Prout, T.and Barker, J.S.F. 1993: F statistics in *Drosophila bussatii*: selection, population size and inbreeding. Genetics. 134: 369-375

SAS 1996: SAS-STAT. User´s manual. Ver 6.0. SAS Inc.

Schiller,G., Conkle, M.T. and Grunwald, C. 1986: Local differentiation among mediterranean populations of Aleppo pine in their isozymes. Silvae Genetica 35(1): 11-19

Schiller, G. and Grunwald, C. 1987: Resin Monoterpenes in Range-wide provenance trials of Pinus halepensis Mill in Israel. Silvae Genetica 36(3/4): 109- 114

Schiller, G. and Waisel, Y. 1989: Among-provenance varaiation in Pinus halepensis in Israel. Forest Ecology and Management 28: 141-151

Slatkin, M. 1995: A measure of population subdivision based on microsatellite allele frequencies. Genetics. 139: 457-462

Spitze, K. 1993: Population structure in *Daphnia obtusa*: quantitative genetic and allozyme variation. Genetics. 135: 367-374

Vendramin, G.G., Lelli, L., Rossi, P. and Morgante, M. 1996: A set of primers for the amplification of 20 chloroplast microsatellites in *Pinaceae*. Molecular Ecology. 5: 111-1-111-4.

Weir, B. S. 1996: Genetic data analysis II. Sinauer Associates, Inc. Publishers. Massachusetts. 445 p.

Weir, B.S. and Cockerham, C.C. 1984: Estimating F-statistics for the analysis of population structure. Evolution 38: 1358-1370

Weinstein, A. 1989: Geographic variation and Phenology of *Pinus halepensis*, *P. brutia* and *P. eldarica* in Israel. Forest Ecology and Management 27: 99-108

Wright, S. 1951: The genetic structure of populations. Ann. Eugenics. 15: 323-354

Yang, R. C., Yeh, F. C. and Yanchuck, A.D. 1996:A comparison of isozyme and quantitative genetic variation in *Pinus contorta* ssp. *latifolia* by F_{ST}- Genetics 142: 1045-1052

Geographical variation of gene diversity of *Pinus pinaster* Ait. in the Iberian Peninsula

S.C. GONZÁLEZ-MARTÍNEZ[1,2], L. SALVADOR[3], D. AGÚNDEZ[1], R. ALÍA[1] AND L. GIL[2]

[1]*Forest Research Centre, CIFOR-INIA, 28080 Madrid, Spain,* [2]*Unit of Anatomy, Physiology and Genetics, ETSIM, Ciudad Universitaria, 28040 Madrid, Spain,* [3]*University of Huelva, Ctra. Palos de la Frontera, 21071 Rábida-Palos de la Frontera, Spain.*

Key words: geographical variation, allozymes, cline, *Pinus pinaster*, Iberian Peninsula

Abstract: The geographical variation of 25 native populations of maritime pine (*Pinus pinaster* Ait.) was studied using ten allozymes encoded by 16 loci. A phylogenetic tree based on Nei's unbiased genetic distance and a Mantel's randomization test were made. Geostatistical analysis was used for describing in more detail spatial patterns of gene frequencies. The phylogenetic analysis found a clear clustering in northwestern populations and a complex pattern, with two differentiated groups, in the southeastern and eastern regions. Clinal trends of variation and fine-scale spatial structure were detected by geostatistical analysis. Directional clinal variation was found from south to north and, though to a lesser extend, from east to west. This pattern was stronger when populations were separated by more than 250 km. The long-distance founder events during postglacial spread and the orientation of the main mountain ranges are proposed to be the main causes of the geographical structure of this species.

Introduction

The knowledge of the geographical variation of gene diversity is fundamental to optimise strategies in breeding and conservation of forest genetic resources. The geographical structuration in forest trees has been studied by means of multivariate methods (Yeh *et al.* 1985; Lagerkrantz and Ryman 1990; Toumi and Lumaret 1998), correlation between genetic and geographical variables (Zanetto and Kremer 1995; Villani *et al.* 1999; Salvador *et al.* 2000) and geostatistical analysis (Monestiez *et al.* 1994; Degen and Scholz 1998; Gömöry *et al.* 1998; Le Corre *et al.* 1998; Bucci and Vendramin 2000). This last approach was first developed in the field of geology and it is useful to detect complex wide-range patterns. In addition, this method does not require the statistical independence of samples improving the classical descriptive methods (Le Corre *et al.* 1998). A complete technical description of the methods used in geostatistics can be found in Journel and Huijbregts (1978) and Cressie (1993).

Müller-Starck, G. and Schubert, R. (eds.). Genetic Response of Forest Systems to Changing Environmental Conditions. Vol. 70, (For. Sci.), 161-171 *© 2001 Kluwer Academic Publishers, Dordrecht, Boston, London*

The Iberian Peninsula has been reported as a glacial refugia for numerous species (Hewitt 1996; Comes and Kadereit 1998) and it is one of the most important native areas of *Pinus pinaster* Ait.. Previous wide-range studies have shown a complex pattern of variation using terpenes (Baradat and Marpeau 1988), chloroplast SSRs (Vendramin *et al.* 1998; Ribeiro *et al.* 2001) and allozymes (Salvador *et al.* 2000), and a correlation among different types of molecular markers (Petit *et al.* 1995). The aim of this work was to study the causes of the geographical variation of *P. pinaster* in the Iberian Peninsula. This objective was achieved by means of intensive sampling of native populations and the application of specific methodologies of spatial analysis.

Material and methods

Population sampling

A set of 25 native populations covering the mainland range of *P. pinaster* in the Iberian Peninsula was analysed (Figure 1).

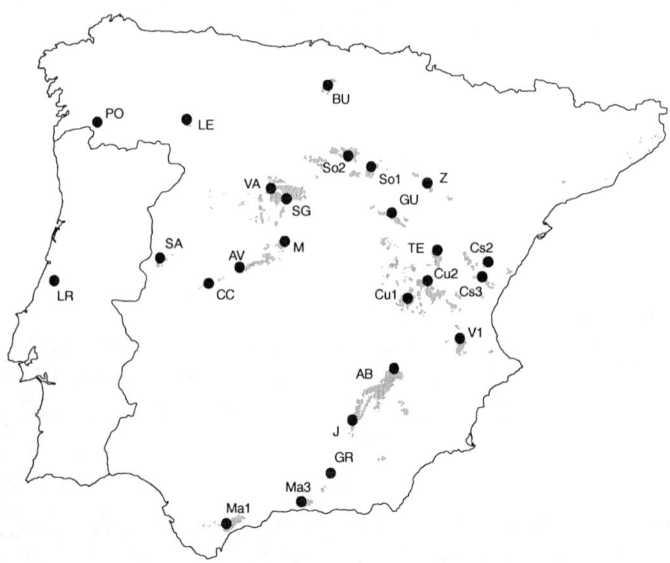

Figure 1. Location of the analysed populations. The shadowed area represents the range of Iberian populations whose native origin has been documented (Alía *et al.* 1996).

The populations used in this study included 9 populations among the 12 analysed by Salvador *et al.* (2000) (3 marginal populations were discarded). In each stand, 2-3 cones were collected from 80 trees, at least 50 m apart from each other. The material analysed was either 70-80 megagametophytes per population (15 populations) or megagametophytes and embryos of 35-40 seeds per population (10 populations). Diploid material (embryos) was analysed to detect departures from Hardy-Weinberg proportions.

Allozyme markers

Seeds were soaked on filter paper at room temperature for 24 h and germinated in a Petri dish in an incubator. Enzymes were extracted from 3-4 mm long radicles, following Conkle *et al.* (1982). Horizontal starch-gel electrophoresis of ten enzymes encoded by 16 loci was performed. The enzyme systems and the scored loci were: 6-phosphogluconate dehydrogenase (*6Pgd-1* and *6Pgd-2*; EC 1.1.1.44), isocitrate dehydrogenase (*Idh*; EC 1.1.1.42), malate dehydrogenase (*Mdh-1*, *Mdh-2*, *Mdh-3* and *Mdh-4*; EC 1.1.1.37), phosphoglucose isomerase (*Pgi-1* and *Pgi-2*; EC 5.3.1.9), acid phosphatase (*Acph*; EC 3.1.3.2), glutamate dehydrogenase (*Gdh*; EC 1.4.1.3), catalase (*Cat*; EC 1.11.1.6), glutamate-oxaloacetate transaminase (*Got-1* and *Got-2*; EC 2.6.1.1), phosphoglucomutase (*Pgm*; EC 2.7.5.1) and leucine aminopeptidase (*Lap*; EC 3.4.11.1). Hardy-Weinberg disequilibrium was only present four times over all populations (16 loci $\times$ 10 populations with diploid data assayed) and in different allozyme loci (*Pgm* in AV, *Mdh-3* in BU, *Acph* in GU and *Pgi-2* in Ma3). Genetic interpretation of enzyme systems and staining methods can be found in Castro (1989) and Salvador (1997). All loci included in this work showed to be neutral under the conditions of the Ewens-Watterson neutrality test (Manly 1985).

Data analysis

A phylogenetic tree based on Nei's unbiased genetic distance (Nei 1978) was obtained by the UPGMA method. In addition, the level of geographic structure was tested using the Mantel's randomisation analysis. This test involves measuring the association between the elements in two matrices, and determining if that association is significant by comparison with the distribution of the statistic found randomly reallocating the order of the elements in one of the matrices (Manly 1997). In our case, one of the matrices includes spatial data (Euclidean distance) and the other genetic distances (Nei's unbiased genetic distance). The correlation found between the two matrices (r) and the strength of the association after 1000 permutations were reported.

A more detailed analysis of the pattern of variation was made by means of geostatistical analysis. Both total and directional variograms were constructed. Only the most common alleles at polymorphic loci, i.e. loci with overall frequency of the most common allele less than 95% (Table 1), were considered (Le Corre *et al.* 1998).

Table 1. Allele number and frequency of the most common allele. In italics polymorphic loci at 95 % level.

Loci	Allele Number	Frequency of the most common allele
6Pgd-1	3	0.9988
6Pgd-2	3	0.9936
Idh	2	*0.9539*
Mdh-1	2	0.9992
Mdh-2	2	*0.8951*
Mdh-3	3	*0.9159*
Mdh-4	2	0.9984
Pgi-1	2	1.0000
Pgi-2	4	*0.8430*
Pgm	3	0.9684
Lap	3	*0.8657*
Got-1	2	0.9976
Got-2	3	*0.8773*
Acph	2	*0.7513*
Cat	1	1.0000
Gdh	2	0.9827

In order to combine the information provided by different loci, a multivariate approach based on principal components analysis was made. In geostatistical analysis, the first two principal components (*F1* and *F2*), that accounted for 37.69% and 34.58% of the total variance, respectively, were used. Geostatistical analysis was based on the description of variograms (i.e. the semivariance of the difference between points separated by a given distance in space) as a function of distance. Variograms were calculated for 8 classes of distance ranging from 0 to 500 km. A wide range of models including linear and exponential models were fitted using weighted least squares. The model with the lower sum of squared residuals was retained. Fitting of data was made with GSTAT program (Pebesma and Wesseling 1998, see also http://www.geog.uu.nl/gstat).

Results

The phylogenetic tree identified two main clusters, dividing the distribution into a N-W and S-E range. Populations from the N-W range showed a clear clustering (Figure 2). In contrast, eastern populations were transitional (note that Cu1 clearly clustered with the N-W group) and populations from southeastern Spain were included into two different subgroups, one of them (GR and Ma3) closer to eastern populations than the other. The Mantel's test found a high correlation between geographic distance and Nei's unbiased genetic distance ($r = 0.4961$ p: 0.001).

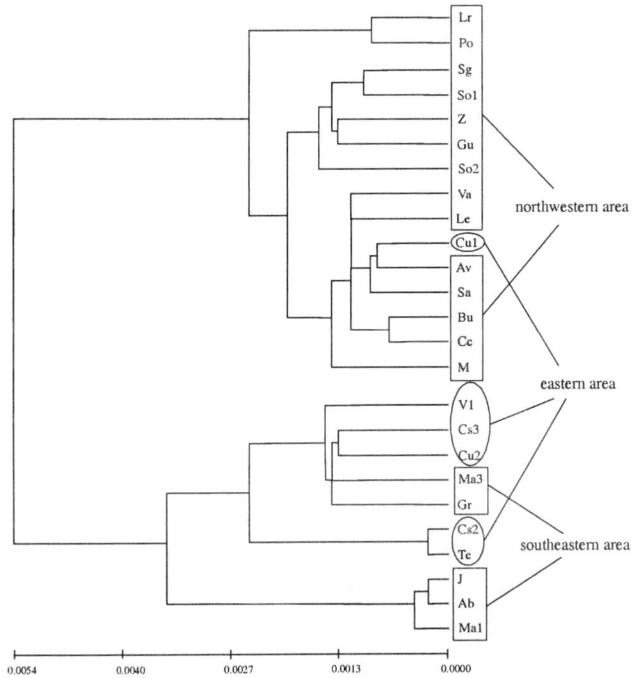

Figure 2. Phylogenetic tree obtained by the UPGMA method.

With respect to the geostatistical analysis, the most common pattern was the clinal one. This pattern is suggested when a linear function is found to be the best fitting of the variogram. The first principal component, *F1*, was mostly related to loci with geographical variation (*Idh, Lap, Acph*) and showed a clear clinal structure in the south-north direction. This pattern was stronger when populations were separated by more than 250 km. In the east-west direction the *nugget effect* (i.e. the discontinuity at the origin) was higher and the clinal structure was less clear (Figure 3). In contrast to the *F1*, the second principal component (*F2*) showed no geographic structure (Figure 4). This result was expected because the *F2* component was correlated with loci without geographic variation (*Pgi-2*, *Mdh-2*, *Mdh-3* and *Got-2*).

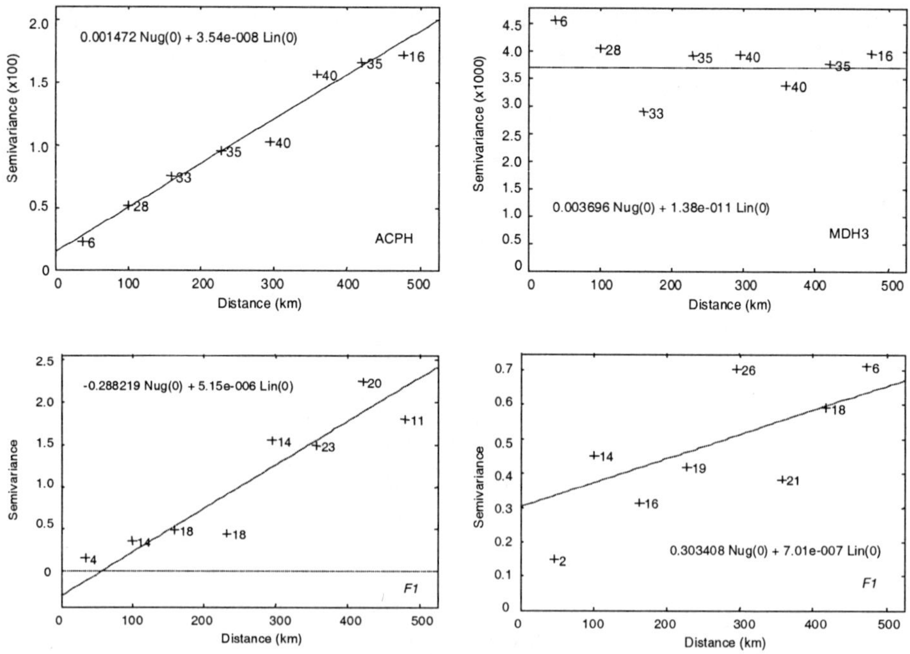

Figure 3. Unilocus variograms showing a linear fitting (*Acph*; top left) and only a nugget effect (*Mdh-3*; top right), and directional variograms for the first principal component (*F1*): from south to north (bottom left) and from east to west (bottom right). The best-fitted model is indicated in each figure.

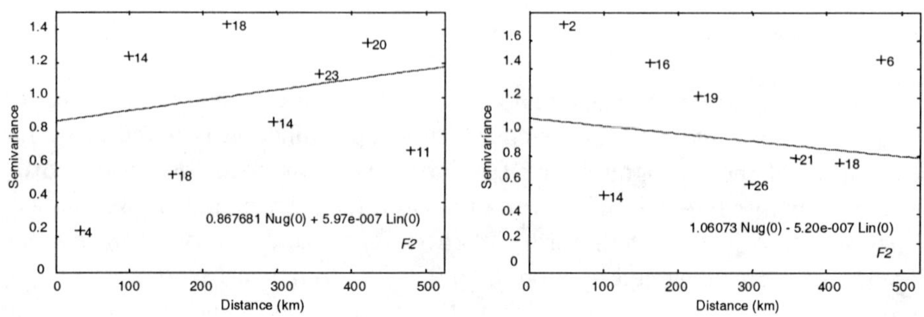

Figure 4. Directional variograms for the second principal component (*F2*): from south to north (left) and from east to west (right). The best-fitted model is indicated in each figure.

Discussion

Wide-range studies of the pattern of variation in maritime pine have shown a complex geographical structure (Baradat and Marpeau 1988; Vendramin *et al.* 1998; Salvador *et al.* 2000). In the Iberian Peninsula 7 geographical races and three main groups (Atlantic, Mediterranean and Maghrebian) have been distinguished using terpene analysis (Baradat and Marpeau 1988). In addition, a clinal genetic structure in the southeastern area of the Iberian range of the species and a lack of genetic structuration in the northwestern area have been reported (Salvador *et al.* 2000; Ribeiro *et al.* 2001). The results presented here are in agreement with the high geographical structuration previously found in the Iberian Peninsula and provide an insight about its causes. The phylogenetic analysis found a clear clustering in northwestern populations that corresponds to the Atlantic group in the study made by Baradat and Marpeau (1988). However, the eastern and southeastern regions showed a more complex pattern than previously described.

With respect to the causes of variation, two main patterns have been detected by means of geostatistical analysis. Clinal trends of variation were detected for 3 loci among the 7 analysed and for the first principal component (*F1*). The preferential direction of the clinal variation was found to be from south to north. This clinal variation may result from three main causes (Le Corre *et al.* 1998): (1) the directional postglacial range expansion, (2) the contact between populations arising from different refugia and (3) the selection acting along ecological gradients. Although the adaptive meaning of allozymes cannot be totally rejected, the studied allozymes are neutral as shown by the Ewens-Watterson neutrality test, and an explanation based on the postglacial evolutionary events seems more reasonable. The colonization of the northwestern area from glacial refugia situated in the Mediterranean Basin and the gene flow between refugia along the coastline could explain the observed pattern. However, the survival of *P. pinaster* during the last glaciation in sheltered areas at low altitude close to the Atlantic Ocean in Portugal has been suggested based on charcoal records (Figueiral 1995) and cpSSR data (Ribeiro *et al.* 2001). The strong clinal trend found when populations were more than 250 km distant, could be related to a limited gene flow beyond this distance. In a large-scale geographical study of Norway spruce (*Picea abies*) over the European natural range using cpSSR markers, Bucci and Vendramin (2000) found some haplotypes to retain genetic structure up to 1800 km while others went unrelated beyond 250-500 km. This pattern of variation was partially expected because of the huge range of the species and the existence of a gradient of variation between European (*P. abies*) and Siberian (*P. obovata*) spruce.

Postglacial migration and founder events have been generally proposed as the cause for the clinal variation of allele frequencies in forest trees (eg. *Quercus petraea*, Zanetto and Kremer 1995; *Fagus sylvatica*, Leonardi and Menozzi 1995; *Pinus halepensis*, Agúndez *et al.* 1999; *Pinus sylvestris*, Prus-Glowacki and Stephan 1994). Gene diversity distribution and palynological records seem to confirm this hypothesis in *P. pinaster* (Carrión and Van Geel 1999; Salvador *et al.* 2000 and references therein). In fact, the highest levels of diversity are found in the eastern and the southeastern populations and an important reduction of gene diversity is observed in the northwestern range of the species. Higher levels of genetic diversity found in refugia may be due to a more stable population dynamics and to larger population sizes, especially in mountainous regions where the risk of a population bottleneck is reduced by altitudinal migration in response to climatic changes (Comes and Kadereit 1998). Stochastic genetic drift in long-distance founder events during postglacial spread has been reported as the main cause of reduced gene diversity in populations near the northern distribution limit of the species (Cwynar and MacDonald 1987; Tomaru *et al.* 1997).

The other main pattern found by geostatistical analysis was a whole *nugget effect* model. The *nugget effect* may cover several sources of variation (Monestiez *et al.* 1994): (1) the experimental error due to inadequate sampling strategies (2) the spatial variation on a finer scale than that of the study and (3) the temporary 'noise' caused by stochastic events such as genetic drift or founder effects. The discontinuity and high altitude of mountain ranges in the Iberian Peninsula contribute to isolation of relatively nearby populations. This fact, in combination with the orientation of the mountain ranges, could partially explain the random component detected by geostatistical analysis. The striking geographical structure found in southeastern Spain may also be related to the distribution of mountain ranges. The two clusters of southeastern populations could be originated by two different gene flow pathways separated by high altitude mountain ranges which are orientated from south-west to north-east in this region (Figure 6).

Other possible sources of noise in wide-range geographical structure analysis are the human action and the fine-scale spatial structure during range expansion. Long-distance colonization events lead to the establishment of small populations in advance of the main invasion front and result in a spatial clustering of genotypes which persists for hundreds of generation (Hewitt 1996; Ibrahim *et al.* 1996). Because of the small effective population size and the relative isolation of each patch, the role of genetic drift in this scenario is highlighted.

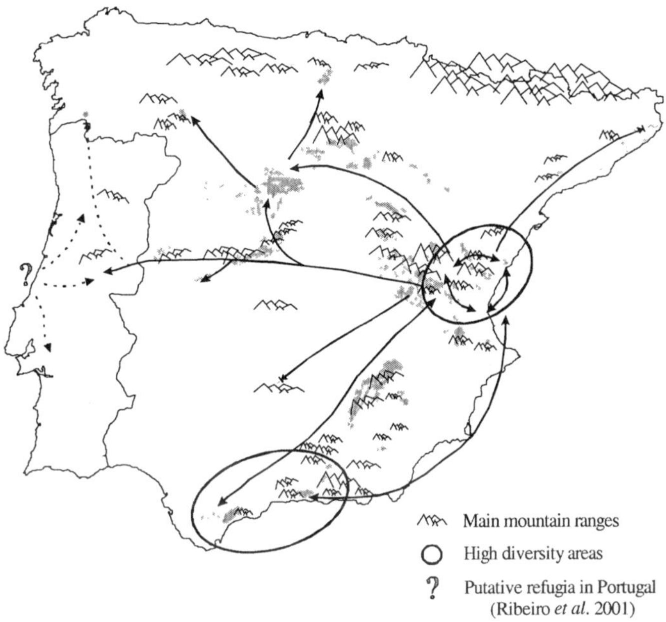

Figure 6. Main mountain ranges and putative migration pathways in *Pinus pinaster* Ait. The shadowed area represents the range of Iberian populations whose native origin has been documented (Alía *et al.* 1996). The existence of a putative glacial refugia in Portugal is discussed in Ribeiro *et al.* (2001).

In conclusion, the data presented here describe the geographical distribution of genetic diversity of *P. pinaster* in more detail than previously reported. This will provide valuable information for the management of forest genetic resources in the Iberian Peninsula.

Acknowledgements

This paper represents a portion of the doctoral research of the first author who is supported by a FPU scholarship from MECD (Ministerio de Educación, Cultura y Deporte). The study was funded by the Cooperation project DGCONA–INIA CC95-0097 and the INIA project SC97-118. Thanks to S. Martín who assisted in the selection and sampling of maritime pine populations, to M.M. Ribeiro who discussed a draft of the manuscript and to P.C. Grant who revised the English language.

References

Agúndez, D., Degen, B., von Wuehlisch, G. and Alía, R. 1999: Multilocus analysis of *Pinus halepensis* Mill. from Spain: Genetic diversity and clinal variation. Silv. Genet. 48(3-4): 173-178

Alía, R., Martín, S., De Miguel, J., Galera, R., Agúndez, D., Gordo, J., Salvador, L., Catalán, G. and Gil, L. 1996: Las regiones de procedencia de *Pinus pinaster* Ait. DGCONA. Madrid

Baradat, Ph. and Marpeau, A. 1988: Le Pin maritime *Pinus pinaster* Ait.. Biologie et génétique des terpènes pour la conaissance et l'amélioration de l'espèce (PhD dissertation). Université Bordeaux I. Bordeaux

Bucci, G. and Vendramin, G.G. 2000: Delineation of genetics zones in the European Norway spruce natural range: preliminary evidence. Molecular Ecology 9: 923-934

Carrión, J.S. and Van Geel, B. 1999: Fine-resolution Upper Weichselian and Holocene palynological record from Navarrés (Valencia, Spain) and a discussion about factors of Mediterranan forest succession. Rev. Palaeobot. Palynol. 106: 209-236

Castro, L.F.T. 1989: Isoenzimas do *Pinus pinaster* Ait. numa perspectiva de aplicaçao ao melhoramento genético da espécie (PhD dissertation). Universidade de Trás os Montes e Alto Douro. Vila Real

Comes, H.P. and Kadereit, J.W. 1998: The effect of Quaternary climatic changes on plant distribution and evolution. Trends in Plant Science 3(11): 432-438

Conkle, M.T., Hodgkiss, P.D., Nunnally, L.B. and Hunter, S.C. 1982: Starch gel electrophoresis of conifer seeds: a laboratory manual. United States Department of Agriculture. Forest Service. General Technical Report PSW-64

Cressie, N. 1993: Statistics for Spatial Data. John Wiley and Sons, Inc. Revised Edition

Cwynar, L.C. and MacDonald, G.M. 1987: Geographical variation of lodgepole pine in relation to population history. Am. Nat. 129: 463-469

Degen, B. and Scholz, F. 1998: Spatial genetic differentiation among populations of European beech (*Fagus sylvatica* L.) in western Germany as identified by geostatistical analysis. Forest Genetics 5(3): 191-199

Figueiral, I. 1995: Charcoal analysis and the history of *Pinus pinaster* (cluster pine) in Portugal. Rev. Palaeobot. Palynol. 89: 441-454

Gömöry, D., Hynek, V. and Paule, L. 1998: Delineation of seed zones for European beech (*Fagus sylvatica* L.) in the Czech Republic based on isozyme gene markers. Ann. Sci. For. 55: 425-436

Hewitt, G.M. 1996: Some genetic consequences of ice ages, and their role in divergence and speciation. Biological Journal of the Linnean Society 58: 247-276

Ibrahim, K.M., Nichols, R.A. and Hewitt, G.M. 1996: Spatial patterns of genetic variation generated by different forms of dispersal during range expansion. Heredity 77: 282-291

Journel, A.G. and Huijbregts, Ch.J. 1978: Mining Geostatistics. Academic Press. London

Lagerkrantz, U. and Ryman, N. 1990: Genetic structure of Norway Spruce (*Picea abies*): concordance of morphological and allozymic variation. Evolution 44: 38-53

Leonardi, S. and Menozzi, P. 1995: Genetic variability of *Fagus sylvatica* L. in Italy: the role of postglacial recolonization. Heredity 75: 35-44

Le Corre, V., Roussel, G., Zanetto, A. and Kremer, A. 1998: Geographical structure of gene diversity in *Quercus petraea* (Matt.) Liebl. III. Patterns of variation identified by geostatistical analyses. Heredity 80: 464-473

Manly, B.F.J. 1985: The statistics of natural selection on animal populations. Chapman and Hall. London

Manly, B.F.J. 1997: Randomization, bootstrap and Monte Carlo methods in biology. Chapman and Hall, 2nd Edition. London

Monestiez, P., Goulard, M. and Charmet, G. 1994: Geostatistics for spatial genetic structures: study of wild populations of perennial ryegrass. Theor. Appl. Genet. 88: 33-41

Nei, M. 1978: Estimation of average heterozygosity and genetic distance from a small number of individuals. Genetics 89: 583-590

Pebesma, E.J. and Wesseling, C.G. 1998: Gstat: a program for geostatistical modelling, prediction and simulation. Computers & Geosciences 24(1): 17-31

Petit, R.J., Bahrman, N. and Baradat, Ph. 1995: Comparison of genetic differentiation in maritime pine (*Pinus pinaster* Ait.) estimated using isozyme, total protein and terpenic loci. Heredity 75: 382-389

Prus-Glowacki, W. and Stephan, B.R. 1994: Genetic variation of *Pinus sylvestris* from Spain in relation to other European populations. Silv. Genet. 43(1): 7-14

Ribeiro, M.M., Plomion, C., Petit, R.J., Vendramin, G.G. and Szmidt, A.E. 2001: Variation in chloroplast single-sequence repeats in Portuguese maritime pine (*Pinus pinaster* Ait.). Theor. Appl. Genet. 102: 97-103

Salvador, L. 1997: Estudio de la variabilidad genética de *Pinus pinaster* en España usando marcadores proteicos e isoenzimáticos (PhD dissertation). Universidad Politécnica de Madrid. Madrid (not published)

Salvador, L., Alía, R., Agúndez, D. and Gil, L. 2000: Genetic variation and migration pathways of maritime pine (*Pinus pinaster* Ait.) in the Iberian Peninsula. Theor. Appl. Genet. 100: 89-95.

Tomaru, N., Mitsutsuji, t., Takahashi, M., Tsumura, Y., Uchida, K. and Ohba, K. 1997: Genetic diversity in *Fagus crenata* (Japanese beech): influence of the distributional shift during the late-Quaternary. Heredity 78: 241-251

Toumi, L. and Lumaret, R. 1998: Allozyme variation in cork oak (*Quercus suber* L.): the role of phylogeography and genetic introgression by other Mediterranean oak species and human activities. Theor. Appl. Genet. 97: 647-656

Vendramin, G.G., Anzidei, M., Madaghiele, A. and Bucci, G. 1998: Distribution of genetic diversity in *Pinus pinaster* Ait. as revealed by chloroplast microsatellites. Theor. Appl. Genet. 97: 456-463

Villani, F., Sansotta, A., Cherubini, M., Cesaroni, D. and Sbordoni, V. 1999: Genetic structure of natural populations of *Castanea sativa* in Turkey: evidence of a hybrid zone. J. Evol. Biol. 12: 233-244

Yeh, F.C., Cheliak, W.M., Dancik, B.P., Illingworth, K., Trust, D.C. and Pryhitka, B.A. 1985: Population differentiation in lodgepole pine, *Pinus contorta* ssp. *latifolia*: a discriminant analysis of allozyme variation. Can. J. Genet. Cytol. 27: 210-218

Zanetto, A. and Kremer, A. 1995: Geographical structure of gene diversity in *Quercus petraea* (Matt.) Liebl. I. Monolocus patterns of variation. Heredity 75: 506-517

Is autochthony an operational concept?

F. N. SCHOPPA AND H.-R. GREGORIUS

Institut für Forstgenetik und Forstpflanzenzüchtung, Universität Göttingen,, D-37077 Göttingen

Key words: autochthony, adaptedness, adaptability, stability, forest history, phylogeny, phylogeography, postglacial re-migration

Abstract: In spite of the high approval and the reference in many forest law and forestry service regulations, the idea of autochthony still seems to lack the definiteness and operationality that at least enables unambiguous specification of its ecological and evolutionary significance. This makes the assessment of autochthony questionable. Explication of the expectations connected with the idea reveals that, when applied to populations, it primarily addresses their adaptedness to the environmental conditions of a special site in combination with the adaptability that can be preserved at this site. Necessary prerequisites for realizing high degrees of adaptedness and adaptability are provided by regular environmental conditions together with balanced gene inflow from genetically heterogeneous and differentiated neighboring conspecific populations. Since successful adaptational processes reduce adaptational stress levels, adaptedness increases the capacity to support the genetic loads that are required for preserving adaptability to temporary as well as to lasting changes. Herewith, adaptability to temporary and lasting changes depends primarily on resident and immigrant genetic information, respectively. The latter requires metapopulation structures with balanced gene flow systems. Because of the complexity of this situation, a comprehensive assessment of autochthony is impossible. However, much of this complexity can be avoided by concentrating on the historical, geographical and genealogical characteristics of populations as separate units of adaptation. It is argued and illustrated by an example, that such an approach to the assessment of autochthony can be realized with the help of phylogeographic methods by testing the hypothesis of positive correspondence between phylogenetic relatedness and spatial distance.

Introduction

In view of the manifold historical and recent impacts on the biodiversity of forest ecosystems, the assessment of the stability and adaptability of the characteristic species becomes increasingly important because of the potential adaptational pressures (e.g. by the global climate change) in the near future. This goes along with some concern that the extent and speed of change of these pressures is out of all proportions with the time required for the evolution of biological diversity. The autochthony of populations has gained considerable interest in this context, since it is believed to warrant stability and adaptive flexibility (see e.g. Behm *et al.* 1997, Marzini 1997).

Müller-Starck, G. and Schubert, R. (eds.). Genetic Response of Forest Systems to Changing Environmental Conditions. Vol. 70 (For. Sci.), 173-185

Most dictionaries agree in their definitions of autochthonous as "native to a particular place", with the usually tacit assumption that this implies continual existence at this place or site. When applied to biological populations, the requirement of nativeness is of course crucial because of the impossibility to define a time of population initiation. This requirement is therefore replaced by that of existence at a particular site for a sufficiently long time, where the time necessary for reaching adaptedness to the local conditions substantiates the meaning of "sufficiently" (see e.g. Hattemer & Ziehe 1996). Yet, even though adaptedness is a common concept, its practical indication is difficult. Moreover, the reference to a fixed site seems to confine the notion of autochthony to sessile organisms, among which, in turn, many spend only a limited number of generations at a particular site because of their determination as pioneer species, for example. Hence, applicability of the term autochthony even seems to be confined to certain kinds of species.

Classification of the origin of forest reproductive material as autochthonous or non-autochthonous (allochthonous, not originating from the site of occurrence) plays an important role in the respective EC legislation and regulations. Usage of the term autochthony is here largely compatible with the colloquial definition in that it refers to a stand "which normally has been continuously regenerated by natural regeneration" (Council Directive 1999/105/EC, Article 2(d)). No indication as to the general significance or practical value of the distinction is made, and recommendations as to the assessment of autochthony are confined to "historical evidence or other appropriate means" (Council Directive 1999/105/EC, Annexe III(1.)). This usage thus suffers from the same problems concerning definiteness, argued significance and operationality as does the formulation of the colloquial concept.

The present paper therefore attempts to outline an approach that might be useful in addressing these problems with more rigor and helps to appreciate the role of retrospective analyses in the assessment of autochthony. By this, the approach is also intended to account for methods of assessment, which are based on the identification of genetic variants or genetic frequency distributions that are typical of certain areas and ecological conditions. Autochthony is refuted in these methods for populations which do not conform with the genetic characteristics predominating in the area. Examples are allozyme variants typically found in spruce populations at high elevations in the Alps (Ruetz & Bergmann 1989), or low amounts of genetic differentiation and high degrees of clustering in dendrograms for local beech populations (Hattemer & Ziehe 1996). In this paper we will proceed by stating the positive expectations associated with the colloquial definition of autochthony, examine the justification of these expectations and develop a framework which could help to put the assessment of autochthony on a firmer basis.

Conceptual reflections

At the outset it might be helpful to recall the trivial fact that a population existing for a large number of generations at the same site must have managed to adapt to a sufficient degree to the local conditions realized during that period of time. The degrees of adaptedness might have changed but at least never fell below a level that would have lead to extinction. Sufficient adaptability in such a population's history can therefore also be assumed. In fact, particularly high degrees of adaptedness and adaptability are generally expected for autochthonous populations. This does not rule out the possibility that the population experienced persistent and considerable stress (impairment of the vital functions), however, without ever running an acute risk of extinction. Hence, autochthony in its historical, geographical and genealogical dimensions alone does not strictly rule out the possibility of persistently poor adaptedness and low adaptability. In view of a generally variable environment it is, however, very unlikely that such a population could have existed under these conditions for a long time.

The population and its environment

The latter remark directs the attention towards characteristics of populations and their environments which allow or prevent the realization of persistently high degrees of adaptedness and adaptability. Sustained distinctly chaotic or large and fast directed changes of the environment clearly exhaust the adaptive capacity of every population. Therefore, concerning the environmental characteristics, autochthony requires a *regular environment* in the sense of quasi constant or recurrent conditions, both including moderate amounts of chaos and directed change. This situation is frequently referred to as one of a "predictable" environment. The current global climate warming, for example, takes place in some areas as a directional change and in others as an increase in the temperature variance. The concern is that either change may introduce amounts of environmental irregularity in the above sense, which could impair the autochthony of many existing populations.

With respect to population characteristics, the prerequisites for adaptational processes are of primary concern. Such processes exist on the *availability* of genetic variation, where the availability refers to resident as well as immigrant (or mutant) genetic variants. Availability of resident variation depends decisively on sufficient population size, while immigrant variation becomes available only through reproductive contact with genetically heterogeneous and differentiated neighboring populations. The rate of immigration (gene inflow) must, however, be limited in order to avoid the accumulation of excessive genetic load. The latter requirement of autochthony would, for example, not be fulfilled for a plant population that is surrounded by artificially planted conspecific populations showing little genetic differentiation, as is typical for many if not most central European forest tree populations.

Adaptedness and two kinds of adaptability

There is a delicate balance between adaptedness and adaptability in that adaptability decreases adaptedness, and adaptedness may be required for adaptability. Adaptability, as an investment into an uncertain adaptational future requires the maintenance of genetic variants which show inferior adaptedness to the current conditions and by this constitute a genetic load. Adaptability is thus paid for by a certain reduction in average adaptedness. Hence, increasing the adaptedness also increases the capacity of supporting the genetic load that is required for the preservation of adaptability. Autochthony may therefore not be warranted by high degrees of adaptedness (low average impairment of vital functions) alone, since these could have been realized at the cost of adaptability.

Environmental changes may be temporary, followed by a return to the initial conditions (perturbations), or they may be lasting. Adaptability to temporary changes can most effectively draw upon genetic information that regularly occurs and is maintained in the population. In the case of lasting changes, however, adaptability is more likely to be provided by genetic information that prevails in other populations and is thus introduced via gene flow. The latter facet of adaptability poses the question as to the autochthony of neighboring populations and, by this, brings into focus the metapopulation aspect of species.

The effect of gene inflow into a population on its adaptability depends on the rate of inflow, as was emphasized before, and on the kind of genetic information that enters the population's gene pool. The kind of introduced genetic information is, in turn, determined by the adaptational status of the neighboring populations from which it originates. The higher the adaptational status (the adaptedness) of the surrounding populations, the more adaptationally relevant genetic information the population receives from its neighbors. The population could thus extend its adaptive capacity (adaptability) to those environmental conditions to which it is most likely to be confronted with in the event of a lasting environmental change. Consequently, the autochthony of a population is strengthened by the autochthony of its neighbouring populations. This is in agreement with the generally held view that the adaptational differentiation combined with balanced gene flow among populations, as is characteristic of metapopulations, adds to the stability of each individual population (see e.g. Brandon 1990). The adaptational advantages of autochthony are, of course, diminished for lasting environmental changes which distinctly differ from all conditions realized in a metapopulation. Global changes such as concerning climate provide an example of this if they proceed at an evolutionary high speed.

The general idea of autochthony and its characteristics

As a first conclusion from the above elementary reflections it appears that *the idea of autochthony aims at populations showing high degrees of adaptedness and adaptability* and that the basic *prerequisites* for a population to reach this status include

(1) regularity of its environment,
(2) sufficient time for adaptation to this environment by way of reducing the average impairment of the vital functions (stress) over the generations,
(3) balanced reproductive contact with neighboring conspecific populations, and
(4) genetic heterogeneity within and differentiation between these neighboring populations.

While prerequisites (1) and (2) follow directly from the historical, geographical and genealogical dimensions of the common concept of autochthony, the other two make sure that the necessary adaptational processes can take place.

If all four prerequisites apply to the neighboring populations referred to in prerequisite (4) as well, this enhances the positive implications of autochthony of the reference population. This fact as well as the possibility of realizing different extents of adaptedness and adaptability suggests that autochthony may be realized to variable degrees, so that populations can be ranked with respect to these degrees. Note that this ranking of autochthony is not only based on its historical, geographical and genealogical dimensions.

Under the premise that the regularity of a population's environment can be specified in the above sense, the present characterization of autochthony can be applied and is not necessarily restricted to sessile species and their sites of occurrence. Motile species may, for example, effect their environments towards regularity by their capacity to evade unfavorable conditions. Hence, if autochthony is required to refer to a particular site this ought to be explicitly stated in the definition. If this is agreed upon, as it almost always is, it must be clear that many sessile pioneer species may a priori not be amenable to any qualification of autochthony.

If not stated otherwise, we will follow the common practice of applying the term autochthony only to time periods over which a population stays at a specified site. However, to guarantee priority for the primarily relevant aspects of autochthony, the specification of "site" ought to be subordinate to the spatio-temporal prerequisites for adaptation, adaptedness, and adaptability. Thus, the site of a population must, for example, be large enough to allow for the establishment of population sizes, which minimize the chances for losing by drift the genetic variation that is to be maintained for adaptability to perturbations. For the same reason, regular reductions in size, such as in pioneer species during succession, indeed force a population to shift between sites within a restricted area in order to maintain their adaptability. It may be reasonable in these cases to extend the notion of autochthony accordingly.

Problems with the assessment of autochthony

Criteria for ranking populations with respect to their extents of autochthony must combine measures of adaptedness and adaptability. In principle, degrees of adaptedness can be measured by some kind of average impairment of the vital functions (stress symptoms) in the population. When applied to the current status, however, this measure may be misleading, since adaptational processes are characterized by repeating phases of decreasing and increasing adaptedness. In the assessment of a population's adaptability, on the other hand, characteristics of the genetic structure of that population as well as of neighboring populations gain special importance. It is quite obvious that any measure for an assessment of the autochthony of a population that includes all these aspects explicitly ought to be of intractable complexity, if it can be designed at all.

These problems support the common practice to characterize autochthony simply by existence at a particular site for a sufficient number of generations as a surrogate measure. Yet, again there is no generally acceptable criterion for what numbers of generations may be sufficient not mentioning the problems of reliable assessment. A last resort therefore frequently consists in stating autochthony if there are no indications of allochthony (introduced from a different site). This can of course not account for the characteristics of autochthony developed above. However, since these characteristics are highly unlikely to be realized without having existed for a long time at the same site and vice versa, a method is desirable, that considers jointly the historical, geographical and genealogical dimensions of autochthony. The principles of such a method will be outlined and demonstrated in the following.

A phylogeographic method for the assessment of autochthony

Among the four prerequisites for autochthony listed above, the second has a particular position in that it depends on the others and comes closest to the colloquial definition of autochthony. This prerequisite relates the time spent by a population at a particular site to its chances to become adapted to this site, and it therefore focuses on the point in time at which a population arrived at this site. Apart from recent establishments or rare fossile records, this point of time is hard to specify in absolute terms. One therefore mostly depends on methods of reconstruction of a population's history which operate at the comparative level. In the following, we introduce an approach to the assessment of autochthony which makes use of methods established in phylogenetics (as presented in the books of Sudhaus and Rehfeld, 1992, and Kitching *et al.*, 1998) and which is largely in accordance with the principles of plylogeographic analyses as summerized in the recent book of Avise (2000).

The phylogeographic prerequisites

Irrespective of the length of time spent at the same site, a migrational history must taken into account that took place prior to the arrival and establishment of a population at this site. In comparison with other populations, this history includes the possibility of common ancestry that may date back to various degrees. The more recently two extant populations separated from their ancestral population, the more are they expected to be genetically similar (for pure drift, this statement holds by probability). Most recent joint ancestry therefore suggests itself as a means of dating the age of populations as separate units of adaptation.

This coincides with the phylogenetic principle of dating, in which relative evolutionary age is measured by the comparison of three populations (for intraspecific phylogenetics), two of which are more closely related to each other than either is to a third. The third population is then considered older than the other two populations in the sense that the most recent ancestor of all three populations is an ancestor of the most recent ancestor of the first two populations. In terms of monophyletic groups this can be cast into the statement that one population is phylogenetically older than a second population, if the smallest monophyletic group containing the first population comprises (but is not equal to) the smallest monophyletic group containing the second population. This information can be taken from cladograms. In phylograms, which utilize phyletic distances in their reconstructions, age can be directly be measured by the phyletic distance between a population and its most recent ancestor. Yet, age specifications based on monophyletic groups are generally more robust.

The description and analysis of associations between phylogenetic relations among populations and their geographical distribution is a matter of phylogeographic analysis for a discussion of the most commonly applied methods of such analyses see Avise 2000, p. 79ff). Assuming natural mechanisms of dispersal, for which dispersal probabilities decrease with geographical distance, one expects closer phylogenetic relations among neighboring than among distant populations. According to this mode of dispersal, geographic distances between populations should ideally increase with the remoteness of their most recent ancestors so that there is a generally positive association between geographic distance and phylogenetic relatedness. Herewith, the measurement of geographical distance can be required to take place along preassigned routes. For reasons of compatibility with the basic three-population definition of relative relatedness given above, it may also be required to allow only for distance comparisons which follow the same pattern. A positive association can then only be stated if, in each case where populations A and B are closer to each other than to C, these two populations also show higher relatedness to each other than to C.

The positive association between geographic distance and phylogenetic relatedness will be called the basic phylogeographical hypothesis in the following. Expressing phylogenetic relationship in terms of the age of lineages, Avise formulates this hypothesis as "older lineages should be more widespread than younger lineages when dispersal is limited" (Avise 2000, p. 80, last sentence). For methods of phylogenetic reconstruction producing unrooted trees it might be necessary to try several roots in order to find the best approximation to a positive association. Any violation of this hypothesis would hint at "artificial" interferences with the natural dispersal mechanisms such as anthropogenic transfer of reproductive material (forest historical seed transfer).

The phylogeographic test of autochthony

The basic phylogeographical hypothesis plays an important role in analyses of the postglacial remigration routes into Central Europe. These routes are of special interest because the re-establishment of populations provides a primary spatiotemporal reference of autochthony. In fact, those of the extant populations, the history of which is chiefly determined by the natural re-migration routes, are most likely to fulfill the above prerequisites for achieving adaptedness and adaptability. Such populations can also be expected to show genealogical and phylogenetic relations which reflect the branching during the remigration processes.

The existence of such relations can be ascertained with the help of the methods of phylogenetic reconstruction by developing a cladogram or phylogram for the populations under investigation and testing the basic phylogeographic hypothesis. If the hypothesis is not rejected, the populations can be considered as autochthonous. The positions of the populations in the cladogram then allow to define different degrees of autochthony on the basis of which of two populations is the phylogenetically older.

Explicit inclusion of certified knowledge on postglacial re-migration routes into testing of the basic phylogeographic hypothesis requires us to measure spatial distances along the routes and to consider their direction through the temporal progress of migration (usually from South to North). Phylogenetic relatedness of the populations is expected to decrease along these routes, if re-migration took place in a natural (continuous) manner, and this should show up in the cladogram. Assessment of autochthony by testing the basic phylogeographic hypothesis now involves a migrational measure of relative age of populations in addition to the phylogenetic measure. The migrational measure firstly serves in rooting the phylogenetic tree (if required by the method of reconstruction) and by this affects the phylogenetic measure of relative age of populations. Secondly, the supposed re-migration movements define routes, along which geographical distances have to be measured. The test of the basic phylogeograhic hypothesis has then to be performed on the basis of the thus measured distances. As a consequence of this specification of geographic distances, the assessment of autochthony and its degrees can be carried out as described above.

Any additional information about the history of populations should be included into the test of the basic phylogeographic hypothesis, since it might contradict the phylogeographic reconstruction of the temporal and geographic succession of phylogenetic separation of the populations. Such information may be available through documentations, for example, of artificial establishment of populations and origin of the used reproductive material.

To prevent possible misunderstanding it should be emphasized that the phylogenetic reconstruction of course depends on the traits used. It is therefore common usage to accumulate the information obtained for several traits which can be argued with respect to their adaptational function and are known for their modes of inheritance and variability characteristics. Based on this knowledge, potentially undesirable effects on phylogenetic reconstruction such as strong drift and opposing or convergent selection can be compensated for by composing a suit of traits which in combination reflects events of common descent. The above reflections and conclusions take this approach for granted. It is beyond the scope of this paper to discuss the suitability of the various types of genetic traits. However, there is one aspect that may deserve special mention and relates to adaptive differentiation as a major driving force for genealogical divergence. Adaptive differentiation draws chiefly from nuclear genetic variation, and this fact warns against uncritical preference of extranuclear over nuclear genetic traits in phylogenetic reconstruction.

An example of application

The above phylogeographic method of testing autochthony will now be briefly demonstrated with the help of data obtained for seven isozyme loci and six beech populations in Germany by Konnert (1995), Turok (1996) and Müller-Starck (1996). The populations are selected from the totality of populations studied by the authors, since they are roughly distributed along the main post-glacial re-migration routes from the South-East which are strongly suggested from paleobotanic studies (see e.g. Frenzel & Gliemeroth 1995). This is of couse not a reprentative sample of the suggested re-migration routes and it may therefore solely serve the demonstration of the method of testing. The location of the populations superimposed by the phylogenetic recontruction is presented in Figure 1. The reconstruction is based on the pairwise distances between the gene pools (measured by d_0, Gregorius 1974) as characterized by the seven isozyme gene loci. The information on the re-migration routes does particularly not allow to arrange the northern populations along a specific migration route, however, it helps in specifying the root of the reconstructed tree.

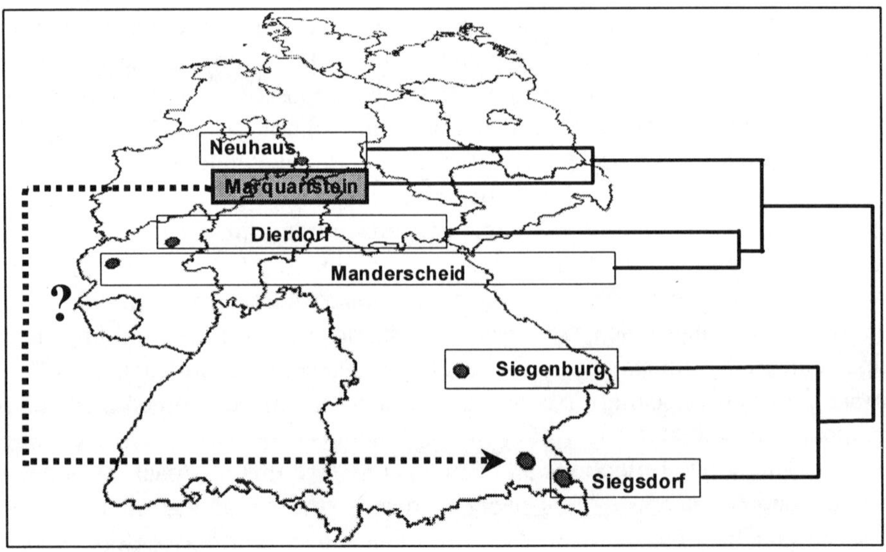

Figure 1. Association between geographical distance and phylogenetic relatedness for 6 German beech populations based on gene pool distances from 7 isozyme gene loci (data source given in the text). Phylogenetic reconstruction is performed according to Fitch & Margoliash (1967). The neighbor-joining method of Saitou & Nei (1987) yielded the same topology. With the exception of the Bavarian population "Marquartstein" the reconstruction is in accordance with the basic phylogeographic hypothesis. The suggested non-autochthony of this population is likely to be due to forest historical seed transfer.

It is seen that there exists a largely positive association between geographic distance and phylogenetic relatedness with the exception of the Bavarian population "Marquartstein". Clustering of this population into the northern array contradicts the basic phylogeographic hypothesis. Since removal of the Marquartstein population from the cladogram restores the validity of the hypothesis, this population is the only one giving rise to doubts on its autochthony. Its phylogeographic position furthermore suggests that it was established by seed that originated from source populations in Central or Northern Germany (perhaps nearby "Neuhaus") one or few generations ago. The doubts on authochthony were indeed confirmed by the Bavarian Forest Service (Konnert, personal communication).

It should be emphasized that the present phylogeographic test of autochthony is probably most suited for the detection of non-autochthony as in the above example. The estimation of degrees of autochthony with the help of relative phylogenetic ages of populations might run into problems when information about new populations is to be integrated into an existing cladogram, since this may change its topology. As a result new monophyletic groups may emerge which change the relative phylogenetic age relations. More conceptual analyses are required here to improve the reliability of the estimates.

Concluding remarks

In conclusion it might be useful to briefly return to the introductory remarks on the use of forest reproductive material obtained from autochthonous populations. The advantage granted to autochthony consists above all in the adaptability of such material when used for the establishment of new populations or for the stabilization of existing populations. According to the above-argued prerequisites for autochthony, regularity of the population's environment is of prime importance. As a rule, this includes only a limited number of local environmental factors such as climatic dynamics and soil characteristics. These regular environmental factors are those to which an authochthonous population has developed high degrees of adaptedness and by this has provided a sound basis for an extension of its adaptability. Hence, reproductive material obtained from autochthonous populations can realize its full adaptational capacity only when grown under the regular environmental conditions of its autochthonous progenitors. It is thus important to identify these conditions with high precision.

Non-regular environmental factors are of lesser significance. Obeying these factors may, however, help to further accelerate the process of reaching the former adaptational status of autochthony. Appropriate use of autochthonous

reproductive material can therefore promote the establishment of populations which are likely to reach a high adaptational status within a comparatively short period of time. Because of the genealogical relation between the reproductive
material and its autochthonous progenitor combined with the similarity in the regular environmental conditions, phylogeographical methods are also likely to detect such kinds of "secondary" autochthony. Explicit consideration of the geographical distribution of regular environmental conditions will help to facilitate the analysis.

Another important practical aspect concerns the present adaptational status of autochthonous populations. Despite realization of the historical, geographical and genealogical characteristics of autochthony, this status may be unsatisfactory due to insufficient population size (only remnants left) or unbalanced gene inflow from artificially established neighboring populations. In the latter case, seed harvested from such populations is unlikely to still harbor the adaptational capacity that enabled the population to survive its past environmental conditions. Hence, statements of autochthony should always be accompanied by statements on the present adaptational status.

Acknowledgements

The authors wish to thank J.W. Kadereit for commenting on an earlier draft of the manuscript. The detailed suggestions of two anonymous reviewers were valuable in focusing our argumentation.

This work was supported by a grant of the German Federal Ministry of Food, Agriculture and Forestry (Bundesministerium für Ernährung, Landwirtschaft und Forsten).

References

Avise, J.C. 2000: Phylogeography – The History and Formation of Species. Harvard University Press

Behm, A., Becker, A., Dörflinger, H., Franke, A., Kleinschmit, J., Melchior, G.H., Muhs, H.-J., Schmitt, H.P., Stephan, B.R., Tabel, U., Weisgerber, H. and Widmaier, Th. 1997: Concept for the conservation of forest genetic resources in the Federal Republic of Germany. Silvae Genetica 46(1): 24-36

Brandon, R.N. 1990: Adaptation and Environment. Princeton University Press, Princeton, New Jersey

Fitch, W.M. and Margoliash, E. 1967: Construction of phylogenetic trees. Science 155: 279-284

Frenzel, B. and Gliemeroth, A.K. 1995: Zur Vegetationsgeschichte Mitteleuropas. Angewandte Landschaftsökologie 4, 15-50. Bundesamt für Naturschutz, Bonn-Bad Godesberg

Gregorius, H.-R. 1974: Genetischer Abstand zwischen Populationen. I. Zur Konzeption der Abstandsmessung. Silvae Genetica 23: 22-27

Hattemer, H.H. and Ziehe, M. 1996: An attempt to infer on the origin of a Beech (*Fagus sylvatica* L.) stand in Rheinland-Pfalz (Germany). Silvae Genetica 45: 276-283

Kitching, I.J., Forey, P.L., Humphries, C.J. and Williams, D.M. 1998: Cladistics. Oxford University Press, Oxford

Konnert, M. 1995: Investigations on the Genetic Variation of Beech (*Fagus sylvatica* L.) in Bavaria. Silvae Genetica 44, 5-6: 346-350

Marzini, K. 1997: Was ist dran an der Autochthonie? Deutsche Baumschule 10: 557-559

Müller-Starck, R. 1996: Genetische Aspekte der Reproduktion der Buche (*Fagus sylvatica* L.) unter Berücksichtigung waldbaulicher Gegebenheiten. PhD thesis, Faculty of Forest Science and Forest Ecology, University of Göttingen. 103 pp

Ruetz, W.F. and Bergmann, F. 1989: Möglichkeiten zum Nachweis von autochthonen Hochlagenbeständen der Fichte (*Picea abies*) in den Berchtesgadener Alpen. Forstw. Cbl. 108: 164-174

Saitou, N. and Nei, M. 1987: The neighbor-joining method: A new method for reconstructing phylogenetic trees. Mol. Bio. Evol. 4: 406-425

Sudhaus, W. and Rehfeld, K. 1992: Einführung in die Phylogenetik und Systematik. Gustav Fischer Verlag, Stuttgart

Turok, J. 1996: Genetische Untersuchungen bei der Buche. Schriftenreihe der Landesanstalt für Ökologie, Bodenordnung und Forsten / Landesamt für Agrarordnung Nordrhein-Westfalen. Band 8: 136 pp

PART 3: GENETIC RESOURCES, REPRODUCTION, MANAGEMENT

Molecular markers in sustainable management, conservation, and restoration of forest genetic resources*

O.P. RAJORA[1] AND A. MOSSELER[2]

[1]*Department of Biology, Life Sciences Centre, Dalhousie University, Halifax, Nova Scotia B3H 4J1, Canada;* [2]*Canadian Forest Service, Atlantic Forestry Centre, Fredericton, New Brunswick E3B 5P7, Canada*

Key words: Molecular markers, conservation, restoration, sustainable forest management, genetic diversity, mating system, conifers, *Pinus*, *Picea*

Abstract: Increased use of forest resources to meet increasing world demand for wood and other forest products threatens forest sustainability and highlights the importance of conservation and sustainable management of these resources. Maintaining well-adapted and productive forests, and conserving natural forest genetic resources are important for sustainable forest management. Over exploitation of species can lead to excessive forest fragmentation and reduction of population sizes to a point that threatens population viability and species existence. In many cases, restoration of genetic resources of threatened species is needed. Molecular genetic markers, combined with population genetic principles and concepts, can greatly facilitate programs in conservation, restoration and sustainable management of forest genetic resources. We have used various biochemical and molecular genetic markers, such as allozymes, microsatellite DNA, random amplified polymorphic DNA (RAPD) and sequence-tagged site markers, to examine effects of alternative silvicultural harvesting and regeneration systems, and forest fragmentation and small population size on genetic diversity, fine-scale population genetic structure, mating system and other population genetic parameters in white spruce (*Picea glauca*), eastern white pine (*Pinus strobus*), and red spruce (*Picea rubens*). The objectives of our studies are to provide genetic benchmarks and indicators for developing guidelines for genetically sustainable forest management practices and scientifically sound strategies for conservation and restoration of forest genetic resources. The results of these studies are discussed, particularly in the context of sustainable management, conservation and restoration of forest genetic resources.

Introduction

Increased use of forest resources to meet increasing world demand for wood and other forest products and decreasing forest land base threatens forest sustainability and highlights the importance of conservation and sustainable

*Contribution FGB0002 (RCGB0002) from the Forest Genetics and Biotechnology Group of Dalhousie University

Müller-Starck, G. and Schubert, R. (eds.). Genetic Response of Forest Systems to Changing Environmental Conditions. Vol. 70 (For. Sci.), 187-202

management of these resources. As a forested nation and the world's largest exporter of forest products (Natural Resources Canada 1999), Canada has an added obligation to meet this increasing demand in a sustainable way. Forest sustainability could be achieved by enhancing the productivity, quality and health of the forests through using genetically improved stock within more intensively managed forests, maintaining well adapted, productive and healthy forests, and conserving natural forest genetic resources. Molecular genetic markers, combined with population genetic principles and concepts provide tools to facilitate tree improvement, conservation, sustainable management, and restoration of forest genetic resources. This paper focuses on applications of molecular genetic markers in sustainable forest management (SFM), conservation and restoration of conifers.

We examine the relationship between genetic diversity, conservation and SFM, and discuss potential genetic effects of forest management practices, forest fragmentation and declining population sizes, and implications for conservation, sustainable management and restoration of forest genetic resources based on results of our case studies involving white spruce (*Picea glauca*), eastern white pine (*Pinus strobus*), and red spruce (*Picea rubens*).

Genetic diversity, conservation and sustainable forest management

Increasing concerns over maintaining biodiversity in our forest ecosystems has focused research attention on impacts of forest practices on biodiversity. Genetic diversity is the basis of all biological diversity, because it provides raw material for adaptation, evolution and survival of species and individuals, especially under changed environment and disease conditions. Since trees are normally the keystone species of forested ecosystems, their genetic diversity has a special significance. Recent studies in Europe (e.g., Bergmann *et al.* 1990; Oleksyn *et al.* 1994; Raddi *et al.* 1994) suggest that reductions in genetic diversity can predispose forests to environment-related decline in health and productivity. Genetic diversity is also the basis for tree improvement. Healthy and productive forests are well adapted to their local environments, and genetic diversity provides the best assurance that they will continue to remain so.

Various evolutionary forces, such as selection, gene migration and mutation that influence genetic diversity also influence survival and fitness. In conserving forest genetic resources, we are primarily concerned with conserving genetic diversity within population of a species, its distribution among populations and the evolutionary processes that maintain genetic diversity (Rajora and Mosseler 2001). Thus, genetic diversity may be viewed as the foundation for forest sustainability and ecosystem stability. The Canadian Standards Association (CSA) has identified genetic diversity as one of the indicators for registration, certification and audit of sustainable forest management systems (CSA 1996a, 1996b). Benchmarking genetic diversity in natural forest tree populations and determining genetic impacts of silvicultural management practices on forest tree genetic diversity, and subsequent monitoring can provide resource managers with an indicator of long-term forest sustainability and ecosystem health (Buchert 1995; Namkoong *et al.* 1996; Mosseler and Rajora 1998).

Genetic effects of forest management practices

Forest management practices relying on natural and/or artificial regeneration systems, including tree improvement, and natural disturbance such as forest fires can significantly impact genetic variability, and various population genetic parameters. The extent of genetic impacts of forest management practices and natural disturbance will depend upon the species' population biology and silvics and the extent of human or natural disturbance. The genetic processes potentially affected by forest harvesting or forest fires include drift, selection, mating system and gene migration. However, genetic impacts of forest management practices are largely unknown for most forest tree species. Forest managers are faced with the challenge of conserving and maintaining genetic diversity while simultaneously maintaining economic viability of the forest industry.

Land managers need silvicultural guidelines, and indicators for measuring, monitoring, and reporting on genetically-sustainable forestry practices. Tree marking and harvesting procedures that maintain population viability and genetic diversity, by ensuring adequate number of residual trees at appropriate spacing, need to be developed. An adequate number of post-harvest residual trees per hectare is that which maintains normal levels of genetic diversity and reproductive fitness. An appropriate spacing of residual trees is one, which maintains or enhances opportunities for cross-pollination among residual trees, and that minimizes inbreeding among groups of related trees.

Silvicultural practices need not cause genetic degradation, and may be used to enhance genetic diversity in forests, especially those with strong family neighborhoods. Shelterwood harvesting in old-growth Douglas-fir (*Pseudotsuga menziesii*) had little effects on genetic variation (Neale 1985) or mating systems (Neale and Adams 1985). No significant differences were found in genetic diversity between virgin forests and naturally regenerated stands of Norway spruce (*Picea abies*) (Gomory 1992), although the planted stands had significantly reduced genetic diversity. In several forest trees, expected heterozygosities were found to be similar between natural stands and seed orchard clones (Savolainen and Karkkainen 1992; El-Kassaby 1995). Genetic diversity was also found to be maintained in advanced generation breeding populations of loblolly pine (*Pinus taeda*) (Williams *et al.* 1995).

Genetic effects of forest fragmentation and declining population size

Forest fragmentation and increased rates of population extinction result in smaller population sizes and increased isolation of populations, particularly among the less common or rare species. Beyond some critical threshold population size, this may lead to increased inbreeding and genetic drift. Whether increased inbreeding has an adverse effect depends on factors such as historical levels of inbreeding, and the inherent genetic load of deleterious genes, a characteristic affected by past levels of outcrossing and inbreeding. For instance, it is generally known that trees characterized by large, extensive populations and high levels of outcrossing tend to have higher genetic loads. When such species experience sudden and strong inbreeding, their reproductive capacity and fitness may be lowered dramatically as deleterious recessive genes, which were once masked in a heterozygous condition, are unmasked as inviable genotypes in the homozygous condition. However, mild protracted inbreeding may purge many of these deleterious genes, reducing genetic load.

Genetic diversity may be maintained by interpopulational gene flow (Slatkin 1985) among a number of smaller, fragmented subpopulations. However, when forest fragmentation reaches the point that such genetic exchange among subpopulations of the larger metapopulation (Hanski and Gilpin 1991) is severely curtailed, then inbreeding and genetic drift may result in losses of genetic diversity within these small populations. As population structure shifts from a series of larger, more extensive populations to a number of smaller, more isolated populations, increased levels of genetic differentiation among populations may result. Such a shift in population genetic structure may not necessarily represent a net loss of genetic variation. However, as populations decline in size, they become more vulnerable to stochastic elimination from the landscape through natural catastrophes (Shaffer 1981). The loss of small, genetically differentiated subpopulations may represent a serious loss of genetic diversity among populations, leaving fewer small populations to represent the entire gene pool.

Molecular genetic markers and genetic diversity measures

Molecular and biochemical genetic markers can facilitate evaluation of genetic resources and the development of strategies for conservation, sustainable management and restoration of forest genetic resources (Millar and Westfall 1992; Buchert *et al.* 1997; Geburek 1997; Rajora *et al.* 1998, 2000a, 2000b; Rajora 1999; Petit *et al.* 1998). However, determining which genetic markers are the most suitable ones presents challenges. Codominant, highly variable and simply inherited genetic markers can be used to determine and monitor genetic diversity and mating behavior, and genetic effects of human and natural disturbance. Allozymes have so far been the most commonly used genetic markers for population and conservation genetic studies in forest trees (Hamrick and Godt 1990; Millar and Westfall 1992; Geburek 1997). However, allozymes have several limitations, including number of loci resolvable, the absence of evident selective value, and potential differences in allozyme expression related to the stage of tissue development. DNA markers, such as restriction fragment length polymorphisms (RFLPs), random amplified polymorphic DNA (RAPD), sequence-tagged-sites (STS), amplified fragment length polymorphism (AFLPs) and microsatellites have also been used in forest trees. Each of these markers has several advantages and disadvantages. Nevertheless, microsatellite DNA (simple sequence repeats (SSR)) markers, due to their favorable features such as codominance, hypervariability, consistency, and suitability for automated high throughput genotyping, are currently considered as one of the most suitable markers for use in genetic resource conservation and management.

Genetic diversity measures are based on allelic richness (e.g., number of alleles per locus), allelic evenness (e.g., heterozygosity) and genotypic richness (e.g., genotype additivity, genotype multiplicity). Allelic richness measures are more useful than allelic evenness measures in genetic resource conservation (Marshall and Brown 1975; Petit *et al.* 1998) and sustainable forest management (Buchert *et al.* 1997; Rajora *et al.* 2000b). Allelic eveness measures, such as heterozygosity, are not as sensitive to changes due to bottlenecks as are allelic richness measures (Buchert *et al.* 1997; Rajora *et al.* 2000b). Genetic drift depletes genetic variance (heterozygosity) very slowly at the rate of $1/2N_e$ per generation, where N_e is the effective population size. Theoretically, two individuals could maintain 50% of the original heterozygosity of a large population. Therefore, populations with a moderate number of individuals could lose little of their expected heterozygosity in one generation. In fact, heterozygosity has even been found to increase in some post-bottleneck populations (Leberg 1992). It should be noted that allelic richness is most likely to contribute to the evolution and development of novel traits, such as resistance to new diseases and pests and adaptation to climate and environment change. Genotypic richness measures are useful and could supplement the allelic richness measures. A combination of allelic and genotypic richness measures could provide appropriate genetic diversity measures from genetic resource conservation, sustainable management, and restoration perspectives.

Case studies: Application of molecular genetic markers in sustainable management, conservation and restoration of forest genetic resources

Genetic effects of alternative silvicultural harvesting and regeneration practices, and phenotypic selection

White Spruce (Picea glauca)

White spruce is a widespread tree species of the boreal forest in Canada, and is one of the most important trees for the production of wood pulp and lumber. Although, white spruce is managed under both artificial and natural regeneration systems, its artificial regeneration after clear harvesting is the primary silvicultrural practice used in Canada. To develop guidelines for genetically sustainable silvicultural practices for white spruce, it is essential to develop benchmarks on genetic diversity inherent in natural forests and measure genetic impacts of alternative silvicultural practices. Very little is known about the genetic effects of silvicultural practices, phenotypic selections and domestication in white spruce. As compared to natural populations, genetic diversity in white spruce phenotypic selections was found to be reduced (Cheliak *et al.* 1988; Stoehr and El-Kassaby 1997), maintained (Cheliak *et al.* 1988) or enhanced (Desponts *et al.* 1993).

The white spruce study (Rajora 1999) was aimed at comparing genetic diversity between pristine old-growth, and both naturally- and artificially-regenerated and phenotypically selected white spruce, and then to determine genetic diversity impacts of silvicultural practices and phenotypic selection, and to provide benchmarks for developing genetic biodiversity indicators and guidelines for sustainable management and conservation of white spruce genetic resources.

Genetic diversity was determined and compared for 51 random amplified polymorphic DNA (RAPD) loci for the adjacent natural old-growth, naturally-regenerated and planted white spruce stands at each of the four sites (30 individuals per stand per site), one 54 year old plantation and open-pollinated progeny of 30 phenotypic tree improvement selections of white spruce (Rajora 1999). Genetic diversity parameters were determined assuming Hardy-Weinberg equilibrium within stands/populations. The old-growth stands had the highest and the phenotypic selections the lowest genetic diversity. The genetic diversity of the natural regeneration was comparable to that of the old-growth, whereas the genetic diversity of the plantations was comparable to that of the selections. (Rajora 1999; Table 1) On average, the genetic diversity of the old-growth and natural regeneration was significantly higher than that of the plantations and selections (Table 1).

Table 1. Overall means* of RAPD genetic diversity parameters for different stand types of white spruce. P, percentage of loci polymorphic; A, mean number of alleles per lcous, Ae, mean effective number of alleles per locus; H, mean expected heterozygosity, SI, Shannon's index

Stand type	P	A	Ae	H	SI	
Natural old-growth		88.7A	1.89A	1.69A	0.381A	0.548A
Natural regeneration	83.8A	1.84A	1.62A	0.349A	0.506A	
Plantation	72.2B	1.72B	1.53B	0.297B	0.431B	
Tree improvement selections	66.7B	1.67B	1.46B	0.259B	0.381B	

*Means for each genetic diversity parameter followed by the same letter are not significantly different at $P<0.05$ according to Duncan's multiple range test

This study indicated that plantations and phenotypic tree improvement selections sampled have significantly reduced genetic diversity, thus have a narrower genetic base, as compared to old-growth and naturally-regenerated white spruce. Since genetic diversity of the natural regeneration was comparable to that of the old-growth stands, it could be suggested that genetic diversity in white spruce might be maintained by a natural regeneration silvicultural system. It is generally known that white spruce does not regenerate well after clearcuts. Therefore, it is primarily regenerated artificially by plantations following harvesting. Natural regeneration under properly designed shelterwood or seed tree systems may not only help in regeneration of white spruce but may also assist in maintaining genetic diversity.

We are using microsatellite DNA markers to investigate genetic diversity effects of harvesting and forest fires of different intensities in white spruce.

Eastern white pine (Pinus strobus)

Eastern white pine is one of the most important softwood timber species in eastern Canada and northeastern United States. This species has been an integral part of the history and economic development of eastern North

America. Once a predominant tree species in forest landscapes of the Great Lakes region of North America, eastern white pine has undergone fundamental changes in population structure and dynamics over 150 years of heavy exploitation (Buchert 1994). Extensive stand harvesting and high grading for timber, and conversion of forest land to agricultural uses has resulted in forest fragmentation and reduction of population sizes (Buchert 1994). Only a small fraction of pristine old-growth white pine forest remains in central Canada. Such intact, pristine old-growth white pine stands provide a unique opportunity to develop benchmarks for the levels of genetic diversity inherent in natural forests and assessing genetic effects of harvesting. Eastern white pine is naturally regenerated under the shelterwood and seed-tree-cut harvesting systems. An understanding of the impact of harvesting on local gene pools will facilitate the monitoring of such practices to ensure their effectiveness in maintaining long-term ecosystem productivity and health. If genetic diversity is substantially reduced by current silvicultural practices, it will be necessary to develop alternative practices which ensure maintenance of genetic diversity in regenerated stands.

The objectives of the white pine study (Buchert *et al.* 1997; Rajora *et al.* 2000b) were to 1) determine genetic diversity, population structure, and genetic effects of harvesting in old-growth white pine, 2) compare informativeness of different genetic markers for this research, and 3) develop genetic benchmarks for developing guidelines and strategies for genetically sound silvicultural practices, conservation and sustainable management of white pine genetic resources.

The study was conducted in two virgin old-growth white pine stands. A partial cutting system close to a seed tree cut was prescribed with the objectives of extracting timber, encouraging natural regeneration and maintaining old-growth structural characteristics in residual stands (Buchert *et al.* 1997). After each stand was marked for harvest, all white pine trees (132 in stand A and 106 in stand B) were numbered and mapped. Foliage samples were collected from all the trees in each stand, and cone samples were collected from all cone-bearing trees in each stand (95 trees in stand A and 96 trees in stand B). Therefore, this study was based on a total census of the individuals within each stand. After harvest, 30 of the original trees were retained in stand A, whereas 23 of the 106 trees were retained in stand B. These residual trees were identified and used for postharvest gene pool analysis. However, allozyme analysis for the postharvest gene pool from stand A was based on 25 trees that produced seeds. Genetic diversity of the preharvest and postharvest gene pools and genetic effects of harvesting were determined by allozyme and microsatellite DNA analyses (Buchert *et al.* 1997; Rajora *et al.* 2000b).

Both preharvest stands had high allozyme and microsatellite DNA genetic diversity (Table 2). Microsatellite DNA genetic diversity at 13 loci was about four-folds that of allozyme genetic diversity at 54 loci. Although the two stands were less than a km apart, there were 13 allozyme and 12 SSR alleles unique to stand A and 20 allozyme and 7 SSR alleles unique to stand B (Buchert *et al.* 1997; Rajora *et al.* 2000b). Genetic effects of harvesting from both allozyme and microsatellite DNA analyses were parallel. There was concurrence in genetic diversity changes in the post-harvest gene pools of the two stands. With removal of about 75% of the trees, allelic diversity was reduced by about 25%, and most other genetic diversity parameters were reduced by 25% to 60% in the postharvest residual gene pools (Buchert *et al.* 1997; Rajora *et al.* 2000b; Table 2). For allozymes, about 40% of the low frequency ($0.25 > p \geq 0.01$) alleles and 80% of the rare ($p < 0.01$) alleles were lost from each residual stand due to harvesting bottleneck. For microsatellites, approximately 18% and 21% of the low frequency and 76% and 92% of the rare alleles were lost from the residual stands A and B, respectively. Latent genetic potential of each stand was reduced by about 40% to 50%. However, heterozygosity was not found to be significantly reduced by harvesting bottleneck. For certain genetic diversity parameters (e.g., heterozygosity), these reductions followed the theoretical expectations, whereas for certain parameters (e.g., number of alleles) the reductions were higher than theoretically expected according to the drift theory.

Table 2. Measures of genetic diversity in preharvest and postharvest gene pools of old-growth white pine stands

Diversity parameter	***Stand A***		***Stand B***	
	Preharvest	***Postharvest***	***Preharvest***	***Postharvest***
Allozymes				
Number of trees	96	25	95	23
Total number of alleles (54 loci)	121	91	128	96
Mean number of alleles per locus	2.24	1.69	2.37	1.78
Number of polymorphic loci	40	28	41	30
Percent loci polymorphic	74.1	51.9	75.9	55.6
Mean heterozygosity (observed)	0.125	0.121	0.126	0.143
Mean heterozygosity (expected)	0.149	0.146	0.157	0.155
Fixation index (F)	0.161	0.171	0.197	0.077
Latent genetic potential	52.4	24.4	58.6	27.6
Genotype additivity (observed)	155	109	160	113
Genotype multiplicity (observed)	2.18×10^{20}	3.43×10^{12}	1.20×10^{21}	2.17×10^{13}
Microsatellites				
Number of trees	120	30	102	22
Total number of alleles (13 loci)	125	93	120	89
Mean number of alleles per locus	9.62	7.15	9.23	6.85
Number of polymorphic loci	12	12	12	12
Percent loci polymorphic	92.3	92.3	92.3	92.3
Mean heterozygosity (observed)	0.505	0.485	0.538	0.535

Mean heterozygosity (expected)	0.599	0.587	0.615	0.585
Fixation index (F)	0.157	0.174	0.125	0.085
Latent genetic potential	66.9	40.3	57.7	39.3
Genotype additivity (observed)	276	135	265	100
Genotype multiplicity (observed)	8.94×10^{10}	4.31×10^{9}	6.91×10^{10}	3.08×10^{9}

Whether and to what extent these genetic diversity reductions are passed on to the post-harvest natural regeneration has yet to be determined. These stands have not regenerated well to allow adequate sampling of the postharvest natural regeneration. However, both our allozyme and microsatellite DNA studies suggest that in the absence of substantial gene migration that could ameliorate genetic losses following harvesting, the ability of the postharvest white pine gene pools to adapt to changing environmental and disease conditions may have been reduced. In order to maintain the genetic integrity of remaining intact old-growth gene pools, white pine stands must be carefully regenerated before any significant genetic erosion due to harvesting activity occurs. Our studies (Buchert *et al.* 1997; Rajora *et al.* 2000b) suggest that silvicultural practices should ensure that the gene pools of remaining pristine old-growth stands are reconstituted via some advanced natural regeneration prior to harvesting.

Genetic effects of forest fragmentation and small population size

Eastern white pine (Pinus strobus)

The rapidly declining numbers and sizes of eastern white pine populations in Newfoundland have become an urgent conservation issue because of threats to its survival as a naturally occurring species on the island. This decline resulted from forest harvesting for sawlogs at the turn of this century, followed by the introduction of the white pine blister rust disease (*Cronartium ribicola*) in the 1930s from Europe to mainland North America and then to Newfoundland. Prior to these events, white pine was widely, albeit diffusely, spread across central and western Newfoundland largely as a minor component of a mixedwood, fire-origin forest. Healthy 'old growth' white pine is no longer a common feature of the island's landscape. The extant white pine now consists of a few small surviving patches of trees of 40 to 70 years in age. These remaining isolated patches often consist of less than 30-50 individuals, creating a situation where inbreeding and genetic drift could begin to affect genetic diversity, structure, and reproductive success.

The objectives of this study (Rajora *et al.* 1998) were to develop benchmark information on genetic diversity and structure of disjunct eastern white pine populations from Newfoundland, and to determine the extent of their genetic differentiation that may have arisen following geographic isolation from ancestral mainland populations from Ontario, in order to monitor the genetic effects of population decline and to plan restoration efforts. We used 20 allozyme loci to determine genetic diversity and structure of six small isolated eastern white pine populations from two different regions in Newfoundland and of three populations from the Ottawa Valley region in Ontario. We also compared the levels of genetic diversity and differentiation within and among populations from the two provinces. On average, 47.8% of the loci were polymorphic, the number of alleles per locus was 1.75, and the observed and expected heterozygosities were 0.215, and 0.195, respectively. The Newfoundland populations were as genetically variable as those from Ontario. Generally, all populations exhibited slight excess of heterozygotes at most loci. The genetic distances among the populations within a province or region were as great as genetic distances among populations between the provinces or regions. Neither provincial nor regional or geographic gradient-related patterns of population variation and differentiation were apparent.

This study indicated that geographic isolation and 8,000 years of post-glacial evolution since white pine re-entered the island of Newfoundland have had little detectable effect on genetic differentiation between Newfoundland and ancestral mainland white pine populations. Newfoundland white pine was found to be as genetically variable as that from central Ontario. Population decline and associated processes of inbreeding and genetic drift apparently have had little or no detectable effect on genetic variability at the biochemical level in the remnant white pine population of Newfoundland. Furthermore, neither clinal, nor provincial, regional, or marginal vs. central patterns of genetic variability and differentiation were evident, and isolation by distance has not played a detectable role in population differentiation at the molecular level. This study provided some genetic benchmarks for monitoring the genetic effects of small population size in eastern white pine following a dramatic decline in both population sizes and numbers. Genetic information from this study may be important for conservation and restoration of white pine - and perhaps other rare and/or marginal species - in Newfoundland by providing a basis for developing conservation strategies. However, it should be noted that allozymes are often considered as selectively neutral genetic markers and the adaptive nature of allozyme variation is controversial in forest trees. In the absence of any information on adaptive trait variation, seed introduced from mainland seed sources as far away as the centre of the geographic range in Ontario may be genetically acceptable for the restoration of white pine in Newfoundland, considering the high genetic similarities and common origin of the Ontario and Newfoundland white pine suggested from our study and assuming their adaptability under Newfoundland conditions.

Red spruce (Picea rubens)

Red spruce is a characteristic species of the late successional "old growth" forests of eastern Canada because of its high tolerance for shaded conditions, its requirement for high levels of atmospheric moisture, and a longevity that ranges between 300 to 400 years under relatively undisturbed conditions. While red spruce has historically been a characteristic component of the Acadian Forest Region of eastern Canada, it has experienced a substantial decline across large portions of its former range (Gordon 1996). Its western-most extension in Canada occurs as a disjunct population of small, widely scattered and isolated stands associated with the Algonquin Highlands of south-central Ontario. This disjunct metapopulation is separated both by geography and the prevailing winds from the center of its distribution in the northeastern U.S.A., southern Nova Scotia, and southeastern New Brunswick. The present distribution of red spruce in Ontario probably represents a relatively recent remnant of a more continuous distribution in the past (Gordon 1996). Red spruce is now recognized as a rare species in Ontario and there is growing interest in its conservation, management, and restoration. In many areas, planting will be necessary to restore red spruce. Successful restoration will depend on proper seed source selection and a good understanding of genetic variation patterns.

Measures of genetic diversity and mating system can serve as indicators of genetic status for assessing and monitoring the viability of populations at risk due to small population size, low within-population densities, and population isolation resulting from forest fragmentation (Mosseler and Rajora 1998). Genetic diversity, population genetic structure, outcrossing rates in the filled seeds, and actual inbreeding levels were characterized in five small, isolated, remnant red spruce populations from the disjunct northwestern limits of its range in Ontario, and five populations from the larger, more extensive Maritime populations in Nova Scotia and New Brunswick to determine genetic and reproductive status; to provide some benchmarks for monitoring genetic changes due to isolation and restricted population sizes; and to assist developing restoration and conservation strategies (Rajora *et al.* 2000a). Thirty-seven allozyme loci were used for genetic diversity assessments, and six of the most polymorphic loci were used for mating system determination. On average, 29.1% (95% criterion) of the loci were polymorphic, the number of alleles per locus was 1.60, and the observed and expected heterozygosities were 0.097, and 0.100, respectively (Rajora *et al.* 2000a). The Ontario populations were comparable to or slightly less genetically variable than those were from the Maritimes. The Maritime populations were genetically less differentiated from each other than were those in Ontario. With the exception of three Maritime populations clustering tightly in one group, there was no clear separation of Ontario from Maritime red spruce populations from genetic distance as well as canonical discriminant analyses. The average multilocus (t_m) and single-locus (t_s) population outcrossing rates were, 0.595, and 0.558, respectively, indicating a comparatively high tolerance for inbreeding up to the filled seed stage in red spruce (Rajora *et al.* 2000a). The Ontario populations, on average, showed higher outcrossing rates (t_m=0.654, t_s =0.641) than the Maritime populations (t_m=0.535, t_s=0.475). Individual family outcrossing rates were similar to their respective population outcrossing rates and no significant difference among families within populations was observed for the multilocus estimates. When such high levels of inbred filled seeds were combined with the high proportions of empty (post-pollination aborted) seeds (Mosseler *et al.* 2000), it appears that actual inbreeding levels varied from 48 to 86% (Rajora *et al.* 2000a), with the highest inbreeding levels occurring in the smallest, most isolated Ontario populations, and in those populations most likely to have been adversely affected by poorer pollination conditions. The results indicate that in the short term, extant remnants of Ontario red spruce have maintained their genetic diversity (and integrity). For artificial restoration of red spruce in Ontario, local seed sources could be used without undue concern over losses of genetic diversity. However, over the longer term, genetic drift and inbreeding may be expected to result in further losses of genetic diversity and/or fitness if population sizes, numbers, and distribution continue to decline.

Conclusions

Molecular genetic markers, in conjunction with population genetics concepts and principles, and wherever possible in conjunction or relation with adaptive traits, can greatly facilitate development of genetic indicators, guidelines, and strategies for genetically sound silvicultural management practices, and conservation and sustainable management of forest genetic resources.

Acknowledgements

The studies summarized in this paper were supported by several funding agencies, including Canada-Saskatchewan Partnership Agreement in Forestry, the Prince Albert Model Forest Association Inc., Ontario Ministry of Natural Resources, Natural Sciences and Engineering Research Council (Individual Research and Strategic Project grants), and the Canadian Forest Service. A number of people provided input or assistance in the studies reported. Their assistance is thankfully acknowledged.

References

Bergmann, F., Gregorius, H.-R. and Larsen, J.B. 1990: Levels of genetic variation in European silver fir (*Abies alba*) - are they related to the species' decline? Genetica 82: 1-10

Buchert, G.P. 1994: Genetics of white pine and implications for management and conservation. Forestry Chronicle 70: 427-434

Buchert, G.P. 1995: Genetic diversity - an indicator of sustainability. In Proceedings of the Workshop, Advancing Boreal Mixedwood Management in Ontario, October 17-18, 1995, Sault Ste. Marie, Ontario

Buchert, G.P., Rajora, O.P., Hood, J.V. and Dancik, B.P. 1997: Effects of harvesting on genetic diversity in old-growth eastern white pine (*Pinus strobus* L.) in Ontario, Canada. Conservation Biology 11: 747-758

Cheliak, W.M., Murray, G. and Pitel, J.A. 1988: Genetic effects of phenotypic selection in white spruce. For. Ecol. Manage. 24: 139-149

CSA 1996a: A Sustainable Forest Management System: Guidance Document. Z808-96

CSA 1996b: A Sustainable Forest Management System: Specification Document. Z809-96

Desponts, M., Plourde, A., Beaulieu, J. and Daoust, G. 1993: Impact de la selection sur la variabilite genetique de l'epinette blanche au Quebec. Can. J. For. Res. 23: 1196-1202

El-Kassaby, Y.A. 1995: Evaluation of the tree improvement delivery system: factors affecting genetic potential. Tree Physiol. 15: 545-550

Geburek, T. 1997: Isozymes and DNA markers in gene conservation of forest trees. Biodiversity and Conservation 6: 1639-1654

Gömöry, D. 1992: Effect of stand origin on the genetic diversity of Norway spruce (*Picea abies* Karst.) populations. For. Ecol. Manage. 54: 215-223

Gordon, A. G. 1996: The sweep of the boreal in time and space, from forest formation to genes, and implications for management. Forestry Chronicle 72: 19-30

Hamrick, J.L and Godt, M.J.W. 1990: Allozyme diversity in plant species. Pp. 43-63 in Brown, A.H.D., Clegg, M.T., Kahler, A.L., and Weir, B.S. (Eds.). Plant population genetics, breeding, and genetic resources. Sinauer Associates Inc., Sunderland, MA.

Hanski, I. and Gilpin, M. 1991: Metapopulation dynamics: brief history and conceptual domain. Biol. J. Linn. Soc. 42: 3-16

Leberg, P.L. 1992: Effects of population bottlenecks on genetic diversity as measured by allozyme electrophoresis. Evolution 46: 477-494

Marshall, D.R. and Brown, A.H.D. 1975: Optimum sampling strategies in genetic conservation. Pp. 53-80 in Frankel, O.H., and Hawkes, J.G. (Eds.) Crop genetic resources for today and tomorrow. Cambridge University Press, London

Millar, C.I. and Westfall, R.D. 1992: Allozyme markers in forest genetic conservation. New Forests 6: 347-371

Mosseler, A. and Rajora, O.P. 1998: Monitoring population viability in declining tree species using indicators of genetic diversity and reproductive success. Pp. 333-344 in Sassa K (Ed) Environmental Forest Science. Kluwer Academic Publishers, Dordrecht, The Netherlands

Mosseler, A., Major, J.E., Simpson, D., Daigle, B., Lange, K., Park Y.-S., Johnsen, K.H. and Rajora, O.P. 2000: Indicators of population viability in red spruce, *Picea rubens*. I. Fecundity and reproductive traits. Can. J. Bot. 78: 928-940

Namkoong, G., Boyle, T., Gregorius, H.-R., Joly, H., Savolainen, O., Wickneswari, R. and Young, A. 1996: Testing criteria and indicators for assessing the sustainability of forest management: Genetic criteria and indicators. In Proceedings of the Workshop Project: Assessing the Sustainability of Forest Management: Testing Criteria and Indicators, 8-12 April 1996, West Java, Indonesia

Natural Resources Canada 1999: The state of Canada's forests 1998-1999. Natural Resources Canada, Ottawa

Neale, D.B. 1985: Genetic implications of shelterwood regeneration of Douglas-fir in southwestern Oregon. For Sci 31: 995-1005

Neale, D.B. and Adams, W.T. 1985: The mating system in natural and shelterwood stands of Douglas-fir. Theor. Appl. Genet. 71: 201-207

Oleksyn, J., Prus-Glowacki, W., Giertych, M. and Reich, P.B. 1994: Relation between genetic diversity and pollution impact in a 1912 experiment with east European *Pinus sylvestris* provenances. Can. J. For. Res. 24: 2390-2394

Petit, R.J., Al-Mousadik, A. and Pons, O. 1998: Identifying populations for conservation on the basis of genetic markers. Conservation Biology 12: 844-855

Raddi, S., Stefanini, F.M., Camussi, A. and Giannini, R. 1994: Forest decline index and genetic variability in *Picea abies* (L.) Karst. Forest Genetics 1: 33-40

Rajora, O.P. 1999: Genetic biodiversity impacts of silvicultural practices and phenotypic selection in white spruce. Theor. Appl. Genet. 99: 954-961

Rajora, O.P., DeVerno, L., Mosseler, A. and Innes, D. 1998: Genetic diversity and population structure of disjunct Newfoundland and central Ontario populations of eastern white pine (*Pinus strobus* L.). Can. J. Bot. 76: 500-508

Rajora, O.P. and Mosseler, A. 2001: Conservation of forest genetic resources: challenges and opportunities. Euphytica 118(2): 197-212

Rajora, O.P., Mosseler, A. and Major, J.E. 2000a: Indicators of population viability in red spruce, *Picea rubens*. II. Genetic diversity, population structure, and mating behaviour. Can. J. Bot. 78: 941-956

Rajora, O.P., Rahman, M.H., Buchert, G.P. and Dancik, B.P. 2000b: Microsatellite DNA analysis of genetic effects of harvesting in old-growth eastern white pine (*Pinus strobus*) in Ontario. Molecular Ecology 39: 339-348.

Savolainen, O. and Karkkainen, K. 1992: Effect of forest management on gene pools. New Forests 6: 329-345

Shaffer, M.L. 1981: Minimum viable population sizes for species conservation. BioSci. 31: 131-134

Slatkin, M. 1985: Gene flow in natural populations. Ann. Rev. Ecol. Syst. 16: 393-430

Stoehr, M.U. and El-Kassaby, Y. 1997: Levels of genetic diversity at different stages of the domestication cycle of interior spruce in British Columbia. Theor. Appl. Genet. 94: 83-90

Williams, C.G., Hamrick J.L. and Lewis, P.O. 1995: Multiple-population versus hierarchical conifer breeding programs: a comparison of genetic diversity levels. Theor. Appl. Genet. 90: 584-594

Sustainable treatment of resources: The genetic basis

H.-R. GREGORIUS

Institut für Forstgenetik und ForstpflanzenzüchtungUniversität Göttingen, D-37077 Göttingen

Key words: concept of sustainable treatment, adaptation, feedback, self-regulation, self-regeneration, mechanisms of the genetic system, indication of sustainable treatment

Abstract Any treatment of a resource is considered sustainable, if it diminishes neither this resource nor any other irreversibly. A conceptually consistent and practicable presentation of the principles of sustainable treatment of resources is given using a system-oriented approach. In this approach the resource appears as a system that is characterized by the capacity to deal to a certain degree with exogenous impacts without losing its identity and integrity. Realization of this capacity rests on the existence of mechanisms of self-regulated regeneration with their characteristic feedback relations. For a treatment of a resource to be sustainable, it is thus necessary to guard the intactness of the resource's own mechanisms of self-regeneration as well as those of all affected systems. Many anthropogenic forms of resource treatment violate even this basic prerequisite of sustainability by disconnecting essential feedback loops. Populations are the most elementary biological resources capable of indefinite self-regeneration, and certain specified mechanisms of the genetic system realize this capacity. Special emphasis is put on the two-fold function of the environment as modifying these mechanisms and providing the conditions for their operation. On this basis it is argued that any treatment of a population can be sustainable only if it obeys the following three prerequisites: (*i*) intactness (operability) of the mechanisms of the genetic system, (*ii*) realization of the mechanisms' external conditions for operation (operational conditions), and (*iii*) availability of genetic variation for alteration of these mechanisms. Indication of these prerequisites is argued to be most effective when the system analytical approach is applied. In essence, this approach consists in modeling the potentially affected mechanisms and their operational conditions, such that they fulfill prerequisites (*i*), (*ii*) and (*iii*). Incompatibility between the model predictions and observations then leads to rejection of the hypothesis of sustainable treatment. The advantage of this approach to indication over an approach involving mere comparative interpretation of genetic processes and indices of genetic variation is demonstrated.

Introduction

It is understood that a form or kind of treatment of a resource is to be considered sustainable if, by application of the treatment, neither this resource nor any other is irreversibly diminished. The forces and conditions that enable resources to withstand certain treatments but not others are therefore the central subjects of studies of sustainability. In essence, this idea underlies most of the rules and recommendations for a sustainable development stated in the

Müller-Starck, G. and Schubert, R. (eds.). Genetic Response of Forest Systems to Changing Environmental Conditions. Vol. 70 (For. Sci.), 203-222

resolutions and conventions following the Rio Declaration of 1992. The term "treatment" is given preference to "development" in order to emphasize the non-anthropocentric orientation of the following derivations and also to point out that any organism in some sense or other "treats" its resources (for an elaboration of this perspective, see e.g. Svirezhev & Shvirejeva Hopkins 1998; for a critical review of the main directions of biological sustainability concepts, see e.g. Callicott & Mumford 1997).

In order to define the position of genetics in the realm of sustainable treatment of resources, the present paper will focus on a system-oriented presentation of the principles of such treatment. Considering a resource as a system requires specification of the general structural and functional characteristics in such a way that the resource's capacity to maintain its identity and integrity under exogenous impacts (the treatment) becomes apparent. It will be shown that populations are the most elementary biological resources that realize these characteristics, and the involved structural and functional genetic components will be described. Unsustainable forms of resource treatment then appear as impairment of these components. Hence, the indication of sustainability requires specification of the components that are among the most affected and that show observable features which can be related to the degree of impairment of the components.

Even though the practical value of a system-oriented approach to sustainability seems to be recognized (see e.g. Schellnhuber & Kropp 1998) and is obvious for the specification of criteria and indicators, explicit consideration of its principles seems rare. This may be due to the complexity of biological systems, particularly ecosystems, which is thought to inhibit sufficient characterization of the structural and functional properties of ecosystems (see e.g. De Leo & Levin 1997). Yet, as indicated above, appropriate generalization of the problem on a system-oriented basis may help to identify surrogates of complexity that are observable and suitable for the indication of sustainability of treatments. Ludwig et al. (1997) provide a useful summary of the approaches relevant to sustainability from a perspective of stability and resilience, as defined in classical dynamical systems theory via equilibrium states and their various attractivity characteristics. In this theory, aspects of (eco-)system identity and integrity are largely identified with equilibrium characteristics rather than system structure and function (cf. De Leo & Levin 1997).

Such structural and functional aspects are in the foreground in the paper of Namkoong et al. (1996), where an attempt is made to base the genetic criteria and indicators of sustainable management on the intactness of the genetic processes that govern changes in genetic structures. Some of the central ideas of this paper will be taken up here and generalized to a degree that allows development of an operational concept of the indication of sustainable treatment of genetic resources. The utility of this concept will be demonstrated with the help of paradigms relating to distinctly different scales of indication, such as concerning mechanisms and conditions of self-regeneration in forestry and the preservation of the adaptability of populations.

Some general principles of sustainable treatment of resources

In a general context, resources are viewed as supplies available (not only to man) for consumption, so that a resource's persistence requires occasional replenishment. From this simplistic point of view, any treatment is sustainable if it ensures replacement of what has been removed, in other words, if it ensures regeneration. This has to be extended to all resources which are directly or indirectly affected by the treatment, so that, in effect, all of these resources are considered to be treated. As a result, complex reactions may occur even in systems with smaller numbers of subsystems if they show high degrees of reticulation. In order to reduce the danger of instability (failure to regenerate after perturbation) associated with complexity, self-regulation of the system or its subsystems is required.

Since self-regulation effects state changes as a function of the preceding system states, this type of regulation involves feedback. In fact, in order to control the effect of a regulatory action, the effect must necessarily be reported back to the regulatory mechanism. Therefore, self-regulation can be conceived of as a control of state dynamics by a feedback mechanism. When regeneration takes place by self-regulation it is commonly referred to as self-regeneration.

In biological systems, feedback loops are abundant, and it is generally recognized that they play a leading role in the organization and stabilization of these systems. In stable ecosystems, for example, the food net is organized by consumption relations which ensure internal (re)cycling of all essential nutrients at balanced rates and modes. Impairment of feedback mechanisms results in unbalanced rates in terms of matter, energy or information feedback per unit of time, which includes the possibility of a change of the mode from negative to positive feedback. Discontinuation of feedback and consequently disintegration of the system may then ultimately result. Self-regulating feedback mechanisms therefore serve the maintenance of the identity (characteristics) and integrity (intactness) of their underlying systems.

In this system-oriented concept, treatments of a resource appear as either perturbations or external regulations. Perturbations are exogenous state changes which do not affect the feedback mechanisms, whereas external regulations do. Since external regulations act on the internal mechanisms of the system, any alteration of these regulatory forces affects the mechanisms of self-regulation, which is obviously more likely to result in unsustainability of the treatment than would any temporary perturbation. In fact, while state perturbations are the rule in all open systems such as biological systems, changes in externally regulating forces are the exception, since they can easily destabilize the system.

Considering the example of harvesting a plant species, the removal of individuals or parts of them in moderate amounts constitutes a perturbation that can be compensated by the species' regenerative capacity. However, if the method of harvesting entails removal of most of the reproductive structures (flowers or fruits), such that regeneration occurs only via vegetative propagation, this constitutes a measure of external regulation since a basic mechanism of self-regeneration (self-regulated regeneration) is affected. In a similar fashion, silvicultural management methods that reduce the harvesting (exploitation) age of trees below the natural age distribution constitute external regulations rather than perturbations, since they change the generation time and by this the natural reproductive feedback rate.

Inherent to this system-oriented representation of resource treatment and its consequences is the following approach: Specification of a principle for the treatment (sustainability), a criterion for achievement of the principle (regenerability), indicators for assessment of the criterion (mechanisms of self- and external regulation), and verifiers for ascertainment of observance of the principle (intactness of the mechanisms). Figure 1 summarizes this with the left side of which referring to regenerating resources as considered above. The right side of the figure extends the principle of sustainable treatment to non-regenerating (non-renewable) resources. Such resources are usually abiotic. As becomes apparent from the diagram, the criterion of regenerability can be fulfilled for non-regenerating resources only by securing the option of restoration of the resource. This can be achieved by recycling the resource products according to self-regulated feedback mechanisms. The establishment of self-regulation requires a temporary or even permanent phase of external regulation in order to stabilize the recycling procedure (the design of Figure 1 and its underlying concept were developed in cooperation with JRG Kleinschmit, F Schoppa, and H von Werder).

In summary it can be stated that for a treatment of a resource to be sustainable, it is necessary to guard the intactness of its own mechanisms of self-regeneration as well as those of all affected systems.

The genetic components of sustainable treatment

The smallest biological unit that is capable of indefinite self-regeneration is the population. It therefore constitutes the most elementary biological resource.

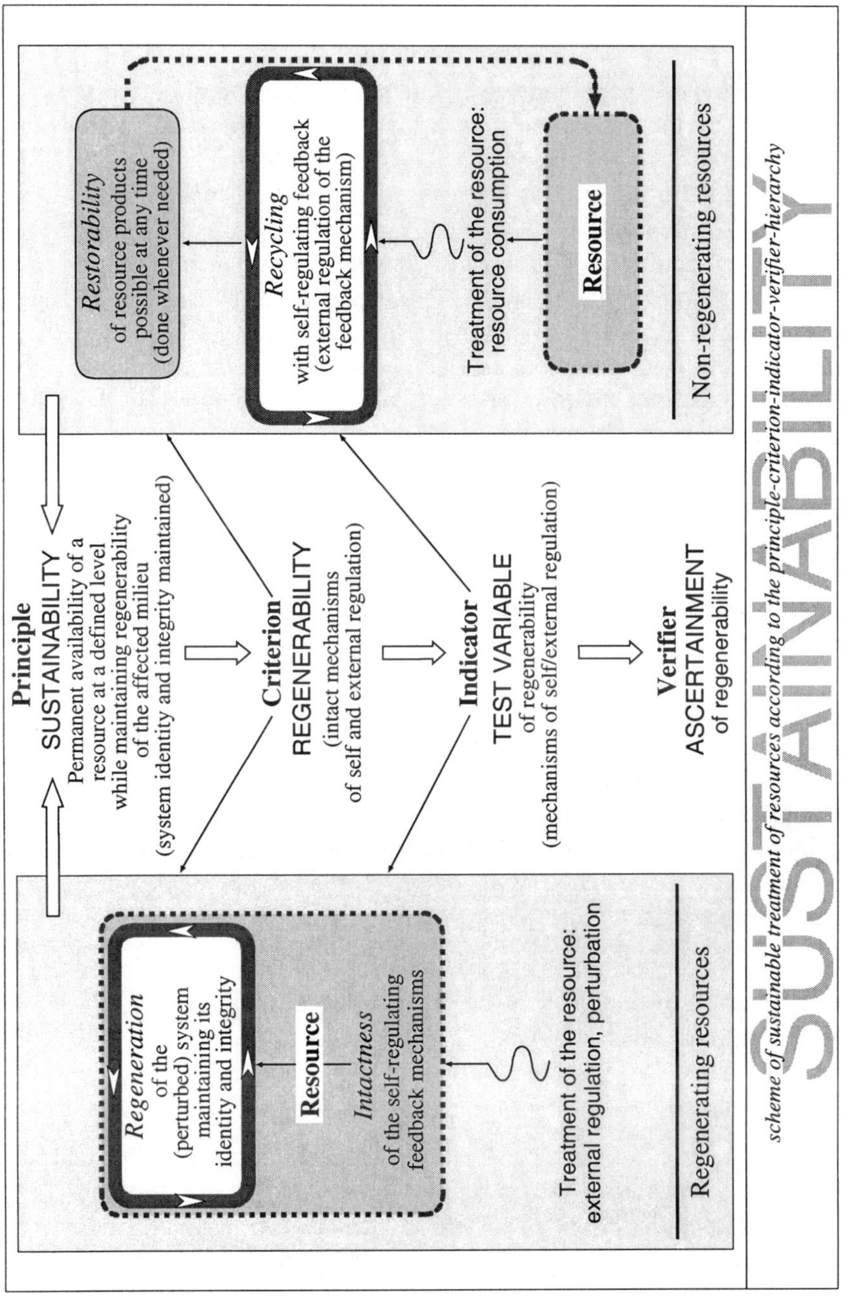

scheme of sustainable treatment of resources according to the principle-criterion-indicator-verifier-hierarchy

Feedback in populations occurs through reproduction, which guarantees the continuation of genetic information. Feedback rates are determined by the reproductively effective population size, which limits the amount of information transmitted, and by the generation time, which governs the speed of feedback. The mechanisms of self-regulation are aimed at adaptation to the prevailing environmental conditions and at the preservation of adaptability to future changes of the environment.

Adaptation is achieved by changes in the genetic composition of a population such that information coding for adapted types becomes available in a sufficient number of population members. Adaptability to further conditions is preserved by preventing complete adaptation, so that genetic variants always remain that code for types with inferior adaptedness to the currently prevailing conditions. Thus, preservation of adaptability amounts to preserving genetic variation that is not required for adaptation to the currently prevailing conditions. The totality of mechanisms and factors involved in the realization of adaptational processes makes up the adaptational system (see the system-oriented representation of the adaptational system in Figure 2, the individual components of which and their relations will be introduced in the following).

The main constituent of a population's adaptational system is its genetic system. In terms of its functions, the genetic system is defined by all mechanisms that serve the organization (e.g. on chromosomes, in molecules), expression (e.g. regulation of gene activity and translation), reproduction (e.g. replication and transmission, different forms and portions of generative and vegetative reproduction), combination (e.g. by meiosis and systems of mating), and distribution over space and time of genetic information. It follows that the basic prerequisites for successful operation of the adaptational system of a population or species are

- (i) intactness (operability) of the mechanisms of ist genetic system,
- (ii) realization of the mechanisms' external conditions for operation (operational conditions), and
- (iii) availability of genetic variation for alteration of these mechanisms.

As to prerequisite (i), it must be taken into account that the development of a mechanism is governed by the genetic information and the environmental conditions of the population members. Hence, mechanisms may be modified by genotype and environment. Environmental components with this effect are referred to as modifying environmental conditions. Each modification represents a variant or state of the mechanism (see Figure 2). In extreme cases, these modifying environmental conditions may not allow certain or even any of the genotypes to develop any mechanism state in operable form. For example,

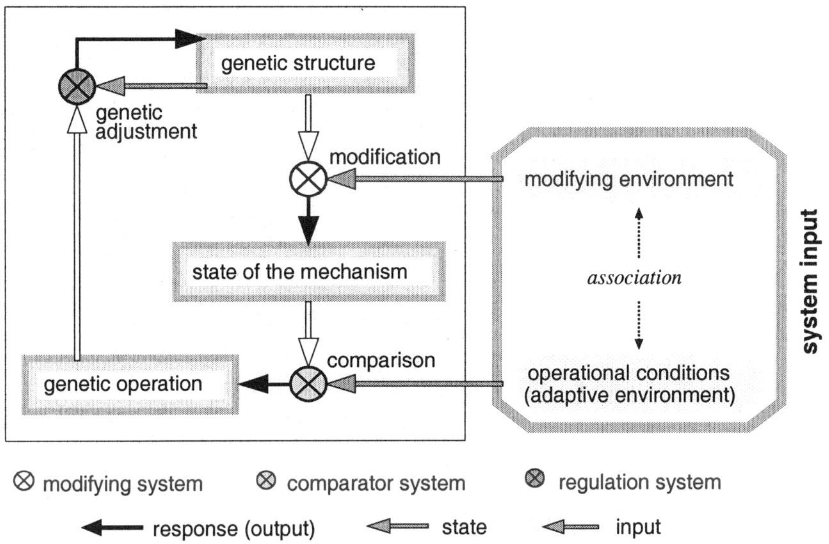

Figure 2: *The adaptational feedback system with its three constituent subsystems (after Gregorius 1997): (a) modifying, (b) comparator, and (c) regulation system. The subsystems are defined according to their roles in (a) specifying the state of the mechanism of the genetic system, in (b) producing the genetic operations that are induced by the operational conditions of this state, and in (c) adjusting the genetic structure through the genetic operations.*

many cell toxins constitute modifying environments that destroy constituent components of the mitotic mechanism. A less extreme example of a modifying environment is provided by mating incompatibility mechanisms, in which the strictness of the incompatibility relations may decline with increasing temperature. In terms of the above general representation of resource treatment, the modifying environmental conditions of a population apparently constitute its external regulation factors.

Prerequisite (*ii*) concerns those environmental conditions of a population that enable the mechanisms to accomplish their special tasks or destinations. These operational conditions may be of a very basic nature in that their absence may imply inoperability of the mechanisms. For example, all structures required for pollen dispersal by certain animal vectors may be well developed, however, the vectors may be absent; the dispersal mechanism is intact here, but its operational conditions are not realized.

On the other hand, the operational efficiency of a (state of a) mechanism may change with alterations of its operational conditions. In principle, such alterations may extend over a range that implies everything from complete inoperability to optimal operation of the mechanism. Hence, since a mechanism may appear in different states, these states may be distinguished by the operational efficiencies which they realize under the same operational conditions. The operational conditions can be viewed to play the role of adaptive environmental conditions because they decide upon the operational efficiency of a particular state of a particular mechanism of the genetic system (see Figure 2). Changes in the adaptive conditions constitute system perturbations in the above general framework of resource treatment.

Modifying and adaptive environmental conditions can be associated to various degrees, including the extremes of independence and identity of both functions. Cases of strong associations of both functions may complicate distinction of their effects, which, in turn, may complicate the evaluation of the effects of environmental changes in terms of perturbations and external regulations. In trees, for example, temperature dynamics acts as an essential modifying environmental condition in the timing of new growth (shooting). Late frost, on the other hand, is part of the temperature dynamics, but it acts as an adaptive condition in that it leads to a die-back of the shoots. Hence, temperature dynamics act as a modifying as well as an adaptive condition on the trait "timing of shooting". Moreover, the trait entails constitutional adaptations to this condition, since damage can be avoided by delaying the timing of shooting, and since this is not a response to the adaptational challenge.

In some trees species such as oak, however, the die-back of primary growth due to frost is frequently followed by an induction of secondary growth (Lammas shoots). This represents a remedy of an adaptational failure showing in the die-back. As part of the temperature dynamics, late frost functions here as a modifying condition for secondary growth. Consequently, the same environmental factor (temperature dynamics) acts as modifying and adaptive condition on one trait (timing of shooting), while the adaptive part of this factor acts simultaneously as a modifying condition on a second trait (secondary growth). Being modified by the adaptational challenge, this second trait constitutes a restituting regulatory adaptation to late frost.

Availability of genetic variation

Prerequisite (iii) relates the genetic capacity for adaptation to variable operational conditions. This capacity is inherent in the existence of genetically distinct states of a mechanism, which differ in their operational efficiency in response to the same operational conditions. The availability of genetic variation therefore determines the adaptability of a population.

Genetic variation can be "available" (for adaptation) in two basically different ways. One way concerns the regular presence or generability of genetic variation within a population. The other concerns the potential for acquiring genetic variation that resides in other populations. The latter way is, of course, only meaningful, if sufficient genetic differentiation among populations is realized. Hence, availability of genetic variation implies in this case the existence of dispersal mechanisms, realization of these mechanisms' operational conditions, and genetic differentiation among those populations which can be connected by (moderate) gene flow. To actually allow the inflow of genes from other populations, a population may need to extend the period of ovule receptivity and lower existing cross-incompatibility barriers (usually in response to adaptational stress).

The first way, in which genetic variation can be available, again contains two aspects. One relates to the potential for generating genetic variation within a population when required for adaptational processes. These mechanisms respond to adaptational stress by, for example, lowering population-internal incompatibility barriers or increasing the mutation rate or recombination frequency. However, genetic variation cannot always be generated in appropriate quality and quantity when demanded by adaptive changes. Therefore, in the long run, availability of genetic variation within populations is ensured by conditions serving the preservation of genetic variation. These conditions may be of an environmental nature (spatial and temporal heterogeneity of operational conditions), of a demographic nature (sufficient population size, balanced age class structure), or they may be inherent in the mechanisms of the genetic system (heterosis, sexual asymmetry, self-incompatibility, limited gene flow, etc.). Both ways of preserving genetic variation, within and between populations, serve the preservation of adaptability.

These considerations reveal that besides the actual presence of genetic variation its availability is largely determined by special mechanisms of the genetic system and their operational conditions. Consequently, prerequisites (*i*) and (*ii*) also play a decisive role in the fulfillment of prerequisite (*iii*).

Genetic operations

The forces that directly effect changes of the genetic structure will be addressed as genetic operations. Among the most important operations (called "genetic processes" by Namkoong et al. 1996) are selection, mating, gene flow, mutation and genetic drift. They are realized by the operations of the mechanisms of the genetic system and are therefore determined by the state of the mechanism and its operational conditions (see Figure 2). Namkoong et al. (1996, p.2, right column) refer to the effects of mechanisms and their operational conditions on genetic operations as "the structures necessary to maintain those processes". Most significantly, genetic operations are the direct forces of adaptation in that they adjust genetic structures according to their performance under the modifying and adaptive environmental conditions (see Figure 2).

While selection and genetic drift by definition depend on the presence of genetic variation, mating relations may depend on genetic variation in some cases (S-alleles in incompatibility systems) but not in others (random mating, partial selfing). Genetic variation is no prerequisite for gene flow and mutation.

Selection is the primary agent of adaptation. It acts by removing those states of a mechanism of the genetic system that are operationally less efficient under the prevailing operational (adaptive environmental) conditions. Operational efficiency is chiefly determined by the mechanisms involved in maintaining the vital functions, including fertility, and involved in mating. If the states of the mechanism are heritable they can be replaced over the generations, and adaptation can take place on the population level. This does not rule out the possibility that selection may fail to restore adaptedness, since the operational efficiency achieved after selective replacement may still not suffice to maintain a population. Mating, gene flow and mutation are genetic operations that produce genetic changes irrespective of the mechanisms for which these changes code. However, if these changes code for different states of a mechanism, they can be selected for later in the reproductive cycle.

The ways in which the mechanisms and their operational conditions bring about genetic operations are obvious for selection, mating, gene flow and mutation, but they appear less self-explanatory for drift. Genetic drift is commonly referred to as random changes in genetic structures, and as such it is not actually a genetic operation resulting from a response of some mechanism to its operational conditions. In fact, the randomness leading to genetic drift is restricted to the action of environmental conditions both in their modifying and adaptive components. In this way, selection may take place at random since the operational conditions do, and the same can be stated for the modifying conditions evoking mutation.

Since all of the four (primary) genetic operations, selection, mating, gene flow, and mutation, are affected by environmental randomness, genetic drift can be conceived of as a secondary operation that results from randomness in any of the four primary genetic operations. Randomness in the primary genetic operations comprises the reduction of populations to small sizes. This is one of the most frequently named causes for genetic drift. Drift may accelerate as well as impair adaptational processes, which leaves us with the four primary genetic operations as the direct forces of adaptation.

Two elementary levels of genetic information

By definition, the characteristics of a species' genetic system must have been evolutionarily stable. Thus, the genetic basis of these characteristics has been maintained despite the continuing challenges of adaptation and the entailed changes in genetic structure. This suggests distinction of two levels of genetic information, at one of which the mechanisms of the genetic system are encoded; this information ought to be relatively conservative. The other level concerns adaptability to changing environmental conditions, and it requires variable genetic information. In essence, the adaptational processes serve to preserve the characteristics of the species' genetic system and its integrity. Herewith, each characteristic can be equated to a trait with variable states.

The latter clearly corresponds to genes and their allelic states, which thus represent the two levels of genetic information. The function of each particular enzyme, for example, is usually indispensable for the realization of a number of vital processes of an individual. It defines a particular mechanism of the genetic system. Moreover, many enzymes are known to exist in different states that are coded for by the alleles of a single gene locus, and the catalytic activity of these states could be modified differently by temperature. Optimal realization of the enzyme's function could then be achieved under different temperature environments by different enzyme states (allozymes). The existence of allelic variation would thus allow for adaptation of the enzyme's function to different temperature environments. By this the intactness of the mechanism defined by the enzyme's function would be guaranteed.

In general, one could state that a gene codes for products, the functions of which determine mechanisms of the genetic system. The states of a gene (the alleles) encode modifications of the products that do not change the functions but ensure their performance (operability) under variable environmental conditions. Alleles thus represent states of the mechanism.

Indication of sustainable treatment

It follows from the above considerations that changes of the genetic structure (brought about by genetic operations) alone are not sufficient for the indication of endangerment of a population. Such changes may be signs of ongoing adaptation, in which case they signify intactness of the mechanisms of the genetic system and availability of genetic variation. They may, however, also be signs of an impairment of these mechanisms and their operational conditions with the result of losing adaptational capacity. If, for example, previously rare but persistent genetic variants suddenly further decrease in frequency and probably vanish, this may be due to disrupted inflow of the rare genes from other populations, which, in turn, may be a consequence either of an impairment of the dispersal mechanisms or of a change in their operational conditions. Hence, any indication of the sustainability of a resource treatment rather has to concentrate on all of the above-listed three prerequisites for successful operation of the adaptational system in order to meet the criterion of regenerability (see Figure 1). Consequently,

> any treatment of a population can be sustainable only if it jointly guards (i) the intactness of the mechanisms of the genetic system, (ii) the realization of their operational conditions, and (iii) the availability of genetic variation.

The numbering reflects an obvious priority: a mechanism must first be developed in an operable state before its operational efficiency under the prevailing operational conditions can be specified. In effect, however, fulfillment of prerequisites (*i*) and (*ii*) must be evaluated jointly, since only in this way can the operational efficiency of a mechanism in response to its operational conditions be quantified. The adaptedness of a treatment should therefore be rated on this basis. Moreover, in combination, both prerequisites determine the capacity to preserve and to acquire the genetic variation required for adaptation to variable operational conditions. Hence, joint indication of prerequisites (*i*) and (*ii*) is indispensable for an evaluation of the sustainability of a treatment. Finally, to make sure that a resource treatment does not impair the adaptability (to environmental changes), prerequisite (*iii*) must be indicated.

Indication of the prerequisites

The possibility of broad-scale effects usually is of primary concern in the assessment of a treatment's sustainability. Indication is focused here on types and degrees of self- and externally regulated regeneration. As mentioned before, this may involve an estimate of feedback rates as they are determined by the reproductively effective population size (amount of information transmitted) and by the generation time (speed of feedback). A detailed indication following the above prerequisites (*i*), (*ii*) and (*iii*) may not be necessary at this level.

On finer scales, effects of treatments on special mechanisms of the genetic system or on their operational conditions, and thus the three prerequisites, become relevant. If these effects are observable, then direct indication is possible without further complication. Otherwise, if these effects cannot be directly verified, the system analytical approach to indication must be taken. In essence, this approach consists in modeling the potentially affected mechanisms and their operational conditions and in comparing the model predictions with actual observations (model validation). Incompatibility between model predictions and observations leads to rejection of the model (for a compact presentation of this approach and its statistical implications see Gregorius 1998). In the present context, rejection of the model amounts to the indication of violations of the sustainability prerequisites (*i*) or (*ii*) as applied to the modeled mechanisms and their operational conditions. Moreover, a non-rejected model can be used to predict capacities for preserving or acquiring genetic information, as is required for the indication of prerequisite (*iii*). Many population genetic models are suitable for this approach.

Provided that its direct deduction from basic principles is not possible, the choice or development of a model must be preceded by a study of mechanisms and their operational conditions in populations, the present appearance and history of which justify the assumption of adaptedness and adaptability (usually autochthonous populations, compare the paper of Schoppa & Gregorius in this volume). Such populations form the reference for developing models of intact mechanisms and their operational conditions. The appropriateness of the model is again tested according to the system analytical approach by comparing its various predictions with the corresponding observations in the reference populations. The choice of the mechanisms to be modeled depends, in turn, on the system components which are likely to be affected by a treatment (Namkoong et al., 1996, for example, provide a list of genetic operations which are likely to be affected by a number of management and utilization practices that are typical of tropical forestry). Thus, indication of the sustainability of a treatment is impossible without knowledge of intact reference systems, which confirms the frequently claimed indispensability of protecting recognizably stable ecosystems (see e.g. Mader 1991).

Given this prior information on conditions for optimal operational efficiency, the system analytical approach for the indication of prerequisites (*i*) and (*ii*) can be summarized in the following steps:

(1) specify mechanisms and operational conditions potentially affected by the treatment,
(2) model these mechanisms and their operational conditions,
(3) determine ranges of model parameters that optimize the operational efficiency,
(4) determine model characteristics which allow the prediction of observable traits,
(5) calibrate the model so as to fit the model predictions to the actual observations,
(6) reject the hypothesis of sustainable treatment either if the parameters of the calibrated model do not belong to the range that

optimizes the operational efficiency or if the calibrated model predicts loss of genetic information.

The last statement in step (6) refers to the verification of prerequisite (*iii*). Loss of genetic variation could result from changes in operational conditions which imply ultimate selective removal of all but a few genetic variants (directional selection) or which exert such strong selective pressures that population size is reduced at least temporarily below a threshold that entails considerable genetic drift. Directional selection, for example, can be suspected to take place, if rare alleles continue to decrease in frequency. As was argued earlier, the potential for such violations of prerequisite (*iii*) can, however, be recognized more comprehensively by verification of prerequisites (*i*) and (*ii*) for those mechanisms and operational conditions which serve the preservation and acquisition of genetic variation. Verification of the actual existence of genetic variation, however, is subject to experimental survey.

Genetic structures as indicators

It was argued above in detail that the indication of violations of the sustainability prerequisites (*i*) and (*ii*) have priority, even though they are not directly concerned with genetic variation. However, as became apparent in the example of gene flow, there are characteristics of the genetic structure that may be utilized to indicate an impairment of the operational conditions of special mechanisms of the genetic system. Another standard example is provided by the relative frequency of homozygotes in the zygotic or some other early developmental stage. The mating system and its operational conditions usually result in typical homozygote frequencies, so that (by application of the system analytical approach) deviations from these typical frequencies may indicate an impairment of the mating mechanisms or their operational conditions.

These examples demonstrate that genetic traits can be utilized in two ways for the indication of the sustainability of a treatment. On the one hand, genetic traits, among others, serve as tools for indicating the operational efficiency of the mechanisms of the genetic system in response to their operational conditions. In special cases, such as assortative mating for a particular genetic trait, a genetic trait is directly involved in the specification of a mechanism and therefore provides the opportunity for direct indication.

As a rule, however, the genetic traits potentially involved in the specification of a mechanism are not known. In such cases of indirect indication, a large number of different genetic markers may be available for studying a particular mechanism in combination with the effects of its operational conditions. The power of indication may then vary with the genetic marker used, depending for example on its degree of polymorphism and its selective significance. Since indirect indication usually involves the system analytical approach, the chosen genetic marker must appear as an observable variable in the pertinent model.

On the other hand, many if not most genetic traits indicate the availability of genetic variation for future adaptational processes. When used for this purpose, it must be remembered that the indication always depends on the particular genetic trait studied. The results thus may not be generalizable to other traits in the same populations. This situation is one of direct indication, even though the mechanisms coded for by the genetic information behind the genetic trait are not known. A more comprehensive compilation of characteristics of genetic structures that are useful for direct and indirect indication of various adaptational aspects is provided by Gregorius (1996).

A typical example of application of the system analytical approach

There are several mechanisms of the genetic system with the capacity to effect adaptational processes that approach stable *polymorphic* equilibrium states. These states optimize the trade-off between adaptedness (and thus operational efficiency) and the preservation of adaptability. The forces enabling such equilibria are to be found in special mechanisms of the genetic system and their operational conditions. One (but not the only) important mechanism is heterosis, i.e. adaptive superiority of phenotypes as the result of interaction between different allelic states in heterozygote individuals. In fact, under the comparatively mild assumption of consistent ranking of average genotypic fitnesses and moderate degrees of homotypic mating, heterosis is necessary for the maintenance of allelic polymorphism.

The observation of a major polymorphism at a gene locus (at least two alleles with high frequencies) in combination with moderate or no homozygote excess therefore suggests that the polymorphism was maintained by heterotic effects in the past. Hence, if after a treatment of the population heterotic effects should not be detectable, continuation of the treatment is likely to endanger the polymorphism. The treatment is thus unsustainable in that it caused changes either in the modifying conditions for the heterozygotes or in the adaptive conditions with the effect of losing genetic variation. Hence, prerequisite (*iii*) is violated through violation of prerequisite (*i*) or (*ii*).

How can the suspected heterosis be verified? A direct approach would be feasible under the reasonable assumption that the heterotic effects are confined to viability components (step (1) of the system analytical approach), and if genotypic frequencies could be scored within a cohort at a stage close to the zygotic and at the adult stage. This situation can be represented by a simple viability selection model (step (2)), in which the relative frequencies of genotypes before and after viability selection are observable (step (4), step (3) has already been taken by the specification of heterosis as optimizing the operational efficiency) and the genotypic viabilities (proportions of survivors) are free parameters to be used for model calibration.

For this simple model, the calibration (step (5)) simply results from division of the actually observed frequency of a genotype at the adult stage by that at the zygotic stage, since this yields a value that is proportional to the genotype's viability. The obtained genotypic viabilities are compared with the hypothesis of heterosis. If it turns out that heterozygotes do not show the highest viabilities, the hypothesis is rejected (step (6)), and, as shown above, the unsustainability of the treatment is indicated.

This direct approach is not realizable for trees because of their longevity. The near zygotic and adult stages can only be observed in the reverse order by scoring parental trees and their seed offspring. This problem can be solved by the assumption of genetic equilibrium, since then the same ontogenetic stages show the same genotypic structure in successive generations. Looking forward or backward in time yields the same result under equilibrium. Therefore, genotypic viabilities can again be estimated by division of genotype frequencies among the parental trees by these frequencies in the seed production. The same procedure as before can now be taken for testing the heterosis hypothesis.

Since this procedure is legitimate only if the equilibrium assumption can be justified, means of testing the assumption are desirable. In fact, there exists a simple method for such a test, which follows directly from the absence of all selection except viability selection. In this case genetic equilibrium implies identity of the allelic frequencies among parents and their seed offspring. If the test yields significance, the equilibrium model is not applicable and more sophisticated models must be induced to verify sustainable treatment.

It should be recalled that the system analytic approach is always aimed at the falsification of hypotheses on characteristics relating to sufficient operational efficiency of special mechanisms. There are of course mechanisms other than heterosis which are capable of maintaining genetic polymorphisms. Negatively frequency dependent selection is one example. This is why emphasis must be put on the verification of the special mechanism (producing heterosis in the above example) considered to be responsible for the observed phenomenon (genetic polymorphism). In this way the mechanisms and their operational conditions can be identified which are affected by a treatment in an unsustainable manner.

Sustainability of some basic forms of utilization of forest trees

Indication of the operational efficiency of special mechanisms of the genetic system in response to their operational conditions is meaningful in cases where treatments do not fundamentally affect the mechanisms of reproduction and thus of self-regeneration. The number of such situations is conceivably very large, and their indication usually requires application of the system analytical approach, as was argued earlier and demonstrated for an example in the last section. However, many if not most of the actual types of forest resource utilization do in fact interfere fundamentally with the mechanisms and conditions of self-regeneration. As was explained earlier, in such situations of highly complex consequences, it is meaningful to concentrate indication of sustainable treatment on the more general principles of self- and externally regulated regeneration. The basic forms of forest utilization addressed in the following are meant to exemplify such fundamental interferences.

Plantations

Plantations are resources that are not designed for self-regeneration. Their sustainable management depends on the self-regenerating resources from which they are derived and on the environment in which they grow. Plantations thus consist of two components, the source population and the production population. Sustainable management of this resource requires that the source population regenerates by self-regulation, and that stable nutrient cycles can be established in the plantation through external control. The latter demand accounts for the requirement that all affected resources be included in the design of a treatment. The external control can be a factor of instability, since it decreases the degree of self-regulation. The question of an optimal relationship between the sizes and structures of areas reserved for source and production populations is essential. Sizes, numbers and distribution of source populations in space need to fulfill the criterion of sustainability, and the possibility of excessive gene flow from plantations into source populations has to be given particular consideration with respect to size, numbers and spatial distribution of production populations.

Seed orchards

Production seed orchards can be viewed as a means of reducing the reproductively effective part of a source population (regenerating by self-regulation) to trees of desired phenotype and to establish certain mating relations among these selected (and possibly cloned) trees. The production populations established from seed of the orchard are expected to reproduce in larger proportions the desired phenotype identified in the source population. Since regeneration of genetic information in seed orchards is limited in time (with occasional compensation planting), sustainable supply with the desired information is again restricted to the source population with its self-regulated regeneration. Production seed orchards are thus subjected to similar principles as plantations.

Breeding

If breeding products cannot be regenerated indefinitely, their source populations must be maintained as units of self-regulated regeneration. On the other hand, if indefinite regeneration is possible, the system of reproduction must be designed such that regeneration occurs essentially in a self-regulated manner. Breeding products such as those resulting from genetic transformation introduce new mechanisms into an integrated system and can therefore not be considered to be derived from source populations which regenerate by self-regulation. Hence, in order to become sustainably manageable, they must at least be shown to allow for the establishment of populations with the capacity of self-regeneration. Since this kind of establishment requires the activity of evolutionary forces, it is highly unlikely that genetically transformed organisms can be developed into a sustainably manageable resource.

Silviculture

Silviculture may be chiefly conceived of as a form of external regulation (silvicultural control methods such as improvement felling) or of perturbation (such as selective logging) of a resource. It becomes the more sustainable the more it succeeds in organizing the external regulation and the perturbations as an obligatory feedback mechanism. By "obligatory feedback" it is understood that the external impacts are inescapably determined by their effects on the regeneration system. Under this premise, the external impacts are actually embodied in a self-regulating process and by this may establish a mechanism of self-regeneration.

Natural regeneration is, by definition, a necessary condition for sustainable silviculture. Control measures such as improvement felling are sustainable to the degree that their consequences react on them. Since such reactions may become apparent after one tree generation at the earliest, the indication of sustainability requires efficient methods of predicting future states from present observations. With the appropriate interpretation, the methods developed and practiced in the increment- and yield-oriented disciplines of forest science could be used to shorten the feedback intervals.

Concluding remarks

According to the above considerations, sustainability of a treatment can be indicated at two levels of specialty corresponding to broader and finer scale effects. While at a broader scale the realization of the basic feedback mechanisms of self-regeneration is concerned, finer scale effects refer to the operational efficiency of the individual mechanisms of the genetic system in response to their operational conditions as well as to the availability of genetic variation. Genetic traits play a special role at both levels of scale in that they help to trace descent, reflect features of the genetic background or are directly involved in the determination of genetic operations.

Yet, at any of these levels, traits other than genetic traits could be more efficient in the indication of the general criterion of regenerability, even though mechanisms of the genetic system or availability of genetic variation may be the subject that is being indicated. Moreover, as the examples of forest practices demonstrate, many of the most important methods of forest utilization can be effectively assessed for their sustainability on the lowest level of specialty, namely on the degrees to which they take into account feedback regulation as the elementary condition of self-regeneration.

Frequently, effects of treatments on a resource's genetic operations are considered to be central for the indication of sustainability (see e.g. Namkoong et al. 1996, p.2, top of right column). The present deliberations confirm this in that they identify genetic operations as the primary forces of adaptational adjustments. Yet these adjustments provide no direct information about the actual adaptational success achieved in terms of increased or decreased operational efficiency. Such information can only be obtained from indications of the prerequisites (*i*) and (*ii*). Characteristics of genetic operations are therefore of limited value for the indication of current adaptational success and its sustainability unless they can be related to the forces which bring them about.

If the latter is possible, the system analytical approach to the indication of sustainability can be taken. In this case, observations on genetic operations can be used to predict their potential effects on the preservation of adaptability as required in prerequisite (*iii*). Directional selection, for example, may occur due to the advent of a new selectively superior genetic variant or due to changes in the operational conditions. Any of these events would impair the preservation of genetic variation and thus of adaptability, but they may do so in different ways depending on the mode of inheritance. Hence, information on the mechanisms causing the observed genetic operation (directional selection) would help in specifying the impairment. Moreover, the other genetic operations of mating, gene flow, mutation or drift may interact by accelerating, slowing down or even preventing the selective substitution.

The wide applicability of the system analytical approach to the indication of sustainability helps to avoid the vagueness frequently felt in the comparative interpretation of genetic processes and indices of genetic variation. In fact, the model-based indication allows to utilize the broad range of theories and methods developed in the field of population genetics for the analysis of experimental data.

Acknowledgment

The author is grateful to the colleagues at his institute and especially to Frank Schoppa for their lively discussions of the topic and their suggestions for improvement. The comments of Gene Namkoong and an anonymous reviewer helped to improve the precision of the representations.

References

Callicott, J.B., Mumford, K. (1997): Ecological sustainability as a conservation concept. Conservation Biology 11: 32-40.

De Leo, G.A., Levin, S. (1997): The multifaceted aspects of ecosystem integrity. Conservation Ecology 1(1): Article 3

Gregorius, H.-R. (1996): The contribution of the genetics of populations to ecosystem stability. Silvae Genetica 45(5-6): 267-271.

Gregorius, H.-R. (1997): The adaptational system as a dynamical feedback system. J. theor. Biol. 189: 97-105.

Gregorius, H.-R. (1998): The system analytical approach to the study of hypotheses. URL http://webdoc.gwdg.de/ebook/y/1998/gregorius/sysana.pdf

Ludwig, D., Walker, B., Holling, C.S. (1997): Sustainability, stability, and resilience. Conservation Ecology 1(1): Article 7

Mader, H.-J. (1991): The isolation of animal and plant populations: Aspects for a European nature conservation strategy. Pp. 265-275 in Seitz & Loeschke

Namkoong, G., Boyle, T., Gregorius, H.-R., Joly, H., Savolainen, O., Ratnam, W., Young, A. (1996): Testing criteria and indicators for assessing the sustainability of forest management: Genetic criteria and indicators. Working Paper No. 10, Center for International Forestry Research, Bogor, Indonesia

Schellnhuber, H.J., Kropp, J. (1998): Geocybernetics: Controlling a complex dynamical system under uncertainty. Naturwissenschaften 85: 411-425.

Schoppa, F.N., Gregorius, H.-R. (2000): Is autochthony an operational concept? This volume

Seitz, A., Loeschke, V. (1991): Species Conservation: A Population-Biological Approach. Birkhäuser Verlag, Basel etc.

Svirezhev, Yu M., Shvirejeva Hopkins, A. (1998): Sustainable biosphere: Critical overview of basic concepts of sustainability. Ecological Modelling 160: 47-61.

Genetic diversity and differentiation of individual effective pollen clouds in trees

H.H. HATTEMER[1], M. ZIEHE[1], R. FINKELDEY[2] AND M. FROMM[3]

[1]Institut für Forstgenetik und Forstpflanzenzüchtung, Universität Göttingen , D-37077 Göttingen, Germany; [2]Eidgenössische Forschungsanstalt für Wald, Schnee und Landschaft, CH-8903 Birmensdorf, Switzerland ; [3]Landesanstalt für forstl. Saat- und Pflanzzucht, D-93317 Teisendorf

Key words: Population density, effective pollen clouds, differentiation, adaptation

Abstract: Tree populations suffer from a variety of natural environmental factors and Man's action. The latter mostly leads to reductions not only in population size but also in population density. Reduction of population density generally means the reduction of possible mating contacts among trees. Consequently, the number of mating partners is reduced. Hence, the number of effective pollen parents of the seed produced by single trees is expected to decrease. This condition is responsible for a tendency towards increased genetic differentiation of pollen contributions. The genetic differentiation of individual effective pollen clouds is an important means of inference on reduced effective number of male mating partners. Reduced population density possibly enforces mating preferences for neighboring trees. If neighboring trees are on average more closely related than more distant trees, preferential mating among neighbors results in inbreeding even for completely self-incompatible or dioecious species. This situation is particularly likely to occur in naturally regenerated stands if gene flow through seed and pollen has been restricted and eventually results in reduced adaptedness and adaptability of progenies under heterogeneous environmental conditions. A reduced number of male mating partners of a cosexual tree may also mean restrictions in the relative amount of neighbor tree pollen and, hence, a relative increase of the tree's own pollen present in its crown.

Consequently, the population density is likely to influence the adaptive potential of tree populations. However, it is difficult to define critical population densities which must be kept in order to avoid intolerable losses of adaptive capacities. Such critical densities are an outcome of the genetic system of a species but they also depend on environmental factors including pollen vectors. The critical population density may vary greatly depending on the environment, the physical structure of forests, and the mode of pollen transport. Although there exists ample evidence of long-distance transport of effective tree pollen, the reproduction system of trees is also characterized by a small-scale spatial pattern. This has a variety of consequences and is sensitive to anthropogenic influence.

Müller-Starck, G. and Schubert, R. (eds.). Genetic Response of Forest Systems to Changing Environmental Conditions. Vol. 70 (For. Sci.), 223-237

Introduction

Tree populations are stressed by a variety of natural environmental factors and Man's action. The latter mostly leads to reductions not only in population size but also in population density and may manifest itself in different ways.

There is no doubt that not only the gene pool of populations but also their genetic system is capable of getting adapted by evolutionary change. However, the large scale and the speed of human impact on the genetic system of tree populations has become a matter of concern to forest geneticists.

When considering the flow of effective pollen in a forest, we have to discriminate between two effects which hinder the mating contact between trees. The filter effect is produced by trees of any species standing in between two given conspecific trees by holding back pollen grains with their crowns. In case of zoophilous species, other tree species visited by the same pollinators may distract these animals. The dilution effect is produced by conspecific trees standing in between two given trees by producing pollen themselves.

For the time being, the paternity of all seeds produced by a tree can hardly be reconstructed unambiguously, but only be inferred on the basis of maximum likelihood. Unfortunately, this is particularly true in some tree genera and in large populations. In other genera promising prospects exist for the detection of numerous nuclear microsatellite gene loci possessing several variants in similar frequencies. The precision of the analysis of gene flow systems will then only be a matter of effort rather than shortcomings of the methodology.

As will be discussed in the present paper, heterogeneity among individual effective pollen clouds as measured by differentiation may indicate that trees possess only a restricted number of mates.

Methods

Estimation of pollen cloud structure

Genotypic structures of single tree progeny arrays can easily be determined if seeds are harvested from single seed trees and carefully kept separate. It is more difficult to estimate only the allelic contributions of male gametes to single tree progeny arrays, which are defined as pollen allele frequencies or genetically effective pollen clouds.[1]

---

[1] This term is illustrative in wind-pollinated species but not necessarily so in zoophilous species. In the present paper, it is still maintained for the sake of brevity.

It is frequently possible to unambiguously identify the male contribution to a progeny if the genotype of both the progeny and its mother tree at a particular gene locus are known. However, this does not hold in case of identical heterozygosity of the progeny and the mother tree. The occurrence of a haploid, maternally inherited, tissue in the seeds of most gymnosperms allows to differentiate between male and female contributions even in this case (Müller[-Starck] 1976, Gibson and Hamrick 1991), but no comparable tissue is available in the seeds of angiosperms. However, it is still possible to obtain maximum likelihood estimates of pollen allele frequencies from heterozygous seed trees if a sample of its progeny is analyzed (Gillet 1997). Regular segregation of both maternal alleles as well as random fusion of gametes is assumed in order to obtain a unique maximum likelihood estimate.

Measurement of differentiation and variation

Genetic differentiation among several subsets of a population is measured by genetic distances of those subsets and their complement (D_j according to Gregorius 1985). This method was later shown by Gregorius and Roberds (1986) to possess other attractive properties. It does not only allow for quantification of the degree of absolute average differentiation δ but also of the contribution of single pollen clouds to the overall differentiation; it is possible to quantify the differentiation among pollen clouds by the measure G_{ST} (Nei 1973, Finkeldey 1994). Genetic variation within pollen clouds or progenies was measured by total differentiation δ_T as derived by Gregorius (1978); this measure assumes values between 0 and unity and has a simple relationship to diversity v, $\delta_T = 1\text{-}1/v$. D_j and δ_T have been discussed in more detail in Hattemer (1991) and other contributions to that volume.

Data sets

Several data sets were taken from various studies of the reproduction system in tropical and temperate tree species with different sexual systems, pollen vectors, and from populations with different densities. All data reported here are based on genetic structures at isoenzyme gene loci, but, depending on the type of available data, different approaches of their interpretation had to be followed..

The African *Pterocarpus angolensis* studied by Rettenberger (1999) and its Asian congener *P. indicus* studied by Finkeldey *et al.* (1999) belong to the family *Fabaceae*. Both of these two species reveal an obviously disomic pattern of segregation and may be considered to be diploid. We may, therefore, assume that the system of incompatibility described for the respective family is active in these species and that there is but a small chance for self-fertilization in the presence of pollen produced by conspecific trees. The reproductive system of teak, *Tectona grandis*, was studied with emphasis on gene flow by Finkeldey (2001).

The small-leaved lime, *Tilia cordata*, is a European member of the predominantly tropical plant family *Tiliaceae*. Nothing is known to the present authors about the existence of incompatibility in this hexaploid species. The reproduction system of European beech, *Fagus sylvatica*, has recently been studied by Ziehe *et al.* (1998) and Ziehe (1999). The male contributions to the seeds of Norway spruce, *Picea abies*, can unambiguously be identified by joint electrophoretic analysis of macrogametophytes and embryos.

All of the species mentioned above have either been reported to produce very little seed following self-pollination or the moderate to low proportion of self-fertilization has been inferred from analyses of their reproduction system.

In this volume, Siregar reports on the differentiation of pollen clouds in different populations of *Pinus merkusii* and their variation.

A frequently encountered situation in degraded populations are few and highly dispersed solitary individuals. Talking about the mating contact of solitary individuals seems a contradiction in itself. Here we consider solitary individuals in the ecological sense which are cross-fertilized to an *a priori* unknown but presumably small proportion.

Results

Differentiation of individual effective pollen clouds

A summary table of the amounts of genetic differentiation among the individual effective pollen clouds in the investigated populations is presented in Table 1. Note that in *Tectona grandis* and *Picea abies* the estimates of G_{ST} differ markedly while the estimates of δ_T are exactly equal.

A natural stand of the Southeastern African species (*Pterocarpus angolensis*) in Botswana has very low and varying density. As all species of the family (*Fabaceae*) this tree is insect-pollinated. In Figure 1, some of the distances between trees are indicated for orientation. Data on genetic diversity of the pollen clouds and their differentiation indicate a rather clear negative relationship between the two parameters. However, the indication of varying numbers of effective pollen parents had obviously very little to do with the position of the individual seed trees. For instance, T1 would have been suspected to be most isolated for reason of its position in the stand. Quite unexpectedly, T3 is most extreme in this respect.

Table 1. Genetic variation in single tree pollen clouds and their differentiation. The table shows gene pool values of 4 to 11 polymorphic enzyme gene loci. Average gene diversity v, maximum value of D_j for any seed tree, δ and G_{ST}.

Species	v	max D_j	δ_T	G_{ST}
Pterocarpus angolensis	2.038	0.416	0.234	
Pterocarpus indicus	1.515	0.170	0.125	0.099
Tectona grandis	1.192	0.068	0.053	0.053
Tilia cordata	1.429	0.300	0.168	
Fagus sylvatica	1.411	0.120	0.062	
Picea abies	1.578	0.078	0.053	0.013

It may be concluded that substantial differences in the flowering times are responsible for deviation of the mating conditions in the stand from randomness. The more or less isolated location of the individual trees does not indicate that the numbers of their mating partners are simply a function of distance.

In *Pterocarpus indicus* no correlation between pairwise spatial distances of the trees and pairwise genetic distances of their effective pollen clouds was detected. Phenological observations suggest that the flowering periods of many trees do not overlap. An isolated road-side tree had the lowest estimated proportion of cross-fertilized seeds; also the maximum proportion of self-fertilization of 34% is at its maximum in this tree. The pollen cloud of this tree is typically most strongly differentiated and the diversity of its seeds is below average.

Results on *Tectona grandis* are presented in more detail elsewhere (Finkeldey 2001). Due to the relatively low level of variation at enzyme gene loci, the average differentiation of pollen clouds is low despite of rather short transport distances of effective pollen.

An isolated beech forest of about 40 ha in size comprises a total of 149 limes (*Tilia cordata)* in various densities. Some of them such as a clump of trees in the southwest were planted, the more scattered individuals are natural. Their age varies between 140 and 250 years. The degree of spatial isolation of the trees varies greatly. Seed from some isolated trees several hundred m apart on agricultural land was also collected.

The lime is pollinated largely by social hymenopteran insects. During its flowering period no major assemblage of other food plants was in bloom.

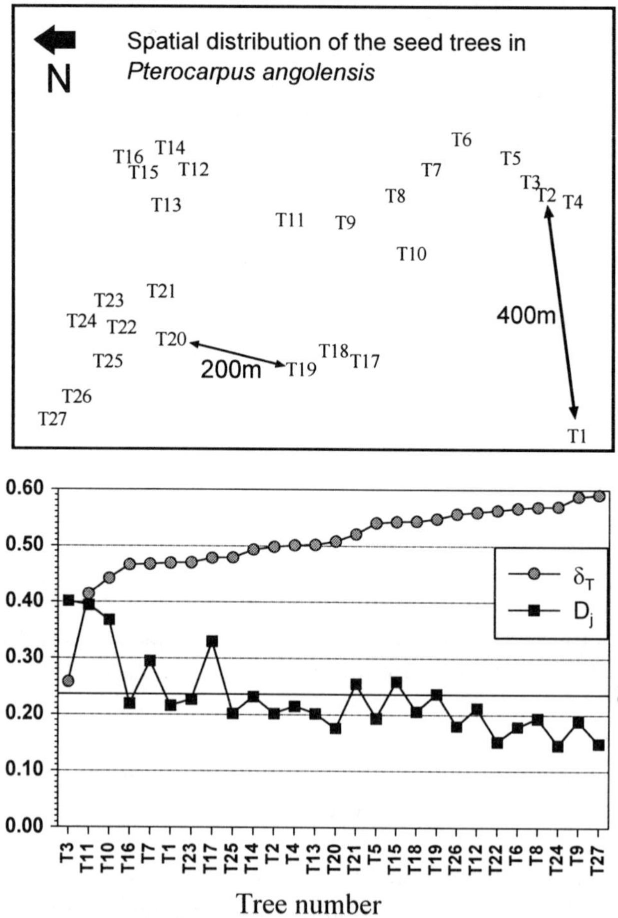

Figure 1. Deviations from random mating in Pterocarpus angolensis in Botswana. Top: Spatial distribution of 27 seed trees in a stand. Only low shrubs grow between the seed trees. Bottom: Variation in and differentiation among the respective 27 effective pollen clouds pooled over 4 enzyme gene loci (from data of Rettenberger 1999).

Figure 2 shows the maximum differentiation value of 0.30 measured for one of the three isolated trees (numbers marked with F). With a value of only 0.04 the least differentiated, i.e. the most representative, was the genic structure of the 149 adult trees on the right.

Some trees with strongly differentiated pollen clouds such as L121, L129 and L134 were represented by only moderate sample sizes. This does not disturb the provisional overall picture of decreasing differentiation with increasing variation.

Single tree progenies of *Fagus sylvatica* have been studied e.g. in the Harz region (Seesen) in a stand of beech trees mixed with spruce. The distances of neighboring beech trees varies considerably. Figure 3 demonstrates for 14 of these beech trees the genetic variation parameters of their estimated effective pollen cloud structures.

According to Figure 3, trees B26 and B31 represent lowest total differentiation δ_T but large D_j for their effective pollen cloud, which confirms the above illustrated relationships for this tree species. On the other hand, a negative correlation between δ_T and D_j is not sufficiently supported by the beech data.

It has to be added that seed tree B26 which according to Figure 3 might be expected to have the smallest number of effective neighbors, is right in the middle of the stand and might, therefore, have had many mating partners. Both of the trees B8 and B10 are more or less isolated and assume contrasting positions in Fig. 3. Tree B4 with low δ_T and large D_j is encircled by spruces. In these instances, the position of a seed tree in the stand is obviously irrelevant. Hence, also the genotypes of the effective neighbors contribute to differentiation and diversity of the pollen clouds. Besides that, wind directions, topology, tree size and selfing-rate modify the relationship between the two parameters. When comparing Figures 1 and 2, the difference between differentiation and diversity is larger in the former. The difference is even more pronounced in Figure 3, where the shear between the two curves is wide open. This difference may have something to do with peculiarities of the pollen transport media and vectors, respectively.

Three old stands of *Picea abies* in western Germany (Sauerland, Nordrhein-Westfalen) were planted between 1850 and 1910. According to historical records, the origin of the planting-stock is in Central Germany (Thüringer Wald). The three populations are located many kilometers apart. On the basis of genotyped samples of 150 to 200 trees, their gene pools have an average differentiation of δ=0.053 (Finkeldey, 1995).

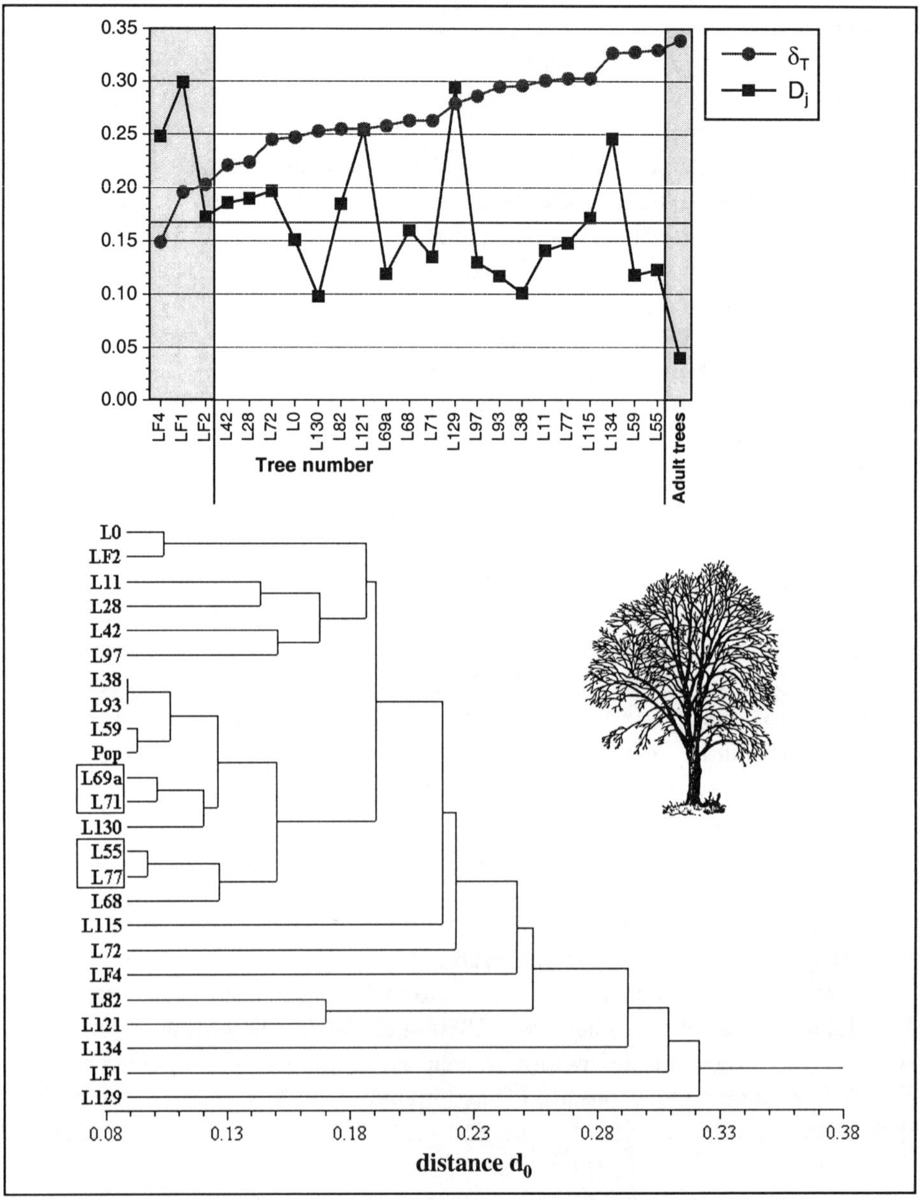

Figure 2. Effective pollen clouds in the small-leaved lime. *Top*: Genetic differentiation and variation of 23 pollen clouds and the set of 140 adult trees (Pop). Trees LF1, LF2, and LF4 are solitary trees several hundred *m* away from the stand. *Bottom*: Dendrogram of the pollen contributions to the seedlots collected from those 23 trees. Two pairs of neighbors are marked by frames. Note the position of the effective pollen clouds of trees LF1 and LF4, and the similarity of those of trees LF2 and L0 (Pop. refers to the pooled pollen cloud of adult trees.).

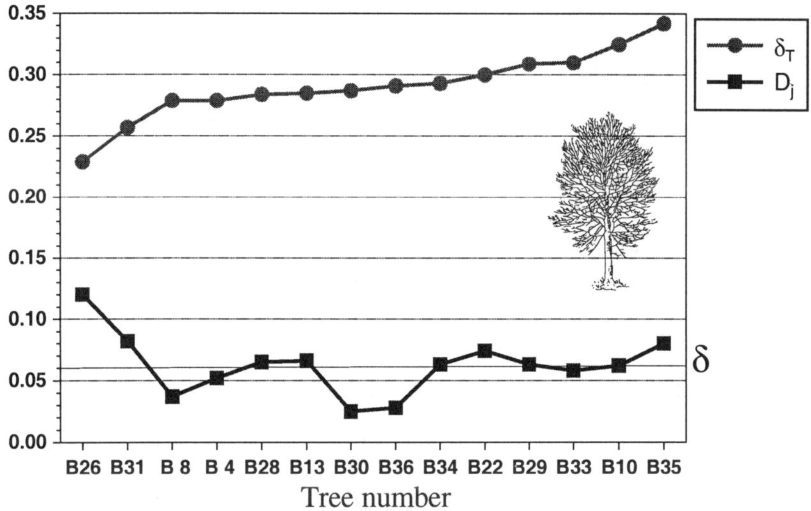

Figure 3. Variation in and differentiation between 14 effective pollen clouds of *Fagus sylvatica* (pooled over 8 enzyme gene loci). Beech nuts were collected from beech trees in the Abt. 210, Seesen. Data were provided by Krabel (unpublished).

100 seeds collected from each of seven trees in every population were investigated at 14 enzyme gene loci. Seven of these were considered eligible for the study of the variation between pollen clouds, since the maximum frequency of the most frequent allele was 95%. The outcrossing rates of the 21 trees were all very high and averaged 0.96, 1.01, and 0.95, respectively, in the three stands. Hence, varying degrees of self-fertilization could not affect the differentiation of effective pollen clouds.

The individual effective pollen clouds of the trees were very similar. In 7 of the 21 tests of homogeneity of allelic structures at single gene loci in the three populations, a significant test statistic was derived. The random association of genotypes in the dense stands of this wind-pollinated tree species was considered to be causal for this result (Finkeldey, 1995).

Discussion and conclusions

Causes of differentiation

The reduction of population density at reproductive age means an increase in distance between mating partners. Wind-carried pollen tends to sink down due to gravity. Although animals travelling between trees transport pollen directionally, the distances of their feeding flights are, nevertheless, steered by the availability of food and their capacity to fly. Therefore, the transport distances of effective pollen are restricted and a reduction in tree density below a critical limit sooner or later inevitably means the reduction of mating contacts among trees. Consequently, the individual tree has fewer mating partners and the number of effective pollen parents of the seed produced by single trees decreases.

The concept of density does, of course, apply only to flowering-tree density (Murawski and Hamrick 1991). In stands with given tree density, the proportion of flowering trees may change considerably during subsequent flowering periods, as Murawski and Hamrick (1992) have shown in a neotropical tree species. Consequently, the genetic implications differ in spite of unchanged tree density.

The concomitant opening of the forest canopy leads to increased air turbulence which may partly compensate for the sinking speed of pollen or may support the feeding flights of less active pollinators. Nevertheless, the individual tree is pollinated by pollen which is produced by a finite or even small number of trees, preferably neighbor trees. Low efficiency of pollen transport enhances the contributions of near neighbors to the pollen clouds. Hence, the pollen fertilizing different individuals is produced by unique groups of other trees. Different individuals share their groups of male mating partners only partly. This is in contrast to the concept of fertilization by a population pollen cloud which is completely homogenized during transport.

This condition has genetic impacts by inducing a tendency towards genetic differentiation of individual effective pollen clouds. It is due to genetic drift.

The hitherto most commonly used indicator of reduced effective number of male mating partners of seed trees is the genetic differentiation of pollen clouds. Nevertheless, it is not the only means. Autocorrelations in the spatial distribution of rare alleles in the seeds are very helpful. Such studies can be considered as a first step towards the construction of curves of pollen distribution in the vicinity of trees carrying unique alleles. Finkeldey (2001) studied the mating preferences of seed trees in *Tectona grandis* for certain male alleles. Any mating preferences are a strong indication of deviation from random mating between genotypes. In particular, changes in the sign of mating preferences may hardly be interpreted but by preferences for gametes produced by trees standing closer.

The differentiation of individual effective pollen clouds is primarily a function of the general level of genetic variation in the population. The eventual clumping of genotypes in the population (Ziehe *et al.* 1998, see also *5.4*) only intensifies the differentiation of the locally produced pollen.

The differentiation of pollen clouds can hardly be expected to be a function of population density alone if effective pollen is considered. The critical population density may vary greatly depending on the environment such as topography, the variation in the speed and directions of winds, the physical structure of forests, the mode of pollen transport, and the sexual system of the species. The variation in flowering phenology in the population may greatly modify the relation between general population density, tree distances and pollen cloud differentiation. This was confirmed by Finkeldey *et al.* (1999). Phenology variation does not only blur the correlation between spatial tree distance and genetic pollen cloud distance, it further reduces the number of trees contributing pollen to fertilization.

Data on the correlation between population density and the differentiation of effective pollen clouds were presented by Murawski and Hamrick (1991), Finkeldey (1995), and Finkeldey *et al.* (1999). However, there are exceptions to this rule as is described in the present paper.

There is no direct relationship between the maximum pollen transport distance and the differentiation of pollen clouds. Neither can much be inferred on the distribution of transport distances. The differentiation of the pollen clouds is, of course, stronger influenced by the genotypes of trees in the neighborhood rather than transport distances of their pollen.

Consequences of increased self-pollination

A reduced number of male mating partners of a cosexual tree may lead to restrictions in the relative amount of neighbor tree pollen and, hence, a relative increase of the tree's own pollen present in its crown.

An increase in self-pollination may have several consequences. If the species neither possesses a system of incompatibility nor self-sterility, self-pollination leads to self-fertilization and inbreeding. The selection against zygotes arising from self-fertilization is sometimes greatly retarded. This contributes to the differentiation of effective pollen clouds as surveyed in the seed stage. A system of incompatibility efficiently prevents inbreeding due to self-fertilization. However, under a system of self-sterility the increased self-pollination is likely to be accompanied by low fruit set or cone set, increased production of empty grains or other types of fertility losses of the population.

In the case of seed production following self-pollination in cosexual species, the mating system contributes to the genetic differentiation of individual effective pollen clouds by the amount of self-fertilization (cf. section 5.3). This does not apply in self-sterile populations where only minor fractions of the pollen produced by the seed trees themselves become effective.

If population density falls short of a certain critical value, it also boosts the homozygosity in the genome. As Farris and Mitton (1984) and Morgante *et al.* (1991) have demonstrated, statements on the critical average distance among the trees are difficult to make, because of the numerous other factors having an impact on the distribution of transport distances of effective pollen.

In a mixed beech stand, Freytag (1998) reported a correlation between the maximum proportion of self-fertilization according to the exclusion principle and the proportion of empty nuts at the level of individual trees. However, this author disregarded her findings on the gene locus *LAP-A* because of the well-known problems in the interpretation of LAP zymograms in this tree species. Since this gene locus happens to be the most variable, it was not possible to further reduce the proportions of seeds that were possibly the result of outcrossing. Nevertheless, the proportion of empty nuts could serve as an indication of the proportion of seeds due to self-fertilization.

However, Siregar (2000) did not find this type of relationship in *Pinus merkusii*. This author also discussed why the two phenomena should not be expected to be correlated. Bendixen (oral communication) neither found this relationship in *Acer campestre*, a species with considerable proportions of empty seeds.

Since a relationship between the density of populations and the proportion of seeds produced by self-fertilization must be expected, the question of the critical density arises. In view of large reported distances of pollen transport in the absence of either the filter effect or dilution effect, the critical density with regard to inbreeding is certainly much lower than commonly supposed. Having in mind the enormous consequences for seed collection, we may only strongly recommend to study this property of self-fertile species carefully.

Self-fertilization does not only contribute to pollen cloud differentiation. If the proportion of self-fertilization does not vary with the distance to the nearest neighbors but with seed tree genotype, the relationship between pollen cloud diversity and local population density must be expected to be blurred. Also the degree of heterozygosity of a seed tree at the surveyed marker gene loci matters greatly to pollen cloud diversity.

Reduced variation of individual pollen clouds

Given that the differentiation of individual pollen clouds is caused by small numbers of male mating partners, the genetic variation of the individual effective pollen clouds is expected to decrease. Some evidence is presented in Figure 1 for *Pterocarpus angolensis,* in Figure 2 for *Tilia cordata*, and in Figure 3 for the wind-pollinated beech (*Fagus sylvatica*).

Since measures of allelic variation such as v and δ_T are functions of the sum of squared allele frequencies, they are most affected by the alleles of medium frequency. In the pollen pool of the population, there is a tendency towards balanced and, therefore, medium frequencies of all alleles. However, the differentiated individual pollen clouds contain more alleles in extreme frequencies and, hence, possess less variation.

Clumping of alleles among reproducing trees

Due to restricted transport of pollen and seed, reduced population density enforces mating preferences for neighbor trees. Low efficiency of both pollen *and* seed transport leads to the clumping of genetically similar if not even related trees.

It is difficult to infer on consanguinity from genetic similarity of the trees in adult stands. Family groups of trees in natural stands regenerated from seed have been detected in some instances (Doúnavi, 2000) but not in others. Since part of those surveys were made in adult stands, also locally heterogeneous viability selection has to be considered as a causal factor of genotype or allele clumping which in conjunction with short effective pollen transport boost the differentiation of effective pollen clouds.

Small-scale spatial pattern of reproduction

Although ample evidence exists that effective tree pollen may be transported over large distances, the reproduction system of trees is also characterized by a small-scale spatial pattern (Freytag 1998). This includes genetic differentiation among pollen clouds depending on the position of the seed tree in the interior of the stand vs. at its margin or depending on the position of fruits or cones picked up from the ground under different parts of the tree crown. This condition can be encountered only in species producing heavy fruits and short distances of their dispersal.

Conclusions

Man's impact on forests does not only influence the condition of tree populations, i.e. their genetic structure. Man has also an impact on the genetic system of tree populations, if the density of tree populations at reproduction falls short of a certain critical bound.

An immediate consequence is the reduced fertility of the population. Genetically more relevant is an increased differentiation of the seeds of trees both in structure and in diversity. This has consequences for the practice of seed collection. Furthermore, there exists a tendency towards more self-pollination and, in self-fertile species, towards more self-fertilization.

The reproductive stress of tree populations can be monitored by the degree of pollen cloud differentiation and some more sophisticated procedures. They represent a set of indicators whether tree populations are really adapted to man-made conditions and are able to produce adaptable progeny.

References

Dounávi, A. 2000: Familienstrukturen in Buchenbeständen (*Fagus sylvatica*). Forstwiss. Dissertation, Univ. Göttingen. (http://webdoc.sub.gwdg.de/diss/2000/dounavi)

Farris, M.A. and Mitton, J.B. 1984: Population density, outcrossing rate, and heterozygote superiority in ponderosa pine. Evolution 38: 1151-1154

Finkeldey, R. 1994: A simple derivation of the partitioning of the genetic differentiation within subdivided populations. Theor. Appl. Genet. 89: 189-200

Finkeldey, R. 1995: Homogeneity of pollen allele frequencies of single seed trees in *Picea abies* (L.) Karst plantations. Heredity 74: 451-463

Finkeldey, R. 2001: Genetische Untersuchungen zur Reproduktion von Teak (*Tectona grandis* L F.) in Thailand. In press

Finkeldey, R., De Guzman, N. and Changtragoon, S. 1999: The mating system of *Pterocarpus indicus* Willd. at Mt. Makiling, Philippines. Biotropica: 31: 525-530

Freytag, T. 1998: Genetische Strukturen ausgewählter Buchen und ihrer Nachkommen in einem Buchen-Eichen-Mischbestand. Forstwiss. Diplomarb. Univ. Göttingen.

Gibson, J.P. and Hamrick, J.L. 1991: Heterogeneity in pollen allele frequencies among cones, whorls, and trees of table mountain pine (*Pinus pungens*). Amer. J. Bot. 78: 1244-1251

Gillet, E.M. 1997: Maximum likelihood estimators of the gametic contributions to single-plant progenies. Biometrics 53: 504-523

Gregorius, H.-R. 1978: The concept of genetic diversity and its formal relationship to heterozygosity and genetic distance. Math. Biosci. 41: 253-271

Gregorius, H.-R. 1985: Measurement of genetic differentiation in plant populations. pp. 276-285 in: Gregorius, H.-R. (ed.) Population Genetics in Forestry. Berlin, Heidelberg etc.: Springer Verlag

Gregorius, H.-R. and Roberds, J.H. 1986: Measurement of genetical differentiation among subpopulations. Theor. Appl. Genet. 71: 826-834

Hattemer, H.H 1991: Measuring genetic variation. pp. 2-19 in: Müller-Starck, G., Ziehe, M. (eds.) Genetic variation in European Populations of Forest Trees. Frankfurt am Main: J.D. Sauerländer's Verlag

Morgante, M., Vendramin, G.G. and Rossi, P. 1991: Effects of stand density on outcrossing rate in two Norway spruce (*Picea abies*) populations. Can. J. Bot. 69: 2704-2708

Müller[-Starck], G. 1976: A simple method of estimating rates of self-fertilization by analyzing isozymes in tree seeds. Silvae Genetica 25: 15-17

Murawski, D.A. and Hamrick, J.L. 1991: The effect of the density of flowering individuals on the mating system of nine tropical tree species. Heredity 67: 167-174

Murawski, D.A. and Hamrick, J.L. 1992: The mating system of *Cavanillesia platanifolia* under extremes of flowering-tree density: a test of predictions. Biotropica 24: 99-101

Nei, M. 1973: Analysis of gene diversity in subdivided populations. Proc. Nat. Acad. Sci. USA 70: 3321-3323

Rettenberger, B. 1999: Untersuchungen zum Paarungs- und Genflußsystem von *Pterocarpus angolensis* in Kasane, Nordbotswana. Magisterarb. Fakultät f. Forstwiss. und Waldökologie, Univ. Göttingen

Siregar, I.Z. 2000: Genetic aspects of the reproductive system of *Pinus merkusii* Jungh. et de Vriese in Indonesia. Forstwiss. Dissertation, Universität Göttingen. Göttingen: Cuvillier Verlag

Ziehe, M. 1999: Anpassungsprozesse auf Populationsebene. pp. 115-134 in: Abschlußbericht 1994-1998 zum BMBF-Verbundprojekt "Veränderungsdynamik von Waldökosystemen", Berichte des Forschungszentrums Waldökosysteme der Universität Göttingen Bd. 61, Göttingen

Ziehe, M., Starke R., Hattemer, H.H. and Turok, J. 1998: Genotypische Strukturen in Buchen-Altbeständen und ihren Samen. Allgem. Forst- u. Jagdztg. 169: 91-99

Microsatellite analysis of small anonymous seedlot samples from pedunculate oak (*Quercus robur*): a promising approach to monitor the number of different seed parents and pollen donors

C. LEXER[1], B. HEINZE[2], S. GERBER[3], S. MACALKA-KAMPFER[1], H. STEINKELLNER[1], B. ZIEGENHAGEN[4], A. KREMER[3] AND J. GLÖSSL[1]

[1]*Zentrum für Angewandte Genetik, Universität für Bodenkultur Wien, A-1190 Vienna, Austria;* [2]*Forstliche Bundesversuchsanstalt Wien, A-1140 Vienna, Austria;* [3]*INRA, Laboratoire de Génétique et d'Amélioration des Arbres Forestiers, F-33611 Gazinet Cedex, France;* [4]*Bundesanstalt für Forst- und Holzwirtschaft, 22927 Großhansdorf, Germany*

Keywords: Oak, *Quercus*, microsatellites, relatedness, seed material, half-sibs

Abstract:We have tested microsatellite analysis to monitor within-stand genetic diversity in seedlot samples from pedunculate oak (*Quercus robur*). We focused on two key criteria of within-stand diversity: the number of seed parents (mother trees) included in the seed harvests, and the number of pollen donors (father trees) contributing to the open pollinated families. Our results show that nuclear microsatellite loci are suitable to detect seed contaminations and to infer the genotypes of the seed parents directly from genotype data of extremely small anonymous acorn samples (five to eight acorns per mother tree), given that the samples from each putative mother tree have been supplied separately. Subsequently, the number of different mother trees for such anonymous seedlot samples can be inferred by comparing the reconstructed maternal genotypes and by calculating F_{ST} pairwise between the families. Furthermore, we have tested an approach to infer the number of different pollen donors directly from open pollinated families by analysing haplotype arrays of closely linked microsatellite markers transmitted from the fathers to the progeny. We selected three closely linked microsatellites from a previously constructed genetic map of *Quercus robur*, then we used simulated data of these linked markers segregating in open pollinated families to test the approach in detail. The procedures presented here are designed for small sample sizes and for situations where no genotype information is available for the parent population. Therefore they are potentially useful for monitoring commercial seedlot samples from *Quercus robur*.

Introduction

Modern forestry management often interferes with the natural process of adaptation, thereby limiting the ability of tree populations to respond to environmental changes. In particular, this may be the case for artificial forest regeneration, as the plant material for afforestation is often not of local origin.

Müller-Starck, G. and Schubert, R. (eds.). Genetic Response of Forest Systems to Changing Environmental Conditions. Vol. 70 (For. Sci.), 239-250

In Europe, the forestry laws of several countries support the use of well adapted plant material, focusing mainly on the use of appropriate provenances as seed sources (Geburek and Heinze 1998).

However, the ability of forest stands to respond to environmental changes also depends on the presence of genetic variability. The importance of genetic diversity and the negative impact of inbreeding have been documented in numerous plant and animal species (Avise and Hamrick 1996). Inbreeding is known to affect populations of naturally outbreeding species by depressing components of reproductive fitness (Frankham 1995). Surprisingly, few attempts have been made to avoid inbreeding and support within-stand genetic diversity in commercial forest seed material through appropriate laws (Geburek and Heinze 1998).

Genetic diversity within forest seed material can be measured using diversity parameters derived from population genetic theory (e.g. Nei 1987, Hattemer *et al.* 1993). However, in practice it may be desirable to express diversity in a simpler way: through the number of seed parents included in seed harvests and the number of pollen donors contributing to the seedlots. A first attempt to include the number of different seed parents in legal regulations about forest reproductive material has been made in Austria. Here, a federal forest seed law regulates the marketing of seed for a variety of conifer and broad leaf tree species (Anonymous 1996 a, b). In the specific case of European oaks (*Quercus petraea* and *Quercus robur*), seedlots in Austria have to be collected from at least 20 seed parents (mother trees) per stand. For any marketable acorn seedlot at least five acorns per mother tree have to be shipped to the Federal Forest Research Centre in Vienna for genetic analysis. The samples of different mother trees are supplied separately; no genetic information is available for the parent populations. In the present study we tested the potential of microsatellite markers to analyse the genetic composition of such anonymous seedlot samples.

Materials and methods

Plant materials and DNA isolation

The experiments presented here are based on maternal half-sib families (=open pollinated families), the basic elements of commercial acorn seedlots. We analysed a model half-sib family harvested from a known mother tree (initial sample size 28 individuals, later enlarged to include 43 individuals) and up to 10 small seedlot samples (sample size four to eight acorns), five of which had been shipped to the Federal Forest Research Centre in Vienna in the course of commercial seed harvest ("anonymous seedlot samples"). A sample of 40 seedlings from one natural *Q. robur* population was used to determine the levels of microsatellite polymorphism within a typical harvest stand. For all of the samples, acorns were sown in the nursery, and DNA was extracted from leaf tissue of germinated seedlings as described elsewhere (Lexer *et al.* 1999).

Microsatellite markers

Nine microsatellite markers isolated from the genomes of *Q. petraea* (Steinkellner *et al.* 1997) and *Q. robur* (Kampfer *et al.* 1998) were used for genetic analysis. All of the markers have been located on a genetic linkage map of *Q. robur* (Barreneche *et al.* 1998). Three of the microsatellites are closely linked on the genetic map (Tab. 1). Microsatellite PCR amplification and electrophoresis were conducted as described elsewhere (Lexer *et al.* 1999).

Table 1. Microsatellite markers used in this study, their species origin, and the observed number of alleles (A_o) as well as expected heterozygosity (H_e) measured in 40 seedlings from one harvest site.

Locus	Species origin	A_O	Gene diversity (He)
ssrQpZAG 1/5	*Q. petraea*[b]	9	0.82
ssrQpZAG 9	*Q. petraea*[b]	10	0.85
ssrQpZAG 15	*Q. petraea*[b]	9	0.75
ssrQpZAG 3/64	*Q. petraea*[b]	18	0.90
ssrQpZAG 110	*Q. petraea*[b]	8	0.60
ssrQrZAG 112	*Q. robur*[c]	15	0.86
ssrQpZAG 36[a]	*Q. petraea*[b]	9	0.82
ssrQpZAG 46[a]	*Q. petraea*[b]	17	0.89
ssrQpZAG 104[a]	*Q. petraea*[b]	15	0.89

[a] closely linked on the same chromosome (θ = 1.1 to 5,6 %)
[b] primer sequences published by Steinkellner *et al.* 1997
[c] primer sequences published by Kampfer *et al.* 1998

Computer simulations

Data of three linked microsatellites segregating in eight open pollinated families were generated by computer simulations. Initially, a diploid population of 10 000 individuals was simulated over 500 generations under the assumption of random mating. The starting allele frequencies for the simulations were taken from the population sample of 40 seedlings (markers ssrQpZAG 46, 36 and 104). The mutation rate was set to 10^{-3} for each locus, using a simple stepwise mutation model. The recombination rates between the markers were set according to the previous mapping experiments (Barreneche *et al.* 1998), where a cluster of five linked microsatellites was identified within one of the linkage groups. After simulating for 500 generations, eight open pollinated families of sample size 40 were created by mating a specified number of pollen donors from that population randomly to one mother tree. This resulted in eight simulated datasets with a fixed number of fathers per family. The simulations were programmed using the C language.

Microsatellite data analysis

General estimates of diversity were calculated for the population sample at each locus in terms of the observed number of alleles, A_0, and expected heterozygosity, $H_E = 1 - \Sigma p_i^2$, where p_i is the relative frequency of the ith allele, according to Nei (1987). For reconstruction of the maternal genotypes we made use of a color code assigning a specific color to each allele within a sample. For each anonymous seedlot sample, the maternal alleles were identified using the following rules based on codominant Mendelian inheritance:

a) An allele can only be of maternal origin if a second allele segregating among the half-sib offspring is present whenever the first one is absent. This step is a simple extension of the general definition of codominance to the genotype data of open pollinated families.
b) An allele that is present in homozygous state in at least one offspring individual must be present in the mother tree.
c) If two different homozygotes are found among the offspring then no other allele can be of maternal origin.

Subsequently, the number of different seed parents (mother trees) was inferred by comparing the reconstructed maternal genotypes. The number of seed parents was also determined by calculating F_{ST} (Weir and Cockerham 1984) pairwise between seedlot samples, using the FSTAT software (Goudet 1995). Seven independent (loosely linked or unlinked) loci were used for this purpose, including ssrQpZAG 104 from the cluster of linked loci marked in Tab. 1. A neighbor-joining tree (Saitou and Nei 1987) was inferred from the F_{ST} data using the PHYLIP software package (Felsenstein, 1989). Alternatively, a principal component analysis was conducted using the SPSS software package (SPSS Inc., Chicago).

Seed contaminations were detected while searching the genotypes of five anonymous seedlot samples for alleles of maternal origin (see above): if at one or more than one locus all of the alleles within a sample were excluded as being of maternal origin (i.e. if all of the alleles were assigned to the pollen contribution), then the sample was assumed to contain contaminations.

Unrelated individuals within the samples were detected as follows: the genetic distance Dps was calculated with the help of the computer program MICROSAT (Minch 1997), using the formula Dps = -ln (ps), where ps is the proportion of shared alleles for pairs of individuals. Nine loci were included in the analysis. Subsequently, a UPGMA cluster analysis was conducted using the PHYLIP software package (Felsenstein, 1989). Moreover, average relatedness within five seedlot samples was computed according to Queller and Goodnight (1989) using the FSTAT software (Goudet 1995).

The number of pollen donors contributing to anonymous seedlot samples was estimated with the help of simulated half-sib families (sample size 40 individuals per family). First, pollen haplotypes were obtained for eight simulated families by subtracting the (known) maternal alleles from the progeny genotypes. Next, the genetic variability of three linked microsatellite loci (listed in Tab. 1) was combined in order to count paternal haplotypes for each family, using the software EXCEL 97 (Microsoft Corporation). The haplotype counts were corrected for rare recombination events between linked markers as described elsewhere (Lexer *et al.* 2000). The chromosome counts were compared to the (known) number of fathers in the datasets by regression statistics, using the SPSS 8.0 software package (SPSS Inc, Chicago).

The potential of the method was tested using the genotype data of a model half-sib family (sample size 43 individuals). Here, the number of paternal chromosomes was counted using the "haplotype approach", then the regression function obtained with the simulated data was used to calculate a lower 95 % confidence interval for the number of pollen donors. The result was validated by calculating genetic relatedness (Queller and Goodnight 1989) between all pairs of pollen gametes that shared a haplotype of three linked markers, using the computer program KINSHIP 1.2 (Goodnight and Queller 1999). Six independent (loosely linked or unlinked) loci were chosen for the calculations (see Tab. 1). The rationale behind this procedure was as follows: if gametes share the same haplotype of linked markers due to identity by descent from the same father, then relatedness among them, based on a set of independent reference markers, will be about 0.5. This will be the case because at each locus the pollen gametes will receive one of the two possible alleles of that father with a probability of ~ 0.5, given that the father was heterozygous. In contrast, if gametes share the same haplotype just by chance, then relatedness is expected to be lower than that. This approach, discussed in more detail elsewhere (Lexer *et al.* 2000), was chosen to obtain basic information about the ability of linked loci to assign a given gamete to a specific father.

Results and discussion

Inferring the number of seed parents from the offspring

Within the present contribution, two different ways were explored to infer the number of seed parents directly from anonymous seedlot samples of *Q. robur*. The first method, based on direct reconstruction of the maternal genotypes, was tested using the genotype data of a model half-sib family (sample size 28 individuals) including the mother tree. Inspection of this model family revealed that first conclusions can already be drawn by a quick look at microsatellite allele frequency distributions in the half-sib offspring. For some highly polymorphic loci it was possible to infer the maternal alleles directly from such allele frequency distributions (Fig. 1A).

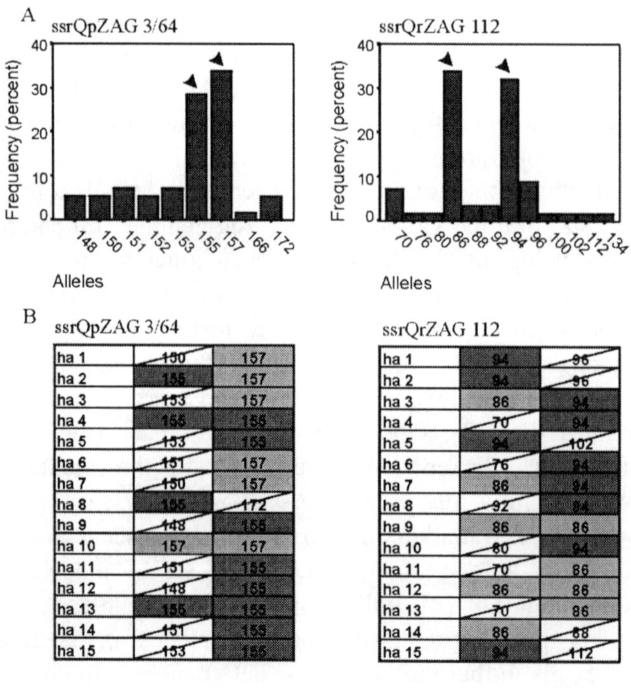

Figure 1. Reconstruction of the maternal genotype at two microsatellite loci in a maternal half-sib family of *Q. robur*. **A** Reconstruction of the maternal alleles based on allele frequency distributions. Alleles (*x-axis*) have been named according to their molecular size in basepairs. The maternal alleles - as inferred from the progeny - are marked by black arrowheads. **B** Reconstruction of the maternal alleles based on genotype matrices. The genotypes of 15 representative offspring are listed. Alleles that were excluded as being of maternal origin have been crossed off the list. The maternal alleles are shaded bright and dark. For details see text.

A more precise approach to reconstruct the maternal genotypes was also identified. This method makes use of the complete genotype matrix instead of allele frequency distributions (Fig. 1B; see Materials and methods). Following this approach, the maternal genotype was reconstructed at all nine loci studied in the model half-sib family, and successful genotype reconstruction was confirmed for each locus by comparison to the microsatellite profiles of the mother tree. Moreover, combining the genetic information of nine microsatellite loci revealed a unique (even if incomplete) maternal genotype for each of the five anonymous seedlot samples studied here (not shown). In other words, the five samples indeed originated from five different seed parents.

In order to explore an alternative way to infer the number of different mother trees, we tested F_{ST} according to Weir and Cockerham (1984). F_{ST} was calculated pairwise between 10 samples, including five anonymous seedlot samples and five samples harvested from the same mother tree. A neighbor-joining tree revealed that F_{ST} among the five samples harvested from the same mother tree was extremely low compared to that between samples originating from different seed parents (Fig. 2A). The result was even clearer when a principal component analysis was used to resolve the pairwise F_{ST} matrix (Fig. 2B). This multivariate technique allowed reduction of the dataset into three variables, the first three principal components, accounting for 56 %, 24 % and 11 % of the total variance in the data, respectively. Separation of the five samples harvested from the same mother tree was extremely weak, especially on the first principal component.

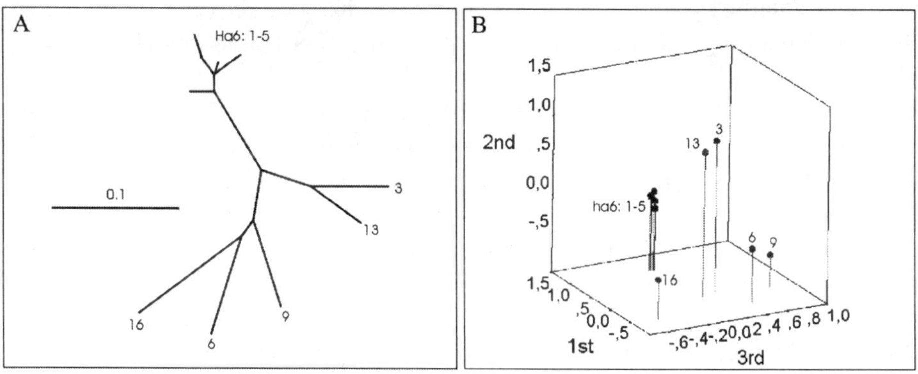

Figure 2. Inferring the number of different mother trees (seed parents) by calculating F_{ST} pairwise between small, anonymous seedlot samples. **A** Unrooted neighbor-joining tree based on pairwise F_{ST} values. **B** The first three principal components of pairwise F_{ST}. *3, 6, 9, 13, 16* indicate five anonymous seedlot samples, *ha 6: 1-5* indicates five samples harvested from the same mother tree.

In the present contribution, F_{ST} according to Weir and Cockerham (1984) was chosen because this type of F-statistics allowed us to compare our results to those obtained for a non-isolated, managed oak stand (Streiff *et al.* 1998); F_{ST} among our seedlot samples 3, 6, 9, 13, 16 was clearly higher than the intra-population values observed by Streiff *et al.* (1998), reflecting the single-tree wise sampling procedure employed during commercial seed harvest (comparison not shown). However, our samples may not fulfill all of the requirements of the sampling model given by Weir and Cockerham (1984), since the genetic structures in our samples arose through the deliberate sampling of family structures rather than genetic drift. Therefore, although the results shown in Fig. 2 were reproducible across a number of genetic distance measures included in the MICROSAT software package (not shown), we encourage the use of model-free approaches in the future (e.g. Gregorius and Roberds 1986).

Detecting seed contaminations within small, anonymous seedlot samples

The genotypes of three anonymous seedlot samples were not consistent with a half-sibling relationship, because at one or more than one locus all of the alleles were excluded as being of maternal origin. (In analogy to Fig. 1B, this is equivalent to all of the alleles within a sample being crossed off the genotype list.) In each sample, one unrelated seedling was identified using a UPGMA cluster analysis based on the genetic distance Dps computed from nine loci; after removing the unrelated plant from each sample, the genotypes of the remaining seedlings were consistent with a half-sibling relationship. Moreover, average relatedness within the five "pure" samples was above 0.25, indicating that the remaining seedlings were on average at least half-sibs (Lexer *et al.* 1999).

It should be stressed that measures of genetic distance between individuals, such as Dps, may yield unprecise results for small numbers of loci, as demonstrated by the following example: if the parents of a full-sib family are heterozygous at all loci under consideration, then two full-sib offspring need not share any allele at all. However, high mutation rates at microsatellite loci and high levels of pollen flow in oaks (Streiff *et al.* 1999) both lead to high levels of microsatellite diversity in natural oak populations (Streiff *et al.* 1998; see also Tab. 1 of the present contribution). Therefore, allele sharing can be expected to be lower for "unrelated" individuals of a population than for closely related ones, and genetic distance approaches such as Dps should reflect these differences, given that a sufficient number of loci are used.

Inferring the number of pollen donors from the offspring

An approach to infer the number of pollen donors contributing to anonymous seedlot samples was developed and tested using simulated data of linked microsatellites segregating in maternal half-sib families. For each of eight simulated families, the number of paternal chromosomes was calculated using the "haplotype approach" (see: Materials and methods). Subsequently, the haploid chromosome count was compared to the simulated number of fathers in each family by regression statistics (Lexer *et al.* 2000). Within this study, three linked microsatellites already provided an acceptable estimate for the number of fathers in the dataset (Fig. 3).

The dashed line in Fig. 3 shows an idealized estimate that would be completely collinear to the number of fathers in each dataset. At the same time, the dashed line reveals the minimum number of paternal chromosomes in each dataset, according to the computer simulations (see Materials and methods); The results suggest that the chromosome estimate is acceptable only when the number of fathers in the dataset is small, since the regression curve drops below the minimum number of chromosomes as the number of fathers exceeds 30.

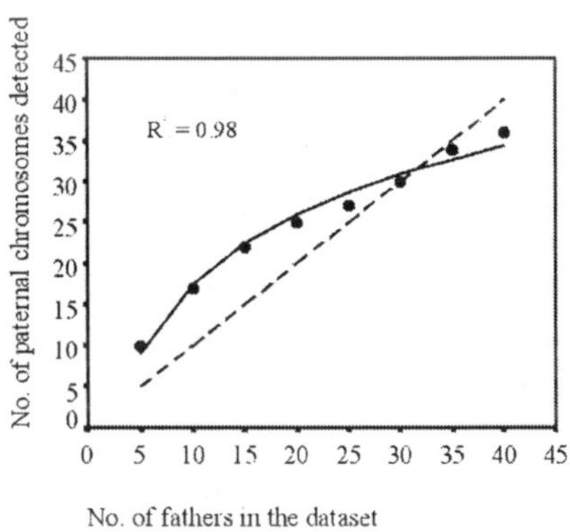

Figure 3. Inferring the number of pollen donors from simulated seedlot samples (sample size 40 individuals per family). Regression curve revealing the relationship between the number of fathers in each simulated sample and the number of paternal chromosomes inferred from the offspring using three linked loci. The *datapoints* represent eight different simulated samples, the *solid line* indicates the logarithmic regression function. The *dashed line* marks perfect collinearity and at the same time reveals the *minimum number* of paternal chromosomes in each dataset, since each of the simulated fathers (x-axis) must contribute at least one chromosome.

Within the present study, the "haplotype approach" was applied to a genotyped model half-sib family with a sample size of 43 individuals and unknown paternal contributions. Three linked microsatellites were used to infer the number of paternal chromosomes from this family. Using the regression curve obtained with the simulated data (Fig. 3), we obtained a lower 95 % confidence limit of 27 fathers, indicating that at least 27 pollen donors had contributed to the progeny.

In order to test the plausibility of the result, we calculated genetic relatedness between all pairs of pollen gametes that shared a haplotype of three linked microsatellite loci (see Materials and methods). Relatedness among those gametes was close to 0.5, suggesting that most of them had originated from one and the same pollen donor, transmitting each of its two alleles at each heterozygous locus to the offspring with a probability of ~ 0.5 (see discussion by Lexer *et al.* 2000).

Practical considerations

The use of polymorphic microsatellites has great potential for genetic analysis of commercial seed material in European white oaks (*Q. robur* and *Q. petraea*). Within the present contribution, suitable methods were presented to detect seed contaminations within small anonymous seedlot samples, and to infer the number of different mother trees as well as the number of different pollen donors directly from such samples. The methods tested within this study have several features in common: they are suitable for small to moderate sample sizes, they do not require any genotype information of the parent population, and they rely on small to moderate numbers of PCR based genetic markers. Hence, they are well suited to practical situations with limited financial resources and a need for quick answers.

However, in order to establish microsatellite analysis of commercial acorn seedlots in practice, more data are required on the effect of sample size (i.e. the number of seeds per family) and the effect of sampling the genome (i.e. the number of markers chosen from the genetic map). A "trade-off" exists between these two parameters, because lowering the sample size will automatically lead to an increase in the number of markers required, and vice versa. Although important optimisation work remains to be done, we have demonstrated that microsatellite analysis of anonymous seedlot samples is a suitable means to monitor within-stand genetic diversity in commercial acorn seedlots.

Acknowledgements

We wish to thank Elizabeth Gillet, University of Göttingen, and an anonymous referee for helpful comments on the manuscript. This work was supported by the Biotechnology Research Programme of the European Commission, DG XII, project ERB-BIO4-CT960706.

References

Anonymous 1996a: Bundesgesetzblatt für die Republik Österreich. 1996/419. Bundesgesetz über forstliches Vermehrungsgut (Forstliches Vermehrungsgutgesetz)

Anonymous 1996b: Bundesgesetzblatt für die Republik Österreich. 1996/512. Verordnung: Forstliches Vermehrungsgut

Avise, J.C. and Hamrick, J.L. (Eds.) 1996: Conservation Genetics. Chapman and Hall, London

Barreneche, T., Bodenes, C., Lexer, C., Trontin, J.F., Fluch, S., Streiff, R., Plomion, C., Roussel, G., Steinkellner, H., Burg, K., Favre, J.M., Glössl, J. and Kremer, A. 1998: A genetic linkage map *of Qercus robur* L. (pedunculate oak) with RAPD, SCAR, microsatellite, isozyme and rDNA markers. Theor. Appl. Genet. 97: 1090-1103

Felsenstein, J. 1989: PHYLIP - Phylogeny Inference Package. Cladistics 5: 164-166

Frankham, R. 1995: Conservation Genetics. Annu. Rev. Genetics 29: 305-327

Geburek, T.H. and Heinze, B. (eds) 1998: Erhaltung genetischer Ressourcen im Wald-Normen, Programme, Maßnahmen. Ecomed-Verlagsgesellschaft, Landsberg, Germany

Goodnight, K.F. and Queller, D.C. 1999: Computer software for performing likelihood tests of pedigree relationships using genetic markers. Mol. Ecol. 8: 1231-1234

Goudet, J. 1995: FSTAT (Version 1.2): A computer program to calculate F statistics. J. Hered. 86: 485-486

Gregorius, H.-R. and Roberds, J.H. 1986: Measurement of genetical differentiation among subpopulations. Theor. Appl. Genet. 71: 826-834

Hattemer, H.H., Bergmann, F. and Ziehe, M. 1993: Einführung in die Genetik für Studierende der Forstwissenschaft. 2. Auflage. Sauerländer`s Verlag, Frankfurt am Main, Germany

Kampfer, S., Lexer, C., Glössl, J. and Steinkellner, H. 1998: Characterization of $(GA)_n$ microsatellite loci from *Q. robur*. Hereditas 129: 183-186

Lexer, C., Heinze, B., Steinkellner, H., Kampfer, S., Ziegenhagen, B. and Glössl, J. 1999: Microsatellite analysis of maternal half-sib families of *Quercus robur*, pedunculate oak: Detection of seed contaminations and inference of the seed parents from the offspring. Theor. Appl. Genet. 99: 185-191

Lexer, C., Heinze, B., Gerber, S., Macalka-Kampfer, S., Steinkellner, H., Kremer, A. and Glössl, J. 2000: Microsatellite analysis of maternal half-sib families of *Quercus robur*, pedunculate oak: II. inferring the number of pollen donors from the offspring. Theor. Appl. Genet. 100: 858-865

Minch, E. 1997: MICROSAT, Version 1.5b. Stanford University Medical Center, Stanford

Nei, M. 1987: Molecular Evolutionary Genetics. Columbia University Press, New York

Queller, D.C. and Goodnight K.F. 1989: Estimating Relatedness using genetic markers. Evolution 43: 258-275

Saitou, N. and Nei, M. 1987: The neighbor-joining method: a new method for reconstructing phylogenetic trees. Mol. Biol. Evol. 4: 406-425

Steinkellner, H., Fluch, S., Turetschek, E., Lexer, C., Streiff, R., Kremer, A., Burg, K. and Glössl, J. 1997: Identification and characterization of $(GA/CT)_n$ -microsatellite loci from *Quercus petraea*. Plant Mol. Biol. 33: 1093-1096

Streiff, R., Labbe, T., Bacilieri, R., Steinkellner, H., Glössl, J. and Kremer, A. 1998: Within population genetic structure in *Quercus robur* L. and *Quercus petraea* (Matt.) Liebl. assessed with isoenzymes and microsatellites. Mol. Ecol. 7: 317-328

Streiff, R., Ducousso, A., Lexer, C., Steinkellner, H., Glössl, J. and Kremer, A. 1999: Pollen dispersal inferred from paternity analysis in a mixed oak stand of *Quercus robur* and *Quercus petraea* (Matt.) Liebl. Mol. Ecol. 8: 831-841

Weir B.S. and Cockerham C.C. 1984: Estimating F-statistics for the analysis of population structure. Evolution 38: 1358-1370

Fructification and genetic structures of *Fagus sylvatica* mixed stands in upper regions of the Harz mountains

D. KRABEL[1], R. HARTZ[1] AND S. HERZOG[2,3]

[1]*Institut für Forstbotanik und Forstzoologie, Technische Universität Dresden, D-01737 Tharandt, Germany,* [2]*Institut für Forstgenetik und Forstpflanzenzüchtung, Georg-August-Universität, D-37077 Göttingen, Germany,* [3]*Wildökologie, Technische Universität Dresden, D-01737 Tharandt, Germany*

Key words: genetic structures, fructification, *Fagus sylvatica*, mixed stand, mountainous region

Abstract: Whereas beech (*Fagus sylvatica*) forms in most parts of its natural range more or less dense pure or mixed stands, this is not the case in mountainous regions. At the upper limit of its natural range, beech normally forms scattered stands with only few individuals per hectar.
We based our studies on the hypothesis that this different structure of beech stands, *i.e.* their occurence as small groups of trees or even single trees, as well as a short flowering period will have impact on the mating system of this predominantly outcrossing species. As a consequence, influences on the genetic structures were expected.
For the assessment of the genetic structures, beech stands from the Harz mountains (Germany) were studied by methods of isoenzyme analysis. Significant deviations from previously observed patterns of genetic variation have been found. This has been explained by genetic drift rather than by selective forces.
No significant excess of homozygotes, indicating a tendency more to self-fertilization, has been observed.
However, this phenomenon may be explained by the observed high proportion of empty seeds, which may indicate a strong viability selection against homozygotes in early ontogenetic stages. Hence, an excess of homozygote as compared to corresponding Hardy-Weinberg structures can be counterbalanced and disappear.The consequences of genetic drift as well as increased self-fertilization rates are discussed in the context of genetic conservation in mountainous regions.

Introduction

Whereas beech (*Fagus sylvatica*) forms in most parts of its natural range more or less dense pure or mixed stands, the situation in mountainous regions is quite different. Especially at the upper limit of its natural range, beech stands are scattered with only few individuals per hectar and often more or less dense parts of mixed species like spruce (*Picea abies*) in between.
We based our studies on the hypothesis that this different structure of beech stands, *i.e.* their occurence in mixed stands, as small groups of trees or even

Müller-Starck, G. and Schubert, R. (eds.). Genetic Response of Forest Systems to Changing Environmental Conditions. Vol. 70 (For. Sci.), 251-258

single trees, will have an impact on the mating system of this predominatly outcrossing species. As a consequence, influences on the genetic structures, especially an increase of selfing or preferential mating of closely related individuals, was expected.

The efficiency of a pollen cloud depends on several factors of the pollen distribution. Local effects, as for example the position of an individual with respect to its next neighbour, isolation effects caused by mixture with other tree species, and microclimatic or microtopographic effects may influence the mating system or at least the mating success by reduction of the density of the pollen cloud.

In this context, the question as to the amount of selfing or mating of closely releated individuals arises. This may be the consequence of a reduced number of individuals of each species in mixed stands as well as of inhibition of pollen dispersion by neighbouring trees of other species. Additionally, reproductive isolation may not only be due to low population density and filter effects of other species but may be reinforced by different flowering periods of neighbouring trees.

The present paper deals with the question as to the rate of selfing of *Fagus sylvatica* in mixed stands of different structure from submountainous to mountainous regions.

Material and Methods

The present study covers altogether 10 mixed stands of beech (*Fagus sylvatica*) and several other tree species as *e.g* oak (*Quercus spec.*), spruce (*Picea abies*) or Douglas fir *(Pseudotsoga menziesii*) from altitudes between 400 meters and 820 meters above sea level of the Harz mountains (State of Niedersachsen, Germany, see Tab. 1). The highest locations are covered with scattered permanents of formerly more dense beech or mixed stands.

In all stands, amount and proportion of empty seeds of beechnuts from the 1995 mast have been evaluated between August 30th, 1995 and April 29th, 1996 using seed collectors of 49 cm x 50 cm positioned below the crown projection of at least nine different trees per stand.

Table 1. Denomination and location of the stands

Stand	Denomination	Altitude above sea level in meters	Number of seeds (mean of beechnuts/m^2)	Proportion of empty seeds (%)
1	Andreasbach 106	400	631	9.8
2	Knollen 27	420	627	8.5
3	Stöberhai 185	600-700	1120	13.7
4	St. Andreasberg 128	600-700	1075	19.1
5	Sonnenkopf	600-700	761	24.2
6	Königskrug 548 A	700	1616	26.0
7	Königskrug 548 B	700	699	29.8
8	Rehberg 551	720	714	25.9
9	Wurmberg 70	740	785	26.8
10	Wurmberg 67	820	974	45.5

Genetic strutures of three out of these ten stands (Königskrug 548 A, Wurmberg 67, Wurmberg 70) were studied by methods of isoenzyme analysis using nine different gene loci (Tab. 2).
The stand Königskrug 548 (0,6 ha) consists of permanents of a formerly more dense beech stand with a crown density of 0,3. Whereas Wurmberg 70 consists of a total of 8 beeches only few meters of distance to each other, Wurmberg 67 was composed of two solitary beeches enclosed with a pole stand of spruce. These three stands were chosen for genetic studies using electrophoretic methods as described by Müller-Starck (1985), Müller-Strack and Starke (1993) as well as Turok (1996). The number of empty seeds in these three stands was the highest compared to the other stands.

Table 2. Enzyme systems and gene loci

Enzyme system	Gene locus	Number of observed alleles
Isocitrate dehydrogenase	*IDH-A*	4
Leucine aminopeptidase	*LAP-A*	4
Malate dehydrogenase	*MDH-B*	4
	MDH-C	2
Diaphorase	*DIA-A*	4
6-Phosphogluconate dehydrogenase	*6-PGD-A*	3
Phosphoglucose isomerase	*PGI-B*	3
Phosphoglucomutase	*PGM-A*	5
Alcohol dehydrogenase	*ADH-A*	4

Genetic variation was found at all loci. The genetic control of the applied enzyme systems was verified previously using the offspring of controlled crosses (see Kim 1979, Thiebaut *et al.* 1982, Müller-Starck 1985, 1989, Müller-Starck and Starke 1993).

Results

The present data show no clear correlation between fructification and the altitude above sea level of the stand. The proportion of empty seeds shows an increase with altitude from about ten per cent up to about 45 per cent in the upper parts of the study area (Tab. 3).

Table 3. Proportion of empty seeds in per cent (stand number see Table 1)

Stand **Date**	**1**	**2**	**3**	**4**	**5**	**6**	**7**	**8**	**9**	**10**
Sept., 25th	21.6	25.3	68.7	87.9	87.2	79.9	84.1	86.3	92.5	78.1
Oct., 9th	11.9	10.9	29.6	33.8	51.3	57.6	75.1	77.3	45.8	86.3
Oct., 21st	9.1	6.1	10.7	12.4	16.2	16.6	19.1	25.3	20.8	29.3
Nov., 3rd	4.8	10.9	8.8	13.1	14.3	15.6	19.2	21.9	17.2	43.7
Nov., 17th	5.5	2.5	9.5	10.7	14.0	17.3	20.7	15.2	26.6	34.4
April, 29th	63.0	14.6	22.8	17.4	23.2	16.8	38.2	16.5	-	-
Total	9.8	8.5	13.7	19.1	24.2	26.0	29.8	25.9	26.8	45.5

Table 4. Proportion of seeds affected by parasites in per cent (stand number see Table 1).

Stand **Date**	**1**	**2**	**3**	**4**	**5**	**6**	**7**	**8**	**9**	**10**
Sept., 25th	21.6	25.3	68.7	87.9	87.2	79.9	84.1	86.3	92.5	78.1
Oct., 9th	11.9	10.9	29.6	33.8	51.3	57.6	75.1	77.3	45.8	86.3
Oct., 21st	9.1	6.1	10.7	12.4	16.2	16.6	19.1	25.3	20.8	29.3
Nov., 3rd	4.8	10.9	8.8	13.1	14.3	15.6	19.2	21.9	17.2	43.7
Nov., 17th	5.5	2.5	9.5	10.7	14.0	17.3	20.7	15.2	26.6	34.4
April, 29th	63.0	14.6	22.8	17.4	23.2	16.8	38.2	16.5	-	-
Total	**9.8**	**8.5**	**13.7**	**19.1**	**24.2**	**26.0**	**29.8**	**25.9**	**26.8**	**45.5**

On the other hand, the proportion of beechnuts affected by parasites decreases with the altitude above sea level (see Tab. 4).
For the investigations on the genetic structure we chose the stand Königskrug 548 A and groups of trees (Wurmberg 67 and 70), because the proportion of empty seeds in 1995 and also in former years (see Lange 1995) in these stands were the highest compared to the other stands. According to Müller-Starck (1996) the proportion of empty seeds is an indicator for an increased selfing rate.
Table 5 shows parameters of genetic multiplicity as well as heterozygosity.

Table 5. Parameters of genetic variation

Population	Königskrug 548 A		Wurmberg 67		Wurmberg 70	
Parameter	Parents	Offspring	Parents	Offspring	Parents	Offspring
Alleles per locus	2.0	2.1	1.4	2.0	1.7	1.9
Population differentiation δ_T*	0.32	0.32	0.42	0.31	0.38	0.32
Heterozygosity H_a	0.38	0.26	0.28	0.24	0.36	0.36

Table 6 shows pairwise genetic distances d_0 between the offspring of the three investigated stands.

Table 6. Pairwise genetic distances (offspring)

Population	**Königskrug 548 A**	**Wurmberg 67**
Wurmberg 67	0.236	-
Wurmberg 70	0.256	0.141

As Table 7 shows, fixation indices show quite homogenous estimations over the different loci and are mostly negative or slightly positive.

PGI-B exhibits a minor polymorphism with frequencies of rare alleles of less than five per cent: the rare allele B_3 was found in two populations once and three times respectively. Thus, these alleles have not been taken into account when calculating fixation indices. The fixation indices show significant deviations (χ^2 test) from Hardy-Weinberg equilibrium only for the two gene loci *MDH-C* and *6-PGDH-A* for the seeds of the Wurmberg 70 stand.

MDH-B is monomorphic for the Wurmberg 67 stand. Königskrug 548 A shows a frequency of 2 per cent for allele B_1. For the Wurmberg 70 we could record no data for *MDH-B*.

Table 7. Fixation indices over different gene loci.

Gene locus	Königskrug 548 A	Wurmberg 67	Wurmberg 70
IDH-A	-0.03	-0.18	-0.23
LAP-A	-0.35	no data	-0.24
MDH-B	0.00	0.00	no data
MDH-C	0.06	-0.06	-0.21
DIA-A	0.00	0.00	0.00
6-PGD-A	-0.04	-0.05	-0.21
PGI-B	0.00	0.00	0.00
PGM-A	0.04	0.11	-0.13
ADH-A	0.06	0.09	-0.31

Discussion

With respect to previously observed patterns (see Konnert *et al.* 2000 for a detailed genetic inventory on the genetic variation of beech in Germany), the investigated stands show some deviations from these results. For example, Turok (1996) found for a total of 27 stands in Western Germany (State of Nordrhein-Westfalen) a higher number of alleles per locus (range between 2.1 and 2.8 alleles per locus) especially compared to the present results from the two Wurmberg stands.

The diversity values δ_T^* are of comparable magnitude, Turok (1996) observed values between δ_T^*=0.23 and δ_T^*=0.38. This holds also true for the actual heterozygosity H_a. Turok (1996) observed values between H_a =0.22 and H_a =0.35 being of the same magnitude like the present results.

The genetic distances d_0 are relatively highly dependent on the low number of adult trees per stand as well as the relative isolation by mixed species.

However, the allelic frequencies found in the present study are not only different between the single stands from the Harz mountains but also between e.g. the allele frequencies found by Müller-Strack (1993) e.g. for the Fichtelgebirge population, a stand from comparable sea levels.

Thus, the different frequencies of alleles in the present study have been explained by the phenomenon of genetic drift rather than by selective forces.

Although no tendency towards self-fertilization can be concluded from the fixation indices, the proportion of empty seeds increasing with the sea level may be due to an increasing self-fertilization. A strong viability selection against homozygotes in early ontogenetic stages will prohibit them to be detected by genetic methods.
The consequences of genetic drift as well as increased self-fertilization rates are of importance in the context of genetic conservation in mountainous regions. The stands may be differentiated not primarily by selective forces but more by random effects. Thus, we cannot assume that we will find one or few stands representing the genetic structure of a whole region and, as a consequence, genetic conservation becomes more difficult in these regions. On the other hand, adaptations to certain ecological conditions may have led to unique genetic structures justifying the declaration as a gene ressource. This dilemma may be at least partially solved by a clear concept of genetic conservation, including a definition as well as a ranking of the goals that have to be reached by the conservation efforts.

References

Kim, Z. S. 1979: Inheritance of leucine aminopeptidase and acid phosphatase isozymes in beech (*Fagus sylvatica* L.). Silvae Genetica 28 : 68-71

Konnert, M., Ziehe, M., Tröber, U., Maurer, W., Janßen, A., Sander, T., Hussendörfer, E., Hertel, H. 2000: Genetische Variation der Buche (*Fagus sylvatica* L.) in Deutschland: Gemeinsame Auswertung genetischer Inventuren über verschiedene Bundesländer. Forst und Holz 55: 403-408

Lange, V. 1995: Die Fruktifikation der Buche in verschiedenen Höhenlagen des Harzes. Diplomarbeit der Forstwissenschaftlichen Fakultät, Georg-August-Universität, Göttingen

Müller-Starck, G. 1985: Genetic differences between „tolerant" and „sensitive" beeches (*Fagus sylvatica* L.) in an environmentally stressed adult forest stand. Silvae Genetica 34: 241-247

Müller-Starck, G. 1989: Genetic implications of environmental stress in adult forest stands of *Fagus sylvatica* L. In: Scholz, F.; Gregorius, H.-R.; Rudin, D. (editors): Genetic Aspects of Air Pollutants in Forest Tree Populations. Springer-Verlag: Berlin, Heidelberg, New-York, 127-142

Müller-Starck, G. 1993: Auswirkungen von Umweltbelastungen auf genetische Strukturen von Waldbeständen am Beispiel der Buche (*Fagus sylvatica* L.). Schriftenreihe der Forstlichen Fakultät der Universität Göttingen 112, J.D. Sauerländer´s Verlag, Frankfurt am Main

Müller-Starck, G., Starke, R. 1993: Genetic control and inheritance of isoenzymes in beech (*Fagus sylvatica* L.). Journal of Heredity 84: 291-296

Müller-Starck, R. 1996: Genetische Aspekte der Reproduktion der Buche (*Fagus sylvatica* L.) unter Berücksichtigung waldbaulicher Gegebenheiten. Dissertation der Forstlichen Fakultät, Georg-August-Universität Göttingen

Thiébaut, B., Lumaret, R., Vernet, P. 1982: The bud enzymes of beech (*Fagus sylvatica* L.): genetic distinction and analysis of Polymorphism in several French populations. Silvae Genetica 31: 51-60

Turok, J. 1996: Genetische Untersuchungen bei der Buche. Schriftenreihe der Landesanstalt für Ökologie, Bodenordnung und Forsten / Landesamt für Agrarordnung Nordrhein-Westfalen 8, 1996, 1-136

Dispersal of seed and effective pollen in small stands of European beech (*Fagus sylvatica* L.)

K. WANG AND H.H. HATTEMER

Institut für Forstgenetik und Forstpflanzenzüchtung der Universität Göttingen, D-37077 Göttingen, Germany

Key words: *Fagus sylvatica* L., allozymes, parentage analysis, gene flow, seed dispersal, pollen dispersal

Abstract: Two isolated small beech stands were used to study gene flow. Seeds were sampled on the ground under each adult tree. Ten polymorphic enzyme coding gene loci were utilized for multilocus-genotyping trees and seeds. The modified most-likely method was used to assess effective pollen dispersal by inferring paternity of offspring. The difference in LOD scores between the most likely candidate parent and the next-most likely candidate parent (i.e. the statistic Δ according to Marshall *et al.* 1998) was used to evaluate confidence in parentage of the most likely parent. The most likely parent pair was inferred and seed dispersal was estimated. Preliminary results showed that the migration distance of seeds within the isolated beech stand $10A_2$ (70 trees on 0.5 ha) is limited to 50m and 75% of the seeds was dispersed within 20m. Most pollen originated from within 50m of the seed tree, although in stand $34B_1$ (24 trees on 0.2 ha) a low proportion of pollen movement was from as far as 300m. In the stand $10A_2$, 6.1% of the seeds were fertilized by pollen parents occurring at least 140m outside the stand. Two of the trees in stand $10A_2$ had unique allozyme alleles (LAP-A_1 and IDH-A_4, respectively), which were used to directly measure pollen movement away from those trees. The frequency of the unique LAP and IDH alleles in seeds declines as the distance from the source tree increases. The mathematical models of pollen and seed dispersal were fitted and the probability distributions of pollen and seed dispersal were derived.

Introduction

In seed plants, there are two possible mechanisms of gene movement - the movement of pollen between mates and the dispersal of seeds. In some species also vegetative plant parts are a means of gene flow. However, this does not occur in the beech family. Patterns of seed and pollen movement influence the mating system of the population (Waser 1993) as well as the population structure (Wright 1951) and the ability of plants to adapt to particular microhabitats (Waser 1993). Pollen and seed dispersal have large impacts on a population's capacity to maintain genetic polymorphisms (Namkoong and Gregorius 1985; Hattemer 1994). Gene flow between stands is extensive, which

Müller-Starck, G. and Schubert, R. (eds.). Genetic Response of Forest Systems to Changing Environmental Conditions. Vol. 70 (For. Sci.), 259-269

may cause problems in tree breeding. Gene conservation areas and breeding populations should be isolated from production populations (Ledig 1986). An essential requirement for seed orchards is that they should be isolated from unimproved stands, as fertilization by pollen from outside sources reduces expected genetic gains (El-Kassaby and Ritland 1986; Adams and Birkes 1989; Di-Giovanni and Kevan 1991; Hattemer *et al.* 1993).

The distances traveled by plant seeds and pollen have been difficult to measure (Levin 1981). Recently, paternity-exclusion analysis has been used to measure gene flow (Meagher 1986; Meagher and Thompson 1987; Devlin and Ellstrand 1988,1990; Adams et al 1992; Boshier *et al.* 1995; Schnabel and Hamrick 1995; Stacy *et al.* 1996). However, multiple males are often found to be genetically compatible with each offspring tested, even when the probability of excluding an unrelated male is high (Marshall *et al.* 1998). Marshall *et al.* (1998) derived likelihood ratios for paternity inference with codominant markers taking account of typing errors in the genotypes of offspring, mother or alleged father, or in some combination of the three, and define a statistic Δ for resolving paternity.

Genetic studies on gene flow in European beech are still very rare. The high cross-pollination rate in an isolated stand reflected long-distance transport of pollen (Merzeau 1991). The migration distance of both pollen and seeds within a spatially isolated beech stand with low population density is limited to approximately 50m, whereas pollen immigrates from sources more than 500 m away (R. Müller-Starck 1996). Pollen can be transported more than 700m (Ziehe 1998).

We will use and develop the method of Marshall *et al.* (1998) in this tree species, *Fagus sylvatica* L., in order to measure dispersal of seed and effective pollen. In addition, two of the trees in stand $10A_2$ had unique allozyme alleles (LAP-A_1 and IDH-A_4, respectively), which were used to directly measure pollen movement away from those trees. The objectives of this study are to: 1) estimate effective pollen dispersal from individual parent trees which carry unique allozyme alleles, 2) estimate gene flow by pollen with the help of paternity-exclusion analysis, 3) estimate gene flow by seed based on parentage analysis.

Materials and Methods

Seeds of two relatively isolated stands were collected on the ground under each adult tree. Seeds were sampled around each adult tree at a radius of 3 m within three circles, which are 40cm in diameter and are located north, southwest and southeast of the stem at angles of 120°. All of the seeds encountered in each circle were sampled. The numbers of seeds sampled under individual trees varied greatly. All of the trees within the two stands were tagged, and the location of each tree was mapped.

Table 1. Description of the isolated beech stands in the forest district of Escherode, northern Germany

Stand	34B 1	10A 2
Age (years)	150	180
Area (ha)	0.2	0.5
Altitude (m)	530	500
Distance to nearest beech stand (m)	300	140
No. of adult trees	24	70
No. of seeds	52 (in 1994) 624 (in 1998)	1702 (in 1994) 2104 (in 1998)

Isoenzyme electrophoresis and direct estimates of pollen dispersal

Ten polymorphic enzyme-coding gene loci were utilized for multilocus genotyping (GOT-A, IDH-A, LAP-A, MDH-B, MDH-C, MNR-A, 6-PGDH-A, PGI-B, PGM-A and ADH-A). Genetic control and mode of inheritance of the respective enzyme systems were verified previously (Müller-Starck and Starke 1993). Two unique alleles were found in stand 10A 2, LAP-A_1 (tree No. 38, genotype A_1A_2) and IDH-A_4 (tree No. 37, genotype A_3A_4). The presence of the allele in the progeny arrays of other trees was therefore deemed to be a direct indicator of gene flow, measured as frequency and distance from this tree. This procedure hardly leads to misinterpretation due to influx of external pollen. The two alleles have been found to be rare in all studies of beech. They were encountered mostly in seeds close to the respective pollen parents in this study.

Parentage analysis

We used the simple exclusion principle to estimate influx of external pollen. This leads to a lower bound of the proportion of external pollen (Ellstrand and Marshall 1985).

The likelihood ratio measures how much more likely it is that the alleged father rather than an arbitrary male passed its genes to the offspring:

$$L(H_1, H_2| g_m, g_a, g_o) = \frac{T(g_o| g_m, g_a).P(g_m).P(g_a)}{T(g_o| g_m).P(g_m).P(g_a)} = \frac{T(g_o| g_m, g_a)}{T(g_o| g_m)} \quad (1)$$

H_1 implies that the alleged father is the true father, and this is tested against hypothesis H_2 that the alleged father is an unrelated individual selected at random from the population. Let g_m, g_a and g_o represent the genotypes of mother, alleged father and offspring, respectively, at a given locus. $T(g_o| g_m, g_a)$ is the probability of the offspring's genotype given the genotypes of the mother and alleged father, and is based on the Mendelian segregation or transition probability. $T(g_o| g_m)$ is the probability of the offspring's genotype given the maternal genotype. $P(g_m)$ and $P(g_a)$ are the frequencies of the mother's and alleged father's genotype in the population. In cases where the mother's genotype is unknown, the likelihood ratio is different:

$$L(H_1, H_2| g_a, g_o) = \frac{T(g_o| g_a).P(g_a)}{P(g_o).P(g_a)} = \frac{T(g_o| g_a)}{P(g_o)} \qquad (2)$$

Here, $P(g_o)$ is the frequency of the offspring's genotype.
(The equations 1 and 2 are taken from the article written by Marshall *et al.* 1998).

Paternity is assigned to a particular male if the likelihood ratio is large relative to the likelihood ratios of alternative males. If several unlinked marker loci are used in paternity inference, the likelihood ratios derived at each locus may be multiplied together and the natural logarithm ($\log_e$) taken. The LOD score is the logarithm of the product of the likelihood ratios at each locus, or equivalently the sum of the log-likelihood ratios at each locus (Meagher 1986). A positive LOD score implies that the candidate parent is more likely to be the true parent than a randomly chosen individual. The most likely candidate parent is the candidate parent with the highest (most positive) LOD score (Meagher 1986). This method is referred to as the most-likely method (Devlin *et al.* 1988; Adams 1992).

In order to discriminate between non-excluded parents, Marshall *et al.* (1998) defined a statistic Δ as the difference in LOD score between the most likely parent and the next-most likely parent. Simulations are used to assess the significance of Δ values. Typical values are 80% for relaxed confidence and 95% for strict confidence. For each offspring tested, the parentage analysis calculates LOD scores for each candidate parent, finds the two most likely parents and calculates the corresponding Δ value as for simulation, and then evaluates the confidence of the Δ value using the appropriate criteria calculated by the simulation.

Modified most-likely method

We directly use the method of Marshall *et al.* (1998) in paternity analysis when the seed parent is known, but take account of the distance between seed parent and pollen parent. The reason is that the mating success is expected to decrease exponentially with increasing distance from the seed plant but at an unknown rate (Müller 1977; Di-Giovanni *et al.* 1991; Adams 1992; Degen *et al.* 1996). We derive possible pollen parents at increasing distances from the seed parent (in circles with a radius of 30m, 50m, 70m and larger than 70m) in the stand10A $_2$.

Seeds were sampled on the ground under each adult tree. Seeds carrying two alleles at any locus not carried by the adult tree were not considered to be produced by that tree. In those cases, the adult tree was excluded as the seed parent so that neither parent of the seed is known. However, if a seed sampled below a certain tree could be considered to be its offspring, that same tree was also considered to be its seed parent. We use the formula (2), when the mother's genotype is unknown, in order to choose a set of parents which is small in number and has positive LOD score, and then to consider them as the parent pair. The parent standing closer to the location of a seed was considered to be the seed parent.

Models of pollen and seed dispersal

We assumed that pollen and seed flow follow a negative exponential function, where f(x) which is the probability density associated with dispersal decreases exponentially with distance x (Müller 1977; Di-Giovanni *et al.* 1991; Adams 1992; Degen et al 1996):

$$f(x)=a\exp(-bx) \qquad (3)$$

Here a and b are constants. At first we estimate the initial values of a and b by least squares after linear transformation (lg f(x) = lg a – bx), and then we used the iteration method of Gauss-Newton as modified by Marquardt (Marquardt 1963) in order to estimate the parameters a and b by minimizing the residual sum of squares ($SS_e(\alpha)=\Sigma\,(f(x_i) - a\exp(-bx))^2$ =Min.). We also use R^2 (the coefficient of determination) in order to evaluate the nonlinear regression model. R^2 is the proportion of the variation of f(x) attributed to the regression on x ($R^2 = SS_{reg}/SS_Y = 1 - SS_e/SS_Y$ where SS_{reg} is the sum squares due to regression, SS_Y is the total sum of squares). We integrated the formula (3) and derived the distribution function Φ(x) for the continuous variable x:

$$\Phi(x)=1-\exp(-bx) \qquad (4)$$

The value of Φ(x) estimates the probability that a seed (or pollen) travelled a fixed distance x or less.

Results

Estimates of pollen dispersal inferred from new alleles

Within the smaller stand $34B_1$ we found two new alleles LAP-A_1 and PGI-B_3 in seeds, which did not exist in seed parents, and should therefore have come from outside the stand. The minimum frequencies of external pollen with LAP-A_1 was 0.5% (in 1998) and with PGI-B_3 were 2.0% (in 1994) and 0.2% (in 1998), respectively.

Estimates of pollen dispersal inferred from unique alleles

Within stand $10A_2$, the unique alleles LAP-A_1 and IDH-A_4 were found in seeds sampled under 21 and 13 adult trees, respectively, except the pollen donor tree. Pollen with LAP-A_1 can reach the edges of the stand, 75m away from the donor tree but the majority of unique alleles were detected in the progeny of adults within a radius of 40m. Most of the pollen with the unique allele IDH-A_4 were found within a radius of 40m and only one seed was obtained at more than 40m. The dispersal of marker pollen fits to the negative exponential model (Fig. 1). The equations are $f(x)=16.1017\ \exp(-0.02402x)$ (R^2 =0.956) and $f(x)=25.896\ \exp(-0.04153)$ (R^2 =0.937) for pollen with unique alleles at the gene loci LAP and IDH, respectively.

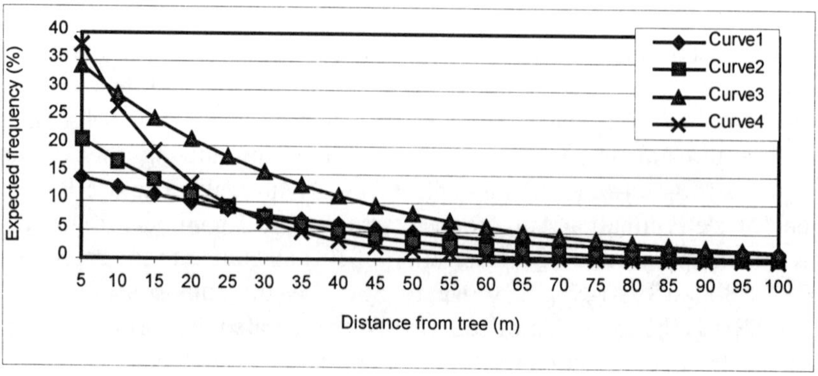

Figure 1. Expected pollen and seed dispersal. Curve 1 represents dispersal of pollen with unique allele LAP-A_1; curve 2 represents dispersal of pollen with unique allele IDH-A_4; curve 3 represents mean pollen dispersal estimated from paternity analysis; curve 4 represents mean seed dispersal estimated from parentage analysis.

Estimates of pollen and seed dispersal inferred from parentage analysis

In this part of the analysis we included only seed trees producing more abundant seeds, i.e. between 28 and 79 seeds encountered in the three circles per tree.

In stand $34B_1$ (24 trees on 0.2ha), most pollen originated from within 50m off the seed tree, although a low proportion of pollen movement (9.6%) was from as far as 300m outside the stand.

In stand $10A_2$ (70 trees on 0.5ha), gene flow data were derived for various circles with a radius of 30m, 50m, 70m, greater than 70m, and outside the stand. We estimated the fractions of seeds which have their pollen parent within a given circle or outside the stand of the total number of the seeds sampled under a tree. Most pollination occurred within the circle with a radius of 50m, but a low proportion of seeds (6.1%) were even sired by males outside the stand.

Among 481 seeds (from 12 seed trees) within the stand $10A_2$ we obtained 123 seeds whose seed parent is known and which possess a significant unique most likely pollen parent. The frequency distribution of pollen dispersal with the distance between the pollen and seed parents was calculated for the distance intervals 5-15m, 15-25m, 25-35m, …, and 105-115m. The pollen distribution fits to the negative exponential equation (Fig. 1). The equation is $f(x)=39.9785 \exp(-0.03191)$; ($R^2=0.994$).

Among 458 seeds (from 15 seed trees) within the stand $10A_2$ we identified 128 seeds (according to parentage analysis), which have an unknown seed parent but a significant most likely parent pair. The frequency distribution of seed dispersal was calculated on the basis of the distances of the location of seeds to their close parent. It was detailed to distance intervals 5-10m, 10-15m, and 55-60m. The seed dispersal showed a typical leptokurtic distribution (see Fig. 1). The exponential equation is $f(x)=53.7801 \exp(-0.06930x)$; ($R^2=0.99$).

Probability distribution of pollen and seed dispersal

Distribution functions of pollen and seed dispersal within stand $10A_2$ were calculated from formula (4) and part of the results were presented in Table 2.

Table 2. Cumulative distributions of pollen and seed dispersal (in percent)

Sources	<20m	<30m	<40m	<50m	<70m	<100m
Pollen with LAP-A $_1$	38.15	51.35	61.74	69.91	81.39	90.95
Pollen with IDH-A $_4$	56.42	7123	81.01	87.46	94.53	98.43
Mean pollen dispersal	47.18	61.61	72.10	79.72	89.29	95.85
Mean seed dispersal	74.99	87.49	93.74	96.87	99.22	99.90

Discussion

Gene dispersal by pollen

The results of analyses based on both of the two unique alleles showed that pollen dispersal decreases with increasing distance from the marker tree. These are similar to the results of some authors (Müller 1977; Adams 1992; Boshier *et al.* 1995; Stacy and Hamrick 1996; Latta *et al.* 1998). Pollen dispersal fits to a negative exponential curve as was found in pine (Müller 1977; Adams 1992) and the coefficient of determination R^2 is high (> 0.93). With the modified most-likely method we can identify on the average 25.6% of the seeds which have one significant most likely pollen parent. The mean pollen dispersal curve, which was estimated from the paternity analysis (when seed parents are known), has the same trend as the analysis on the basis of unique alleles, but slightly higher pollen frequencies at given distances (Fig.1). Both methods (unique allele method and paternity analysis) showed that most pollen were dispersed within 50m. A low proportion (up to 10%) can be from outside the stand (300m outside the stand $34B_1$ and 140m outside the stand $10A_2$).

It was reported that the migration distance of pollen within a spatially isolated European beech stand with low density is limited to approximately 50m, whereas pollen immigrates from sources more than 500m away (R. Müller-Starck 1996) and in mixed stands effective pollen can be transported more than 700m (Ziehe 1998). For two out of 81 seedlings only one possible pollen parent was found at a distance of 29.3 m and 29.7m, respectively, in European beech with RAPD and I-SSR markers (Troggio *et al.* 1996). Hashizume and Sugimoto (1980) investigated structures of Japanese beech (*Fagus crenata*) and found that similar allozyme phenotypes were distributed locally within a circle of 30 to 50m radius around the seed trees. Kawano and Kitamura (1997) studied pollen dispersal in two transects in stands of Japanese beech (*Fagus crenata*). They found that 77% of pollen donors were located within a transect of 55 x 10m in size, and 75% within another transect of 50 x 10m in size. A maximum pollen transport distance of 500m was inferred.

The size of the studied stands is relatively small and defines a priori the maximum distance of pollen and seed dispersal observable within a stand. In the smaller of the two stands studied here, the maximum distance in east-west direction was only 40m. This condition implied also strong edge effects.

The present results are part of a more extensive study. Its goal is the dynamics of small populations including the proportion of external effective pollen. This can efficiently be studied only in small stands which are hardly representative of all beech stands.

In the program CERVUS of Marshall *et al.* (1998) a certain percentage of mislabelled alleles can be considered. A chance of mislabelling alleles due to misinterpretation of the zymograms exists virtually only between the moderately frequent allele A_5 and the rare allele A_4 at the gene locus LAP-A. This chance was minimized by using the buffer system of Tris-Citro. Consequently, possible mislabelling was not accounted for in the present analyses.

Gene dispersal by seed

Seed dispersal is relatively limited. With increasing distance it decreases faster than pollen dispersal (Fig. 1). Our result showed that most seeds (75%) were dispersed within 20m. The migration distance of seeds within a spatially isolated beech stand with low density was found to be limited to approximately 50m by R. Müller-Starck (1996). Our results are compatible with the statement of R.Müller-Starck (1996) that seed of beech can be dispersed up to 20m by wind, and perhaps over 50m by birds and rodents. These diverging distances of pollen and seed transport were used as a priori information in distinguishing between pollen and seed parent of a given seed. In their study of *Fagus crenata*, Kawano and Kitamura (1997) inferred that 95% of the seedlings had seed parents within the two transects mentioned above.

Further research

We have only preliminary results from this study. We are doing more analyses and hope to obtain the results for two more years for comparison.

Acknowledgment

We are grateful to Dr. J. Kleinschmit, Dr. E. Gillet, Dr. M.Ziehe and Dr.T.C. Marshall (University of Edinburgh) for useful help in material collection and data processing. We are also grateful to an anonymous reviewer for his helpful comments. Kesheng Wang is supported by a DAAD scholarship.

References

Adams, W.T. 1992: Gene dispersal within forest tree populations. New Forests. 6:217-240

Adams, W.T. and Birkes, D.S. 1989: Mating patterns in seed orchards. Pp. 75-86 in Proc. 20th South. For. Tree Improv. Conf., Charleston, South Carolina

Adams, W.T., Griffin, A.R. and Moran, G.F. 1992: Using paternity analysis to measure pollen dispersal in plant populations. Amer. Nat. 140:762-780

Boshier, D.H., Michael R.C. and Bawa, K.S. 1995: Population genetics of *Cordia alliodora* (Boraginaceae), a neotropical tree. 3. Gene flow, neighborhood, and population substructure. Amer. J. Bot. 82: 484-490

Degen, B., Gregorius, H.-R. and Scholz, F. 1996: ECO-GENE, a model for simulation studies on the spatial and temporal dynamics of genetic structures of tree populations. Silvae Genetica 45: 323-329

Devlin, B. and Ellstrand, N.C. 1990: The development and application of a refined method for estimating gene flow from angiosperm paternity analysis. Evolution. 44: 248-259

Devlin,B. , Roeder, K. and Ellstrand, N.C. 1988: Fractional paternity assignment: theoretical development and comparison to other methods. Theor. Appl . Genet. 76: 369-380

Di-Giovanni, F. and Kevan, P.G. 1991: Factors affecting pollen dynamics and its importance to pollen contamination : a review. Can. J. For. Res. 21: 1155-170

El-Kassaby, Y.A. and Ritland, K. 1986: Low levels of pollen contamination in a Douglas-fir seed orchard as detected by allozyme markers. Silvae Genet. 35: 224-229

Ellstrand, N.C. and Marshall, D.L. 1985: Interpopulation gene flow by pollen in wild radish, *Raphanus sativus*. Am. Nat. 126:606-616

Hashizume, H. and Sugimoto, S. 1980: A study of the breeding system in natural forests of Bura (*Fagus crenata* Blume) using the peroxidase isozyme technique. Hardwood Research. 1:59-71 (in Japanese with English summary)

Hattemer, H.H. 1994: Die genetische Variation und ihre Bedeutung für Wald und Waldbäume. Schweiz. Z. Forstwes. 145: 953-975

Hattemer, H.H., Bergmann, F. und Ziehe, M. 1993: Einführung in die Genetik. J. D. Sauerländer's Verlag, Frankfurt am Main, 2. Auflage

Kawano, S. and Kitamura, K. 1997: Demographic genetics of the Japanese beech, *Fagus crenata*, in the Ogawa forest preserve, Ibaraki, Central Honshu, Japan. 3. Population dynamics and genetic substructuring within a metapopulation. Plant Species Biol. 12: 157-177

Latta, R.G., Linhart, Y.B., Fleck, D. and Elliot, M. 1998: Direct and indirect estimates of seed versus pollen movement within a population of ponderosa pine. Evolution.52: 61-67

Ledig, F.T. 1986: Heterozygosity, heterosis, and fitness in outbreeding plants. pp.77-104 in : Soule, M.E. (Eds.) Conservation Biology: The Science of Scarcity and Diversity. Sinauer Sunderland, Mass.

Levin, D.A. 1981: Dispersal versus gene flow in plants. Ann. Missouri Bot. Gard. 68:233-253

Marquardt, D.W. 1963: An algorithm for least squares estimation of nonlinear parameters. J.Soc. Ind. Appl. Math. 2: 431-441

Marshall, T.C., Slate, J., Kruuk, L.E.B. and Pemberton, J.M. 1998: Statistical confidence for likelihood-based paternity inference in natural populations. Mol. Ecol. 7: 639-655

Meagher, T.R. 1986: Analysis of paternity within a natural population of *Chamaelirium luteum*. 1. Identification of most-likely male parents. Am. Nat. 128: 199-215

Meagher, T.R. and Thomson, E. 1987: Analysis of parentage for naturally established seedlings of *Chamaelirium luteum* (Liliaceae). Ecology. 68: 803-812

Merzeau, D. 1991: Estimation des parametres du modes de reproduction et des structures génétiques du hêtre (*Fagus sylvatica* L.) Dissertation, Université de Bordeaux,153+37pp.

Müller, G. 1977: Cross-fertilization in a conifer stand inferred from enzyme gene markers in seeds. Silvae Genet. 26: 223-226

Müller-Starck, G. and Starke, R. 1993: Inheritance of isoenzymes in European beech (*Fagus sylvatica* L.). J. Heredity. 84: 291-296

Müller-Starck, R. 1996: Genetische Aspekte der Reproduktion der Buche (*Fagus sylvatica* L.) unter Berücksichtigung waldbaulicher Gegebenheiten. Berichte des Forschungszentrums Waldökosysteme, Reihe A, Bd. 135. Göttingen,103+X II p.

Namkoong, G. and Gregorius, H.R. 1985: Conditions for protected polymorphisms in subdivided plant populations. 2. Seed versus pollen migration. Am. Natur. 125: 521-534

Schnabel, A. and Hamrick, J.L. 1995: Understanding the population genetic structure of *Gleditsia triacanthos* L.: the scale and pattern of pollen gene flow. Evolution 49: 921-931

Stacy. E.A., Hamrick, J.L., Nason, J .D., Hubbell, S.P., Foster, R.B. and Condit, R. 1996: Pollen dispersal in low-density populations of three neotropical tree species. Am. Nat. 148:275-298

Troggio, M., DiMasso, E., Leonardi, S., Ceroni, M., Bucci, G., Piovani, P. and Menozzi, P. 1996: Inheritance of RAPD and I-SSR markers and population parameters estimation in European beech (*Fagus sylvatica* L.). Forest Genetics. 3: 173-181

Waser, N.M. 1993: Population structure, optimal outbreeding and assortative mating in angiosperms. pp. 139-199 in: Wilmsen-Thornhill, N. (Eds.) The Natural History of Inbreeding and Outbreeding: Theoretical and Empirical Perspectives. Univ. of Chicago Press, Chicago

Wright, S. 1951: The genetical structure of populations. Ann. Eugen. 15: 323-354

Ziehe,M. 1998: Anpassungsprozesse auf Populationsebene. pp. 115-141. In: Abschlussbericht 1998 des Forschungszentrums Waldökosysteme. Teil 1. ISSN 0939-1339

Patterns of seed dispersal in a scattered forest tree species (*Sorbus torminalis*) based on multi-scale investigation of population genetic structure for chloroplast DNA

S. ODDOU-MURATORIO[1], D. GUESNET[1], E. OZDEMIR[1], R. J. PETIT[2] AND B. DEMESURE[1]

[1]Conservatoire Génétique des Arbres Forestiers, Office National des Forêts, Campus INRA, F-45160 Ardon, France. [2]Station de Recherches Forestières, Laboratoire de Génétique des Arbres Forestiers, INRA, Pierroton, F-33611 Gazinet cedex France

Key words: cpDNA, phylogeography, seed, gene flow, *Sorbus torminalis* L.

Abstract: The wild service tree, *Sorbus torminalis* L. Crantz, is a post-pioneer temperate forest species, that occurs at low population density. In such species, where populations go through frequent extinction and recolonisation, seed migration is expected to be an important component of gene flow dynamics. To investigate the importance of seed dispersal in the wild service tree, we examined the spatial distribution of maternally inherited chloroplast DNA (cpDNA) markers among populations separated by tens, hundreds or thousands of kilometres.
The survey of cpDNA PCR-RFLP in 1031 individuals distributed among 57 populations allowed the detection of 29 haplotypes. Levels of differentiation as measured by G_{ST} at the different scales are among the lowest observed for cpDNA in forest trees so far. No phylogeographic structure appears across the investigated range. Only a weak pattern of isolation by distance was detected, for distances up to 50 km. These characteristics strongly differ from that of the majority of tree species studied so far. We show that seed-mediated gene flow strongly affects patterns of genetic structure, and also that post-glacial history of *S. torminalis* is much less reflected in the present genetic structure than what has been described for other widespread species.

Introduction

Despite the increasing number of population genetics studies dealing with forest tree species, still little is known about species that combine widespread distribution and low population densities (the so-called "scattered" species). This lack of knowledge may be due first to the low economic impact of these scattered species compared to social forest trees (e.g. oaks or beech) and second to the difficulty to inventory them. Yet they are an important part of the biodiversity in the forest, as they increase the number of different tree species and also the number of animal species which rely on them. Moreover, the few

Müller-Starck, G. and Schubert, R. (eds.). Genetic Response of Forest Systems to Changing Environmental Conditions. Vol. 70 (For. Sci.), 271-280

results acquired on scattered trees allow to identify some general features of the genetic diversity dynamic of these species, both in temperate forests (e.g. Raspe 1998, Demesure *et al.* 2000) and in tropical ones (e.g. Karron *et al.* 1987, Doligez *et al.* 1998). Indeed, as compared to widespread species, scattered forest trees generally show lower genetic diversity within populations and higher differentiation among populations with isozyme markers, and higher within population diversity with cpDNA markers. Beside these specific patterns of diversity, scattered forest trees are often characterised by high colonisation abilities, combined to low competition capacities with other social species. This results in typical population dynamics, characterised by frequent local extinction and recolonisation. Seed dispersal, one of the basic mechanisms of colonisation, is thus expected to be an important component of gene flow dynamics in those species.

Our objective here is to investigate the importance and the pattern of seed dispersal in a scattered forest species, the wild service tree, which is one of the most economically important wild fruit trees in Europe. Its pollen is disseminated by insects, and it produces fleshy fruits which are dispersed by birds and mammals.

In Angiosperm species, for which chloroplast and mitochondrial DNA (cpDNA/mtDNA) are generally maternally inherited, patterns of cytoplasmic diversity result only from seed-mediated gene flow. In various studies at different geographic scales, genetic variation of cytoplasmic DNA has been found to be highly geographically structured. Studies at a regional or local scale often indicate low levels of within-population diversity, together with spatial clustering of haplotypes, that authors usually explain by limited seed dispersal (e.g. McCauley 1994; Petit *et al.* 1997). At range-wide scale, the present distribution of cpDNA diversity is likely to reflect historical events the species went through. Being very long-lived, forest trees in particular have been shown to keep in their genetic structure strong footprints of postglacial recolonisation (e.g. Demesure *et al.* 1996; Dumolin-Lapègue *et al.* 1997). When results of population structure at various scales are accounted for, a much clearer picture of species dynamics may emerge (Le Corre *et al.* 1997).

In order to characterise seed dispersal in *S. torminalis*, diversity was investigated for maternally inherited cpDNA markers (Oddou-Muratorio *et al.* in press). Patterns of gene flow being obviously scale-dependent, genetic structure has been assessed across the European range of the species, among collections separated by tens, hundreds, or thousands of kilometres.

Material and methods

Plant Material

Dormant buds were collected from 1031 *S. torminalis* individuals, distributed among 57 indigenous populations. A population sample consists of 10-58 (18.1 on average) mature wild service trees or young stems separated from each other by 50 m (to avoid sampling of clones) over an area of 20 to 50 ha. The sampling effort was approximately constant (16 individuals per population on average) in most French populations, though the high variability of local *S. torminalis* density sometimes resulted in lower sample size. Sampling design was slightly modified for the Eastern Europe populations. Indeed, for technical reasons, we could not sample as many populations in this area as in France. However, as this study is the first investigation of cytoplasm diversity in the Eastern range of *S. torminalis*, we tended to slightly increase sampling effort within populations. This variance in sampling effort was accounted for in the results interpretation. Sample include:

- 258 individuals (14 populations) from the French region Poitou-Charentes;
- 605 individuals (37 populations) covering most other regions of the French species range;
- 168 individuals (6 populations) distributed from Switzerland to East of Bulgaria.

PCR-RFLP Procedure

Total DNA was isolated from 5-6 buds or from 1-2 fresh leaves per tree following a classical β-Mercaptoethanol / dichloromethane procedure adapted in Oddou-Muratorio *et al.* (in press). Three chloroplast primer pairs described in Demesure *et al.* (1995) were used for the PCR amplifications, and five µl of the PCR products were digested with 5 units of restriction endonuclease. Four primer-endonuclease combinations were used to characterise all Sorbus individuals : trnH-trnK1 (combined to HaeIII), trnK1-trnK2 (combined to ScrfI) and trnSa-trnfM (combined to MseI and DdeI). Restriction fragments were separated by electrophoresis in 6% polyacrylamide gels using tris-borate EDTA buffer (1X) at 300V during 3 hours. The gels were stained with ethidium bromide and scanned under UV light with FM-BIO II scanner (Hitachi Software).

Phylogenetic Analysis

Polymorphic fragments of each haplotype were scored as unordered multistate characters: each fragment was considered as a character and the different states were the different sizes of the fragments. The length variants were noted from 1 to 5, 9 being reserved for restriction site mutation. The distance between two haplotypes was defined as the number of characters presents in different states, i.e. the number of restriction fragments with different sizes. The neighbour-joining tree was obtained with the procedure NEIGHBOR of PHYLIP 3.5 (Felsenstein 1993).

Genetic Diversity Analysis

The frequencies of the haplotypes were used to estimate the level of subdivision of chloroplast diversity (G_{ST} and N_{ST}) and their standard deviations following Pons and Petit (1995, 1996). While G_{ST} estimates the level of subdivision of diversity for unordered alleles, N_{ST} takes into account the similarity between haplotypes, on the basis of the same distance matrix as used to obtain the phenetic tree. Pons and Petit (1996) provide a one-sided U-test that allows to compare N_{ST} and G_{ST} and to test phylogeographic structure.

Populations were grouped according to the mean geographical distance separating them. Three groups were distinguished, corresponding to three spatial scales. A first group (the “finest spatial scale” group) includes the 14 populations from Poitou-Charentes (separated by mean distance of 70 km). Two populations were randomly sub-sampled in Poitou-Charentes and added to the other 37 French populations in order to compose a homogenous “middle-scale” group (mean between-population distance : 370 km). The “largest-scale” collection includes the six Eastern Europe populations, plus three randomly sub-sampled French ones (mean between-population distance: 1300 km).

Description of the Spatial Structure

The existence of a spatial structure was tested with the program SGS (a software produced by Bernd Degen, unpublished), which plots genetic distances and their confidence intervals as a function of geographic distances. Here, pairwise measures of Wright’s F_{ST} were used as differentiation estimates.

Results

cpDNA polymorphism and characterisation of haplotypes

In the total sample of 1031 individuals, 9 restriction fragments were polymorphic and a total of 25 length variants and 4 point mutations were detected (more information on these polymorphism can be obtained from the first author). These variants grouped into 29 haplotypes. The Neighbour-joining tree (Figure 1) allows to distinguish three groups of haplotypes (noted A, B, C), but lineage A is largely dominant in the sample. Inside this lineage, six haplotypes (number 9, 11, 12, 16, 18, 19) are found which are shared by more than 95% of the sampled individuals. These six haplotypes have been observed all over Europe.

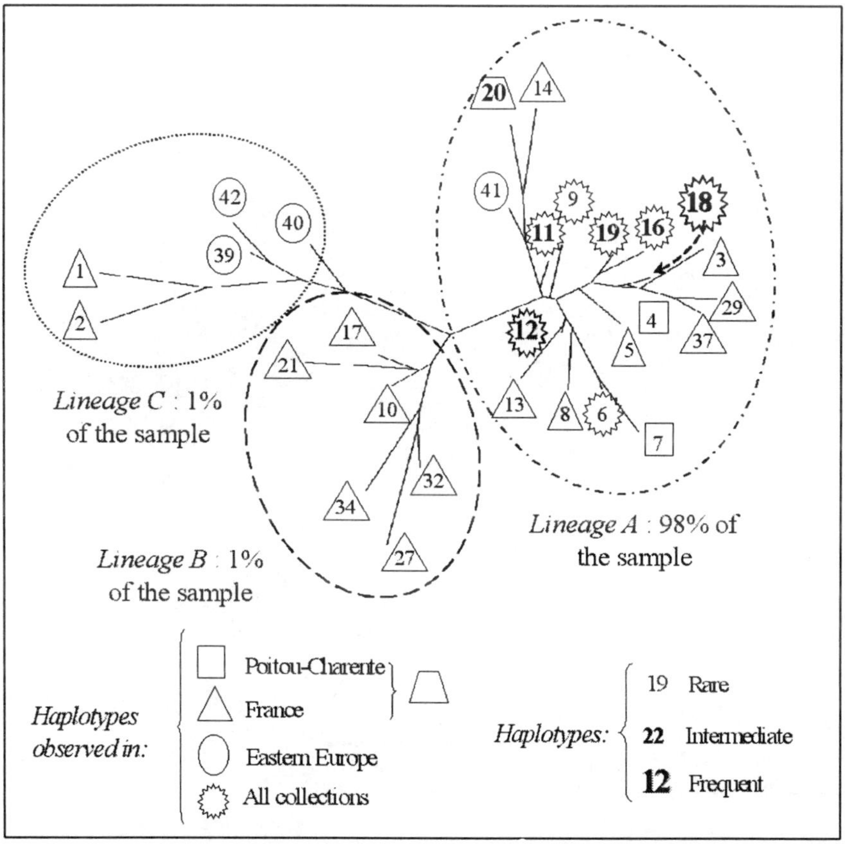

Figure 1. Neighbor-Joining tree of the haplotypes observed in the total sample. A haplotype observed in less than 1% of the total sample is considered as rare, in 1 to 10% of the sample as intermediate, in more than 10% as frequent.

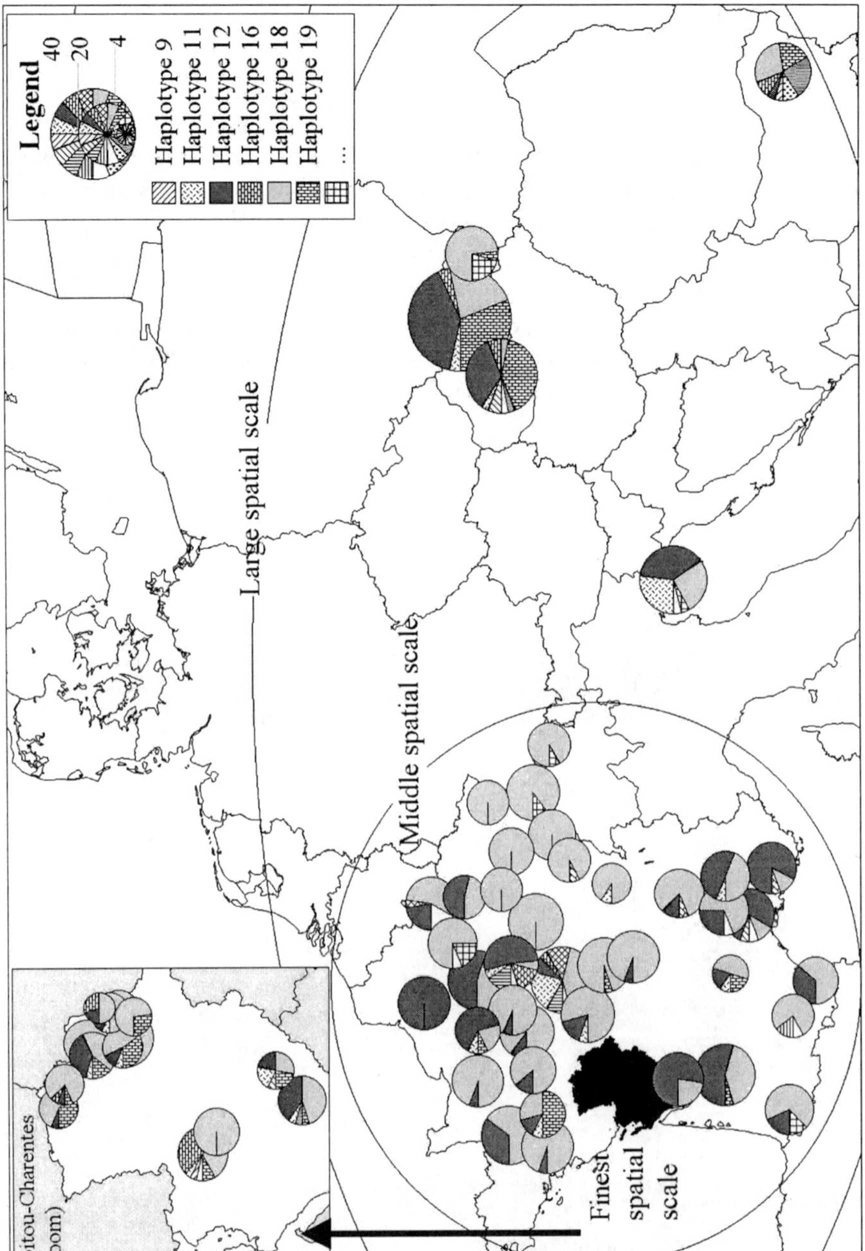

Figure 2. Distribution of cpDNA diversity at the three investigated scales. Each pie chart is a population, and the pie chart size is proportional to the sample size. Left-top image shows a zoom on Poitou-Charentes region (the finest-spatial scale).

Patterns of genetic structure

Figure 2 gives a global overview of the cpDNA diversity distribution. Most haplotypes are observed all over the studied area, and only rare and unique haplotypes are private to a given region. Moreover, populations are rarely fixed for one haplotype (mean of 3 haplotypes per population). The haplotypes 12 or 18 are shared by more than 84 % of the sampled individuals: these two haplotypes are both widespread over the investigated range and dominant within populations. These patterns have been observed at all three scales. Some differences however appear between the eastern and western part of the European range. In particular, haplotype 18 is less frequent in the east than in the west, and within-population diversity seems higher in the eastern populations.

Results of diversity analyses at the three different scales are compiled in Table 1. G_{ST} and N_{ST} never differ significantly from each other in any group. The random sampling of the two Poitou-Charentes populations does not significantly affect parameters estimation at the intermediate scale. G_{ST} is significantly higher in the intermediate-scale group than in the finest-scale one. On the opposite, genetic parameters greatly vary when sub-sampling French populations to be included in the largest-scale group, showing that additional sampling at the largest scale is needed to get precise estimates.

Table 1. Results of the analysis of diversity at different scales. Standard deviations are in parentheses.

Scale	**Nb. of populations**	**Nb of individuals / pop.**	**Nb of haplotypes / pop. (A_0)**	h_S	G_{ST}	N_{ST}
Finest	14	18.4 (5.1)	3.64	0.53 (0.06)	0.13 (0.03)	0.13 (0.04)
Middle	39	15.2 (4.6)	2.54	0.35 (0.04)	0.33 (0.05)	0.33 (0.05)
Largest	9	28 (14.8)	5	0.56 (0.07)	0.29 (0.05)	0.31 (0.04)

Patterns of geographic structure

Distograms computed separately at the three scales do not reveal any significant pattern of isolation by distance. It is only when the two low and middle-scale groups are combined that a slight pattern of isolation-by-distance appears. Differentiation values for cpDNA markers are in this later case smaller than expected by chance in the first distance class [0-50 km] (Fig. 3).

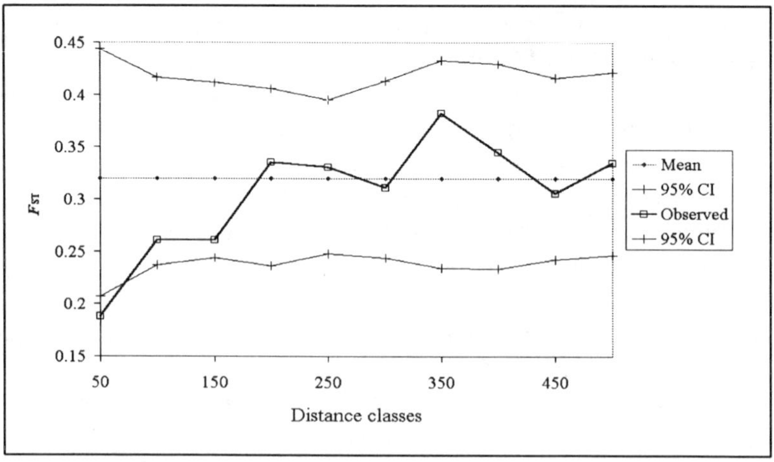

Figure 3. Genetic distogram of cpDNA genetic differentiation (Wright F_{ST}) in ten geographic distance classes (0-500 km), and 95% confidence intervals of F_{ST}, computed by mean of 5000 permutations.

Discussion

Earlier investigations in forest trees based on maternally inherited markers have detected a strong differentiation among populations and low levels of within-population diversity as compared to nuclear markers. Though this trend still holds in *S. torminalis*, patterns of cpDNA diversity strongly differ from what has been observed in widespread species studied so far. Populations are rarely fixed for one haplotype, and the high level of within-population diversity ($0.35 < H_S < 0.56$) indicates that several mothers contribute to the founding of new populations. G_{ST} found in *S. torminalis* at the different scales ($0.13 < G_{ST} < 0.32$) is among the lowest documented so far in plant species for which this genome is maternally inherited (e.g. $G_{ST} = 0.82$ for *Quercus* sp. -Dumolin-Lapègue *et al.* 1997; $G_{ST} = 0.83$ for *Fagus sylvatica* -Demesure *et al.* 1996). Moreover, no strong spatial structure is apparent at the different scales investigated, confirming the results obtained with isozymes (Demesure *et al.* 2000). Distogram analyses reveal a weak genetic structure, as pairwise G_{ST} appears to be lower in the first 0-50 km distance class, but this trend is not strong enough to persist at the different investigated scales. Finally, among-populations differentiation does not differ when measured by G_{ST} or N_{ST}, which indicates an absence of phylogeographic component in the genetic structure (Dumolin-Lapègue at al. 1997).

The usually recognised aptitude of maternally inherited chloroplast markers to reflect post-glacial history of long-lived species is not supported by our results. Yet, this species is found in dry habitats, in the plains at lower latitudes, and is absent at high altitudes. Hence, the climate during the last ice-age probably did not allow numerous small populations to persist, and the hypothesis of glacial refugia is more probable. Though sampling could be increased around the classical Iberian, Italian and Balkan refugia, it is highly unlikely that a picture of recolonisation routes as clear as what has been observed for instance in oaks or beech, or even in *Alnus* (King and Ferris 1998) may emerge. Intensive seed movements could have blurred a preexisting phylogeographic structure, but such a structure might also have never existed, due to a more continuous mode of recolonization.

Nevertheless, cytoplasmic markers appear very useful to provide information on gene flow dynamics. Observed patterns of cpDNA diversity confirm the importance of seed flow in this species characterised by endozoochhorous seed dispersal and by a typical extinction/recolonisation dynamic. Oddou-Muratorio *et al.* (in press) have measured a very low pollen vs. seed dispersal ratio in *S. torminalis*. The present study provides interesting insights concerning the general shape of the seed dispersal curve. Indeed, the observed patterns of genetic structure are obviously scale-dependent. The G_{ST} value observed at the finest spatial scale is significantly lower than at the middle spatial scale. Moreover, the trend to isolation by distance detected when the middle and the finest spatial scale groups are combined does not appear at the finest spatial scale alone. The number of sampled populations differ among scales, but we used resampling procedures to correct for this bias. These observed patterns, which represent the cumulative results of gene flow over numerous generations, show that it is only over distance within one hundred kilometres range that seed dispersal can be considered as limited. The biology of the species, notably the dispersion of fruits by birds and mammals, is consistent with such efficient seed movements. Both the observed pattern of cpDNA diversity and the importance of the seed dispersal process will be major points for the management and conservation of this species` genetic resources. In order to validate our hypothesis of seed dispersal over large ranges, other methods than those presented here will be needed, based for instance on parentage analyses.

Acknowledgements

We thank Berthold Heinze and an anonymous reviewer for stimulating comments, as well as the Office National of Forests technicians and all the helpful colleagues from Eastern Europe who collected the Sorbus samples.

References

Demesure, B., Sodzi, N. and Petit, R. J. 1995: A set of universal primers for amplification of polymorphic non-coding regions of mitochondrial and chloroplast DNA in plants. Molecular Ecology 4: 129-131

Demesure, B., Comps, B. and Petit, R. J. 1996: Chloroplast DNA phylogeography of the common beech (*Fagus sylvatica* L.) in Europe. Evolution 50: 2515-2520

Demesure, B., Le Guerroué, B., Lucci, G., Prat, D. and Petit R.J. 2000: Genetic variability of a scattered temperate forest tree: Sorbus torminalis L. (Crantz). Annals of Forest Science. 57: 63-71

Doligez, A., Baril, C. and Joly H. 1998: Fine-scale spatial genetic structure with non-uniform distribution of individuals. Genetics 148: 905-919

Dumolin-Lapègue, S., Demesure, B., Fineschi, S., Corre, V. L. and Petit R. J. 1997: Phylogeographic structure of white Oaks throughout the European continent. Genetics 146: 1475-1487

Felsenstein, J. 1993: PHYLIP (phylogeny Inference Package), Department of Genetics, University of Washington, Seattle

Karron, J. D. 1987: A comparison of levels of genetic polymorphism and self-compatibility in geographically restricted and widespread plant congeners. Evolutionary Ecology 1:47-58

King R.A. and Ferris C. 1998: Chloroplast DNA phylogeography of *Alnus glutinosa* (L.) gaertn. Molecular Ecology 7: 1151-1161

Le Corre, V., Machon, N., Petit, R.J. and Kremer, A. 1997: Colonization with long-distance seed dispersal and genetic structure of maternally inherited genes in forest trees: a simulation study. Genet. Res. 69: 117-125

McCauley, D. E. 1994: Contrasting the distribution of chloroplast DNA and allozyme polymorphism among local populations of *Silene alba*: Implications for studies of gene flow in plants. Proc. Natl. Acad. Sci. USA 91: 8127-8131

Oddou-Muratorio, S, Petit, R. J., Le Guerroué, B., Guesnet, D. and Demesure, B.: Pollen versus seed-mediated gene flow in a scattered forest tree species. Evolution, *in press.*

Petit, R. J., Pineau, E., Demesure, B., Baccilieri, R., Ducousso, A., and Kremer A. 1997: Chloroplast DNA footprints of postglacial recolonization by oaks. Proc Natl Acad Sci USA 94: 9996-10001

Pons, O. and Petit R. J. 1995: Estimation, variance and optimal sampling of gene diversity I. Haploid locus. Theor. Appl. Genet. 90: 462-470

Pons, O., and Petit R. J. 1996: Measuring and testing genetic differentiation with ordered versus unordered alleles. Genetics 144: 1237-1245

Raspe, O. 1998: Biologie de la reproduction et variation génétique d'un arbre entomophile: *Sorbus aucuparia* L. (Rosaceae: Maloideae). Pp. 111. Thèse de Doctorat de l'Université Catholique de Louvain

Gene flow and mating system in a seedling seed orchard and a natural stand of *Pinus merkusii* Jungh. et de Vriese in Indonesia

I.Z. SIREGAR[1] AND H.H. HATTEMER[2]

[1]Laboratorium Silvikultur, Fakultas Kehutanan, Institut Pertanian Bogor, Kampus Darmaga, Bogor 16001, INDONESIA. [2]Institut für Forstgenetik und Forstpflanzenzüchtung, Universität Göttingen, D-37077 Göttingen, Germany

Key words: *Pinus merkusii*, mating system, isozyme marker, seedling seed orchard, pollen dispersal, gene flow

Abstract: Isozyme gene markers were used to study gene flow and the mating system in a seedling seed orchard and a natural stand of *P. merkusii* in Indonesia. Embryos and megagametophytes of at least twenty seeds per family were analysed electrophoretically in order to determine allele frequencies at eight polymorphic isozyme gene loci (GOT-B, GOT-C, GOT-D, PGM-A, PGM-B, SKDH-A, NDH-A and FDH-A). Outcrossing rates were estimated based on a mixed mating model. In the seedling seed orchard the average single-locus estimate of outcrossing (t_s) was 0.83±0.04, while the multilocus estimate (t_m) was 0.87±0.03. Both estimates were slightly higher (t_s = 0.90±0.05 and t_m = 0.98±0.03) in the natural stand. Estimates of outcrossing rates of single trees were more heterogeneous in the seed orchard than in the natural stand. Individual outcrossing rates in the seed orchard seemed to be related to their ancestral background. Trees belonging to one particular open-pollinated family tended to have homogenous outcrossing rates. Two open-pollinated families consisting of three trees showed lower outcrossing rates than the remaining three. Pronounced variation of outcrossing rates was not detected between individual trees in the natural stand, although comparisons were made between trees growing in contrasting density, i.e. closely together and spatially isolated. The pattern of allelic differentiation among effective pollen clouds was slightly different between the seed orchard and the natural stand. In general, pollen allele frequencies of single seed trees were moderately homogeneous in both populations.

Introduction

Most population genetic studies on the reproduction of plants concentrated on the estimation of outcrossing and selfing rates of populations or individual plants. However, gene flow and the mating system are not fully genetically determined, but depend on many environmental factors such as population density, spatial structure, stand development, flower phenology, and vectors for pollen transport (Finkeldey 1998).

Müller-Starck, G. and Schubert, R. (eds.). Genetic Response of Forest Systems to Changing Environmental Conditions. Vol. 70 (For. Sci.), 281-292

It is possible to characterise gene flow through pollen by an observation of pollen allele frequencies of single seed trees at marker loci. Homogeneity of "genetically effective pollen clouds" of single seed trees was observed in *Picea abies* plantations (Finkeldey 1995). Pollen clouds fertilising single seed trees in a population of *Pinus pungens* were more heterogeneous (Gibson and Hamrick 1991). Thus, gene flow through pollen seems to be more efficient in the investigated *Picea abies* populations. Homogeneity of pollen allele frequencies may be interpreted as the absence of mating preferences of individual seed trees for male gametes, i.e. random mating (G. Müller-Starck and Gregorius 1988). The assumption of homogeneity of the outcrossed pollen pool underlies the mixed mating model of Brown and Allard (1970) for the estimation of outcrossing rates.

P. merkusii is a tropical pine of Southeast-Asia. It occurs naturally in Myanmar, Thailand, Laos, Cambodia, Vietnam, Indonesia, and the Philippines (Critchfield and Little 1966). Outcrossing rates t_m were estimated in natural populations of *P. merkusii* in Thailand by Changtragoon and Finkeldey (1995b). These authors found extraordinarily high proportions of selfing ($0.017<t_m<0.65$) in 9 out of 10 analysed populations. Inbreeding coefficients were accordingly high. However, no information was obtained from plantations or seed orchards.

The establishment of seed orchards is common practice all over the world in order to supply genetically improved forest seeds. A maximum phenotypic similarity between the orchard population and their offspring is aspired (G. Müller-Starck 1991). However, the phenotypic performance of the progenies depends on a large number of factors including the realised mode of reproduction in the seed orchard. It is therefore of great interest to compare the mating system as a component of the reproduction system in a natural population and a seed orchard of *P. merkusii.* No clonal seed orchards of *P. merkusii* have been established in Indonesia. The only available pine seed orchards are progeny tests which were directly converted into seed orchards at the end of the measurement period. The aims of this study are to measure differentiation among single tree pollen clouds, and to estimate the rates of outcrossing in a seedling seed orchard and a natural stand of *P. merkusii*.

Materials and Methods

Natural stand and seed orchard

Seeds were harvested in two separate populations of *P. merkusii*: A natural stand in Aceh, Sumatra (latitude 4°30'N, longitude 96°51'E, altitude 900 m) and a seedling seed orchard in Java (latitude 7°67'N, longitude 113°52'E, altitude 600 m). Distinct differences between the two stands refer to tree dimensions, densities, and the spatial distribution of trees. On average, the trees in the natural stand are larger than those in the seed orchard. On the basis of the distribution of individuals in the natural stand, the population was divided into two parts: 1) trees growing in clumps and 2) trees growing more or less isolated. Four trees growing in clumps and four isolated trees were investigated. Five families, each comprising of three trees, were selected in the seedling seed orchard. The three trees from the same family were left in different replicates of the former field experiment. Hence, they are no direct neighbours. Cones were randomly sampled from the upper part of the crown in order to reduce the chance of natural self-fertilisation. The cones were then exposed to sun drying and the seeds were extracted manually. A sample of 100 seeds per family was soaked in 70% ethanol in order to estimate the percentage of filled seeds.

Electrophoresis

Eight polymorphic enzyme gene loci (GOT-B, GOT-C, GOT-D, PGM-A, PGM-B, SKDH-A, NDH-A and FDH-A) were studied. The genetic interpretation of zymograms was similar to Changtragoon and Finkeldey (1995a). The results on the genetic control of enzyme systems are described in detail by Siregar (2000).

At least twenty seeds per family were analysed electrophoretically. Tissue of the embryo and the megagametophyte was investigated separately in order to determine allele frequencies of maternal and paternal contributions to the embryos at eight enzyme gene loci (e.g. G. Müller(-Starck) 1977). Horizontal starch gel electrophoresis (10.5% starch concentration plus 2.5-3.5% sucrose) was performed according to Feret and Bergmann (1976). Details of laboratory methods are described in Siregar (2000).

Data analysis

Differentiation among "sitpocs" (single-tree effective pollen clouds or pollen allele frequencies: cf. Finkeldey 1995) in the natural stand and in the seedling seed orchard was separately quantified by the measures D_j and δ (Gregorius and Roberds 1986) and the fixation index F_{ST} of Wright (1965), which is equivalent to G_{ST} according to Nei (1973). Heterogeneity among pollen allele frequencies was checked for statistical significance by means of a goodness-of-fit test (*G*-test). Rare alleles were lumped before the *G*-test was performed in order to avoid problems associated with low expected frequencies (Sokal and Rohlf 1981). The computer programs GSED (version 1.1.,Gillet 1998) and BIOSYS-2

(Swofford and Selander 1997) were used to calculate the above genetic measures and to perform statistical tests. Population and single tree outcrossing rates were estimated using the multilocus procedure of Ritland and Jain (1981). The multilocus mating system program (MLTR, version 1.1) of Ritland (1994) was used to calculate outcrossing rates and their standard errors by means of bootstrapping (100 bootstraps). In addition, the paternity exclusion method was used to estimate minimum outcrossing rates (R. Müller-Starck, 1996, pp. 74ff).

Results

Differentiation among single-tree pollen clouds

Tables 1 and 2 show the differentiation among pollen allele frequencies for the two populations. Significant differentiation was observed at four loci in both populations. Two loci, PGM-A and SKDH-A, showed significant differentiation in both stands. The gene pool differentiation δ and fixation index F_{ST} among sitpocs is lower in the seed orchard than in the natural stand. Within the natural stand the clumped trees had only a slightly lower differentiation (δ) than the isolated trees (Table 2).

Estimation of outcrossing rates and its relation to seed production

Single tree multilocus outcrossing (t_m) rates based on the estimation procedure of Ritland and Jain (1981) as well minimum estimates of the outcrossing rate based on paternity analysis (minimum t_m) are presented in Table 3. Large differences between the estimates of t_m and minimum t_m were observed for certain trees. The genotype of the seed tree at the investigated marker loci is likely to be mainly responsible for these differences. The overall outcrossing rates (t_m and t_s) in the seed orchard were 0.87 and 0.83, respectively, while in the natural stand the respective estimates were 0.98 and 0.90.. The average single locus estimates (t_s) were slightly lower than the multilocus estimates (t_m).

In the seed orchard, the single tree estimates of outcrossing rates vary among trees. However, the outcrossing rates of trees of the same families are similar. For example, trees J31, J33 and J34 are from the same family and show low outcrossing rates of 0.73, 0.67 and 0.69, respectively. There is no positive correlation between the level of outcrossing and the percentage of filled seed (see Figure 1).

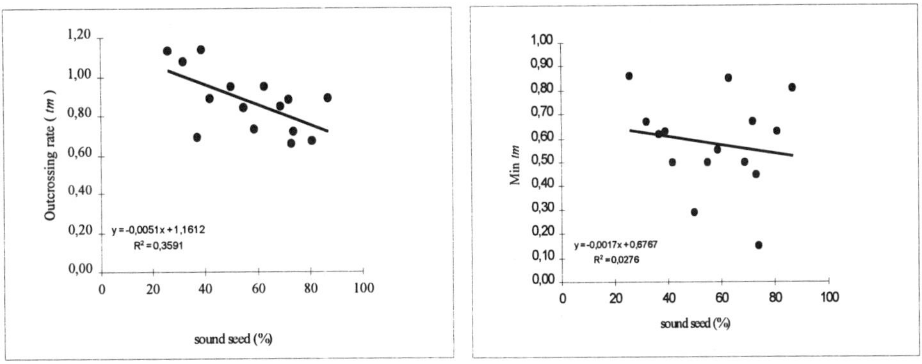

Figure 1. Regression analysis between outcrossing rates and percentages of filled seed. The graph on the left refers to multilocus outcrossing rates based on a mixed mating model (Ritland and Jain 1981), while the one on the right hand side refers to minimum outcrossing according to the paternity exclusion method.

Table 1. Differentiation D_j and δ among single tree pollen clouds (sitpocs), fixation index F_{ST} and results of a test for statistical significance of the heterogeneity (G_h) in a *P. merkusii* seedling seed orchard of Java. N: number of investigated seeds per tree; NS: non significant; (*),(**),(***): significant heterogeneity at 5% , 1% and 0.1% level, respectively.

Gene loci	Family 1			Family 2			Family 3			Family 4			Family 5			δ	F_{ST}	G_h
	J06 (N=24)	J07 (N=24)	J09 (N=24)	J11 (N=22)	J12 (N=40)	J14 (N=26)	J26 (N=62)	J27 (N=58)	J29 (N=22)	J31 (N=22)	J33 (N=26)	J34 (N=29)	J36 (N=40)	J37 (N=32)	J38 (N=32)			
GOT-B	0.021	0.021	0.107	0.107	0.030	0.034	0.057	0.018	0.060	0.107	0.043	0.074	0.081	0.107	0.025	0.087	0.151	NS
GOT-C	0.059	0.145	0.223	0.098	0.121	0.191	0.534	0.079	0.168	0.191	0.159	0.191	0.246	0.111	0.127	0.096	0.219	***
GOT-D	0.071	0.071	0.157	0.087	0.060	0.003	0.166	0.024	0.071	0.087	0.228	0.117	0.029	0.036	0.100	0.176	0.113	NS
PGM-A	0.038	0.079	0.038	0.016	0.012	0.038	0.078	0.007	0.016	0.038	0.097	0.007	0.038	0.038	0.033	0.112	0.160	*
PGM-B	0.076	0.095	0.181	0.298	0.027	0.110	0.142	0.058	0.017	0.169	0.117	0.023	0.052	0.181	0.140	0.038	0.202	**
SKDH-A	0.070	0.102	0.102	0.132	0.002	0.136	0.003	0.049	0.194	0.132	0.092	0.403	0.051	0.156	0.156	0.119	0.216	***
NDH-A	0.097	0.168	0.097	0.074	0.020	0.161	0.009	0.009	0.074	0.122	0.226	0.106	0.089	0.097	0.097	0.090	0.125	NS
FDH-A	0.129	0.042	0.042	0.044	0.146	0.172	0.006	0.080	0.050	0.090	0.044	0.061	0.059	0.172	0.213	0.059	0.135	NS
Gene pool	0.070	0.090	0.118	0.107	0.052	0.106	0.124	0.041	0.081	0.117	0.126	0.123	0.081	0.112	0.111	0.097	0.165	

Table 2. Differentiation D_j and δ among single tree pollen clouds (sitpocs), fixation index F_{ST} and results of a test for statistical significance of the heterogeneity (G_h) in a natural stand of *P. merkusii* at Aceh (Sumatra). N: number of investigated seeds per tree; NS: non significant; (*),(**): significant heterogeneity at 5%, and 1% level, respectively.

Gene loci	clumped trees (*CT*)				Pooled *CT* (ΣN=104)	isolated trees (*IT*)				Pooled *IT* (ΣN=91)	δ	F_{ST}	G_h
	A01 (N=29)	A04 (N=20)	A13 (N=28)	A39 (N=27)		AI08 (N=24)	AI09 (N=24)	AI40 (N=18)	AI41 (N=25)				
GOT-B	0.093	0.053	0.123	0.009	0.070	0.035	0.035	0.006	0.032	0.027	0.058	0.112	NS
GOT-C	0.160	0.113	0.156	0.156	0.146	0.050	0.117	0.198	0.166	0.133	0.191	0.233	NS
GOT-D	0.015	0.313	0.003	0.090	0.105	0.003	0.047	0.172	0.150	0.093	0.093	0.138	NS
PGM-A	0.041	0.041	0.047	0.041	0.043	0.041	0.041	0.205	0.041	0.082	0.066	0.152	**
PGM-B	0.026	0.094	0.117	0.038	0.069	0.073	0.117	0.175	0.199	0.141	0.139	0.130	NS
SKDH-A	0.238	0.172	0.097	0.117	0.156	0.006	0.093	0.272	0.151	0.131	0.193	0.196	**
NDH-A	0.149	0.095	0.147	0.068	0.115	0.009	0.271	0.089	0.229	0.150	0.195	0.205	*
FDH-A	0.195	0.024	0.077	0.078	0.094	0.033	0.077	0.369	0.351	0.208	0.132	0.213	*
Gene pool	0.115	0.113	0.096	0.075	0.100	0.031	0.100	0.186	0.165	0.120	0.133	0.178	

Table 3. Multilocus estimates (t_m) of single-tree outcrossing rates, multi- (t_m) and average single locus (t_s) estimates of population outcrossing rates together with their standard errors, and percentage of filled seeds. Estimation according to the method of Ritland and Jain (1981) for $t_m \pm \Delta t_m$ and paternity exclusion for min t_m.

Tree No.	% Filled Seed	Ritland and Jain Method $t_m \pm \Delta t_m$	Paternity Exclusion Method min. t_m
		Seedling seed orchard	
J06	42.0	0.89 ± 0.32	0.50
J07	72.0	0.88 ± 0.21	0.67
J09	32.0	1.08 ± 0.28	0.67
J11	73.2	0.66 ± 0.14	0.45
J12	74.0	0.72 ± 0.23	0.15
J14	55.0	0.84 ± 0.17	0.50
J26	63.0	0.95 ± 0.04	0.85
J27	50.0	0.95 ± 0.08	0.29
J29	26.0	1.13 ± 0.18	0.86
J31	59.0	0.73 ± 0.15	0.55
J33	81.0	0.67 ± 0.13	0.63
J34	37.0	0.69 ± 0.09	0.62
J36	69.0	0.85 ± 0.13	0.50
J37	87.0	0.89 ± 0.22	0.81
J38	39.0	1.14 ± 0.21	0.63
Population	50.7	0.87 ± 0.03 (t_m) 0.83 ± 0.04 (t_s)	0.59
		Natural stand	
		Clumped trees	
A01	45.7	0.98 ± 0.13	0.86
A04	40.6	1.05 ± 0.47	0.95
A13	65.7	1.03 ± 0.47	0.94
A39	17.7	0.95 ± 0.27	0.53
		Isolated trees	
AI08	17.3	0.92 ± 0.09	0.79
AI09	13.6	1.41 ± 0.31	0.96
AI40	5.5	1.52 ± 0.26	0.89
Ai41	7.0	1.07 ± 0.18	0.88
Population	15.9	0.98 ± 0.03 (t_m) 0.90 ± 0.05 (t_s)	0.85

Discussion

Heterogeneity of single tree pollen clouds

Studies on the measurement of genetic differentiation among single tree pollen clouds in *P. merkusii* have not been reported previously. The values of average allelic differentiation (δ) reported in this study ranged from 0.097 to 0.133, indicating stronger genetic differentiation than in *Picea abies* as reported by Finkeldey (1995). This author found low genetic differentiation (δ ranging from 0.051 to 0.055; F_{ST} inbetween 0.011 and 0.015 in three plantations) and also homogeneity of pooled pollen clouds among three planted stands. The differences between the results of Finkeldey (1995) and the results reported here are much more pronounced if F_{ST} rather than δ is used as a measure of genetic differentiation among "sitpocs". This may be at least partially explained by higher levels of genetic variation at the observed marker loci in the investigated populations of *Picea abies*.

A low level of differentiation of pollen allele frequencies indicates extensive pollen transport within the respective population creating spatial homogeneity in pollen clouds for the population. Considerable differentiation of "sitpocs" was observed both within the seed orchard and the natural stand. This is particularly evident if the differentiation is measured as F_{ST}. High values of F_{ST} were found at single loci even if the differentiation was not significant presumably because of the moderate sample sizes (see Table 1 and 2).

Heterogeneity of "pollen clouds" was observed even within the crown of single trees in *Pinus pungens* (Gibson and Hamrick 1991). Thus, the composition of the successful pollen is likely to show strong variation over short distances for *Pinus pungens* and possibly also for *P. merkusii*, but not for other conifers such as *P. abies* (Finkeldey 1995). The spatial structure of a stand may have an impact on pollen dispersal since pollen of conifers is preferentially transported to near neigbours (G. Müller-Starck 1977). In addition, temporal variation of the pollen composition may have a profound impact on pollen allele frequencies, although it is difficult to prove this hypothesis experimentally. Phenology is likely to be a crucial for pollen allele frequencies of *P. merkusii* since flowering extends over longer periods as compared to temperate pines (e.g. Changtragoon and Finkeldey 1995b). Temporal variation of the pollen pool is likely to be reflected in heterogeneous pollen allele frequencies of trees with differing periods of female receptivity.

Outcrossing rates, mating system and seed production

The outcrossing rates (t_m and t_s) observed in this study are consistent with estimates for other pine species and conifers in general (Ledig, 1998). The slight differences between t_m and t_s are possibly due to inbreeding other than selfing, which depresses the single-locus estimates of outcrossing rates but affect the multilocus estimates less (Ledig 1998). In the present material, cross-fertilisation among closely related trees is unlikely at least in the seed orchard because the trees belonging to the same family were left in different replicates of the former field experiment.

Analysis of the mating system has commonly been restricted to an estimation of parameters of the mixed-mating model which subdivides all mating events into self-fertilisation and random outcrossing. The use of this model in the analysis of mating systems has some drawbacks, such as the violation of one or more assumptions to be made. In this study, for example, the assumption of homogeneity of pollen allele frequencies of the outcrossed pollen is not fulfilled. According to Mitton (1992), such violation of mixed-mating model assumptions are common in forest trees. They frequently cause biologically meaningless estimates of outcrossing rates above 100%, which were also obtained for several seed trees in this study (Table 3).

Seed orchards are established in order to produce large amounts of genetically improved seeds. However, the selfing rate in the investigated orchard is comparatively high (15%) eventually resulting in inbreeding depression and reduced seed quality. Experimental studies on pollen flow and mating systems in seed orchards are useful to improve the design and management of seed orchards. The observed similarity of outcrossing among trees of the same family indicates genetic control of the self-fertilisation rate in individuals of *P. merkusii*. Burczyk *et al.* (1991) reported similar outcrossing rates among the grafts of the same clone in a clonal archive of *Larix decidua*.

Low seed production in the natural stands of Aceh as observed in this study has also previously been reported (Ganefia and Siregar 1995). Frequent abortions of female flowers, high proportion of empty seeds and low seed germination have been also reported for *P. merkusii* in Thailand (Sa-Ardavut *et al.* 1989). Evidence for high inbreeding and selection against inbred individuals during the development of natural stands of *P. merkusii* in Thailand was provided by Changtragoon and Finkeldey (1995b). Cones of pines receiving high proportions of pollen from the same tree contain numerous aborted seeds due to the expression of lethal alleles (Lindgren 1975). Karkkainen *et al.* (1999) suggested that variation between maternal genotypes is another important factor causing abortions due to "bad" maternal genotypes. Cones with many outcrossed seeds usually contain a high proportion of vigorous progenies showing good survival. However, no positive correlation was found in the present study between outcrossing rates and percentages of filled seeds. Comparable results have been reported for Douglas-fir (Omi and Adams 1986), European larch (Burczyk *et al.* 1991), and European silver fir (*Abies alba*; Schroeder 1989). A strong correlation between outcrossing and percentage of filled seed is expected only if self-fertility is uniformly high among all individuals or families (Sorensen 1982). A study on flower phenology and pollination in *P. merkusii* revealed the significance of controlled pollination which at least doubles the average number of seeds per cone in comparison to open pollination (Danarto 1983).

Acknowledgements

Financial support for I.Z. SIREGAR from German Academic Exchange Service (DAAD) is gratefully acknowledged.

References

Brown, A.H.D. and Allard, R.W. 1970: Estimation of the mating systems in open-pollinated maize populations using isozyme polymorphisms. Genetics 66:133-145

Burczyk, J., Kosiński, G. and Lewandowski, A. 1991: Mating pattern and empty seed formation in relation to crown level of *Larix decidua* (Mill.) clones. Silva Fennica 25: 201-205

Changtragoon, S. and Finkeldey, R. 1995a: Inheritance of isozyme phenotypes of *Pinus merkusii*. J. Trop. For. Sci. 8(2): 167-177

Changtragoon, S. and Finkeldey, R. 1995b: Patterns of genetic variation and characterization of the mating system of *Pinus merkusii* in Thailand. Forest Genetics 2:87-97

Critchfield, W.B. and Little, E.L. 1966: Geographic Distribution of the Pines of the World. U.S. Department of Agriculture. Washington

Danarto, S. 1983: Studi Fenologi Pembungaan, Pembuahan dan Penyerbukan Terkendali *Pinus merkusii* Jungh et de Vriese di Sempolan, Jember. Skripsi Sarjana, Fakultas Kehutanan, Universitas Gajah Mada, Yogyakarta

Feret, P.P. and Bergmann, F. 1976: Gel electrophoresis of proteins and enzymes. Pp. 49-77 in MIKSCHE, J.P. (Ed.): Modern Methods in Forest Genetics. Springer, Berlin, Heidelberg, etc.

Finkeldey, R. 1995: Homogeneity of pollen allele frequencies of single seed trees in *Picea abies* (L.) Karst. plantations. Heredity 74:451-463

Finkeldey, R. 1998: An Introduction to Tropical Forest Genetics. Lecture Notes. Institut für Forstgenetik und Forstpflanzenzüchtung, Universität Göttingen. 225 pp.

Ganefia, Y. and Siregar, I.Z. 1995: Vegetative Propagation Technology of Tusam (*Pinus merkusii*) at PT. Tusam Hutani Lestari, Aceh. Research and Development of Kalimanis Group. Jakarta

Gibson, J.P. and Hamrick, J.L. 1991: Heterogeneity in pollen allele frequencies among cones, whorls, and trees of Table Mountain pine (*Pinus pungens*). Am. J. Bot., 78: 1244-1251

Gillet, E.M. 1998: GSED: Genetic Structure from Electrophoresis Data, version 1.1. User's Manual. Institut für Forstgenetik und Forstpflanzenzüchtung, Universität Göttingen Germany. http://www.uni-forst.gwdg.de/forst/fg/index.htm

Gregorius, H.R. and Roberds, J.H. 1986: Measurement of genetical differentiation among subpopulation. Theor. Appl. Genet. 71: 826-834

Ledig, F.T. 1998: Genetic variation in *Pinus*. Pp. 251-280 in Richardson, D.M. (Ed.): Ecology and Biogeography of *Pinus*. Cambridge University Press. Cambridge, UK

Lindgren, D. 1975: The relationship between self-fertilisation, empty seed, and seed originating from selfing as a consequence of polyembryony. Stud. For. Suecica, No. 126, 25pp.

Karkkainen, K., Savolainen, O. and Koski, V. 1999: Why do plants abort so many developing seeds: bad offspring or bad maternal genotypes ?. Evolutionary Ecology 13: 305-317

Müller(-Starck), G. 1977: Cross-fertilization in a conifer stand inferred from enzyme gene-markers in seeds. Silvae Genetica 26:223-226

Müller-Starck, G. 1991: Genetic processes in seed orchards. Pp.147-162 in Giertych, M. and Mátyás, C. (Eds.): Genetics of Scots Pine. Akadémiai Kiadó, Budapest

Müller-Starck, G and Gregorius, H-R. 1988: Analysis of mating systems in forest trees. In Weir, B.S., Eisen, E.J., Goodman, M.M., Namkoong, G. (eds.): Proceedings of the Second International Conference on Quantitative Genetics. Sunderland, Massachussetts: Sinauer. pp. 573-595

Müller-Starck, R. 1996: Genetische Aspekte der Reproduktion der Buche (*Fagus sylvatica* L.) unter Berücksichtigung waldbaulicher Gegebenheiten. Berichte des Forschungszentrums Waldökosysteme, Reihe A, Band 135. Göttingen: Forschungszentrum Waldökosysteme

Mitton, J.B. 1992: The dynamic mating systems of conifers. New Forests 6: 197-216

Nei, M. 1973: Analysis of gene diversity in subdivided populations. Proceedings of the National Academy of Sciences, USA 70: 3321-3323

Omi, S.K. and Adams, W.T. 1986: Variation in seed set and proportions of outcrossed progeny with clones, crown position, and top pruning in a Douglas-fir seed orchard. Can. J. of For. Res. 16: 502-507

Ritland, K. and Jain, S. 1981: A model for the estimation of outcrossing rate and gene frequencies using *n* independent gene loci. Heredity 47:35-52

Ritland, K. 1994: Multilocus mating system program (MLTR version 1.1). Department of Botany, University of Toronto

Sa-Ardavut, P., Chairuangsirikul, T. and Chanpaisaeng, S. 1989: *In situ* gene conservation in northeast Thailand. Pp. 291-297 in Gibson, G.L., Griffin, A.R. and Matheson, A.C (Eds.): Breeding Tropical Trees: Population Structure and Genetic Improvement Strategies in Clonal and Seedling Forestry. Oxford Forestry Institute and Winrock International, Oxford, U.K. and Arlington, U.S.A.

Schroeder, S. 1989: Outcrossing rates and seed characteristics in damaged natural populations of *Abies alba* Mill. Silvae Genetica 38: 185-189

Siregar, I.Z. 2000: Genetic Aspects of the Reperoductive System in *Pinus merkusii* Jungh. et de Vriese in Indonesia. Cuvillier Verlag Göttingen. 147pp.

Sokal, R.R. and Rohlf, F.J. 1981: Biometry. The Principles and Practice of Statistics in Biological Research. Second edition. W.H. Freeman and Co., San Franscisco, Calif., U.S.A. 859pp.

Sorensen, A.E. 1982: The roles of polyembryony and embryo viability in the genetic system of conifers. Evolution 36: 725-733

Swofford, D.L. and Selander, R.B. 1997: BIOSYS-2 : A Computer Program for the Analysis of Allelic Variation in Genetics. User's Manual. Department of Genetics ansd Development, Univ. of Illinois at Urbana-Champaign, Urbana, Illinois. USA

Wright, S. 1965: The interpretation of population structure by F-statistics with special regard to systems of mating. Evolution 19: 395-420

The pattern of genetic variation in *Pinus nigra* subspecies *pallasiana* natural populations from the Kazdağı and Bolkar mountains, Turkey: Implications for *in situ* gene conservation

Z. KAYA, A.A. TOLUN[1], B.N. ÇENGEL[2], E. VELIOĞLU[2] AND G.G. TOLUN[1]

[1]*Department of Biological Sciences, Middle East Technical University, 06531 Ankara, Turkey;* [2]*Forest Trees and Seeds Improvement Research Directorate, 06560 Gazi, Ankara, Turkey.*

Key words: *Pinus nigra* subsp. *pallasiana,* genetic diversity, isoenzymes, genetic distance, *in situ* gene conservation, gene management zone

Abstract: Eleven populations of black pine (*Pinus nigra* Arnold subspecies *pallasiana*) were studied at 17 polymorphic isoenzyme gene loci. The mean number of alleles per locus (A) was 1.6. Observed heterozygosity (H_{obs}) was the lowest in the Gürgendağ population from Kazdağı (0.119) and the highest in the Çamlıyayla population from Bolkar Mountains (0.287). The expected heterozygosities (H_{exp}) ranged from 0.296 (Asar, from Kazdağı) to 0.181 (Cehennemdere, from Bolkar Mountains). F_{ST} was 0.0592 for Kazdağı and 0.0358 for Bolkar Mountains. F_{IS} value was –0.25 in Bolkar Mountains and 0.42 in Kazdağı, indicating considerably less heterozygosity in Kazdağı populations than expected. There were significant deviations from Hardy-Weinberg expected frequencies in some of loci in both Kazdağı and Bolkar Mountain populations. The deviations in Kazdağı populations were due to excess of homozygotes while they were due to excess of heterozygous genotypes in Bolkar Mountains. The estimated genetic distances between populations ranged from 0.003 to 0.666. According to the genetic distance results, populations of Kazdağı form a genetically distinct group compare to the populations of Bolkar Mountains. Based on genetic diversity and distance statistics, three populations from Kazdağı (Asar, Mıhlıdere and Gürgendağ) and two populations from Bolkar Mountains (Ulukışla and Gülekdere) are suggested as potential *in situ* conservation sites.

Introduction

The Anatolian black pine (*Pinus nigra* Arnold subspecies *pallasiana* (Lamb.) Holmb.) is one of the subspecies of European black pine, occurring naturally as a widespread mid-elevation species in Taurus, western Anatolian and northern Anatolian Mountains of Turkey. The range in elevation varies from 250m to 1550m (*Figure 1*). This

Müller-Starck, G. and Schubert, R. (eds.). Genetic Response of Forest Systems to Changing Environmental Conditions. Vol. 70 (For. Sci.), 293-305 © 2001 Kluwer Academic Publishers, Dordrecht, Boston, London

subspecies covers a large area of more than 2 million hectares. It is an important timber species and the first choice for afforestation of the high Anatolian steppes (Kaya and Temerit 1994). *Pinus nigra* subspecies *pallasiana* is therefore of great importance in Turkish forestry (Koski and Antola 1993; Kaya and Temerit 1994). Knowledge of genetic variation in the species is necessary if genetically well-adapted seed sources of Anatolian black pine are to be identified and used to establish plantations. This type of information is also required to make sound decisions on how to protect the genetic diversity of the species and ensure sustainable forest management (Meffe and Carroll 1994; Kaya *et al.* 1997; Ledig 1998).

The Gene Management Zone (GMZ) concept was developed in the context of the project named "*In situ* Conservation of Plant Genetic Diversity in Turkey", which commenced with the collaboration of Turkish Ministry of Forestry, Ministry of Agriculture and Rural Affairs and Ministry of Environment in 1993. This project was supported by a special World Bank fund, Global Environment Facility (GEF). The project aimed to implement GMZ concept as a new type of *in situ* gene conservation program, allowing evolutionary processes to take place in populations of plant and animal species while protecting genetic resources of important tree species as well as wild relatives of crop species in selected pilot sites such as Kazdağı, Bolkar Mountains and Ceylanpınar State Farm. The Kazdağı Region was selected because it is considered a principal meeting place of Euro-Siberian and Mediterranean elements of flora. Surveys indicated that the area is very rich in forest tree species and wild relatives of fruit trees. The Bolkar Mountains Area is well known for its richness in endemic plants as well as in forest tree species. Taxonomic studies have shown that there are about 1700 plant species in the Bolkar Mountains and 350 of them are endemic to Turkey (Gemici 1994).

In both the Kazdağı and Bolkar Mountain regions, Anatolian black pine has a widespread distribution (e.g., wide range in elevation and habitats such as high vs. low elevation, south vs. north facing slopes, coastal vs. inland etc.). They are economically important timber species. Information on the distribution of genetic variation in some or whole range of natural distribution of the species can help develop useful cost-effective conservation strategies. Isoenzymes (multiple molecular forms of enzymes as neutral markers) can be used effectively to determine the magnitude and pattern of genetic variation in forest tree populations (Adams and Joly 1980a; Wendel and Weeden 1989) and be a useful tool for quick decision making process in conservation of genetic resources *in situ* (Ledig 1998).

To reduce the cost and increase the efficiency of *in situ* conservation sites in both regions, a majority of the sampled Anatolian black pine populations were selected from potential GMZ sites based on biodiversity, presence of other target species and accessibility for protection against fire etc. Thus, in this paper, isoenzyme data were pooled from previous studies (Tolun *et al.* 2000, Çengel *et al.* 2001) which were dealt with genetic control of isoenzymes in Anatolian black pine, using open-pollinated seeds, representing optimum (optimum elevation for species distribution, natural stands) and extreme habitats (extreme elevations for species, south facing slopes, highly degraded stands) for the species in the Kazdağı and Bolkar Mountain regions. General pattern of genetic variation in seven populations from Kazdağı and four populations from Bolkar Mountains were explored with data from 14 isoenzyme systems common to both regions. The implications of the results of this study for *in situ* conservation of genetic resources of the Anatolian black pine is also discussed.

Materials and Methods

Open-pollinated seeds of Anatolian black pine were collected from a total of 315 and 190 parent trees (half-sib families) from seven natural populations at Kazdağı and four natural populations in the Bolkar Mountains, respectively, in the fall of 1995 (*Figure 1*). Geographic and topographic descriptions of sampled populations as well as information on sampling of parent trees in each population and populations from each of Kazdağı and Bolkar Mountain Regions are provided by Tolun *et al.* 2000 and Çengel *et al.* 2001.

To study the plant isoenzymes, the procedures described in Conkle *et al.* (1982) were used. All the chemicals used in this study were purchased from Sigma Chemical Company, USA. Eight seeds of each of the randomly selected mother trees were analyzed. The seeds were germinated on a germination table for 2 to 4 days. When radicles reached 2-5 mm long, megagametophytes were separated from the embryos and placed in labeled grinding plates. The extraction and sample preparation procedures are as described in Conkle *et al.* (1982). The gels were run for four hours. For detailed information on sample preparation and isoenzymes procedures see Tolun *et al.* (2000) and Çengel *et al.* (2001). Data from the 14 enzyme systems which were common to both Kazdağı and Bolkar Mountain Regions were included for the data evaluation in the present study. These enzymes systems were; aconitase (ACO); acid phosphatase (ACP); diaphorase (DIA); glutamate dehydrogenase (GDH); glutamate oxaloacetate transaminase (GOT); isocitrate dehydrogenase (IDH); leucine aminopeptidase (LAP); malate dehydrogenase (MDH); menadione reductase (MNR); phosphoglucoisomerase (PGI); phosphoglucomutase (PGM); mannose phosphate isomerase, (PMI); shikimate dehydrogenase (SKDH); superoxide dismutase (SOD).

In order to determine the amount of genetic diversity in a standardized way, allelic richness (A), proportion of polymorphic loci (P), expected (H_e) and observed (H_{obs}) heterozygosities were estimated for the populations (Nei, 1987). Chi Square tests for heterogeneity in allele frequencies among the populations for all polymorphic loci were carried out at $p<0.01$. Chi square tests for segregation of heterozygous mother trees in each population of both regions were also done at $p<0.01$ for all polymorphic loci to estimate whether their segregation was according to the Hardy –Weinberg expectations.

The Wright's F-Statistics using the measure of heterozygosity defines three level of inbreeding such as FIT = Fixation index within population, FIS = Fixation index over the total populations and FST = Reduction in fixation index due differences among populations. These are essential in determining the presence of any deviation from the Hardy-Weinberg expectations in populations and genetic differentiation of populations.

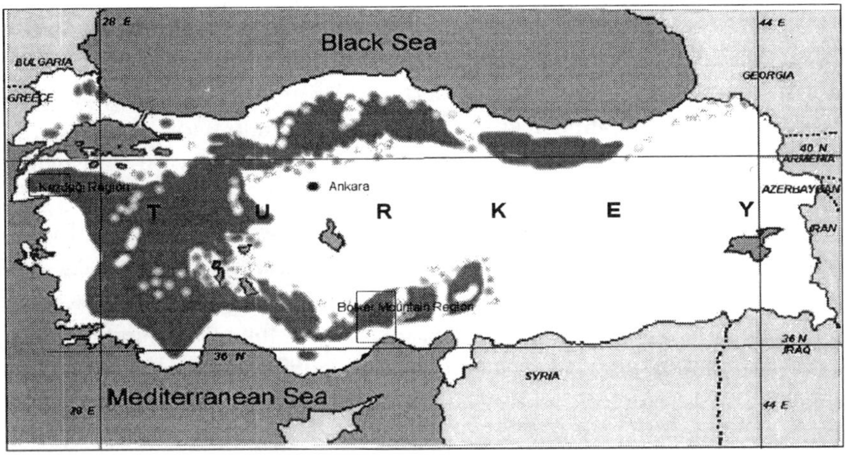

Figure 1. The map showing the natural distribution of *Pinus nigra* Arnold subsp. *pallasiana* in Turkey and the regions (Kazdağı and Bolkar Mountain) where populations were sampled.

The Wright's F-Statistics were estimated for each locus and overall loci including all populations according to Nei (1987).The estimates of Nei's (1972) standard genetic identity (*I*) and standard genetic distance (*D*), unbiased for sample size (Nei 1978) for all pair-wise population comparisons were calculated to show the genetic relationships between studied populations. Data analysis for estimating genetic parameters were carried out by using POPGENE (Yeh *et al.* 1997) and BIOSYS-1 (Swofford and Selander 1989) computer softwares.

Results

Twenty-one zones of activity were resolved consistently for the 14 enzyme systems assayed. For banding patterns see Tolun *et al.* (2000) and Çengel *et al.* (2001). Variants in each zone of activity segregated according to the expected 1:1 ratio in pooled megagametophyte samples from heterozygous mother trees, indicating single locus control (Nazlier 1998; Tolun 1998).

Heterogeneity ($p<0.01$) in allele frequencies among the populations was detected at 17 loci (for frequency distributions see Tolun *et al.* 2000 and Çengel *et al.* 2001). ACO, GDH, GOT1 and PMI1 were monomorphic. Most loci in both regions showed significant deviations from expected Hardy-Weinberg frequencies. In the Kazdağı populations, there were more homozygous genotypes than expected while in the Bolkar Mountain populations, there were more heterozygotes than expected. Also, for almost all loci, Chi square values for heterogeneity tests in both Kadağı and Bolkar Mountain populations were statistically significant at $p<0.01$ (data available on request).

Mean number of alleles per locus varied between 1.5 in Çamlıyayla and Gülekdere populations to 1.7 in Eybekli, Asar, Gürgendağ and Mıhlıdere populations. The average allelic richness for Kazdağı Region was 1.66, for Bolkar Mountains was 1.55 and for all populations was 1.62. No significant difference in average number of alleles per locus appeared within regions and among populations considering the standard errors of estimates (*Table 1*). The percentage of polymorphic loci varied from 42.9% (Çamlıyayla and Gülekdere) to 61.9% (Eybekli, Asar, Gürgendağ and Mıhlıdere). The average percentage of polymorphic loci (using 0.99 criterion) was determined as 59.84% for Kazdağı Region, 46.45% for Bolkar Mountains and 54.97% for overall populations (*Table 1*). Observed heterozygosity was the highest (0.184) in Asar and Katrandağ and the lowest (0.119) in Gürgendağ populations from Kazdağı. For Bolkar Mountains, the highest observed heterozygosity was in Çamlıyayla (0.287) and the lowest was in Cehennemdere (0.212). When all the populations were considered, the highest was in Çamlıyayla (0.287) and the lowest observed heterozygosity was in Gürgendağ (0.119). The expected heterozygosities ranged between 0.296 (in Asar) and 0.243 (in Kapıdağ) in Kazdağı Region and between 0.181 (in Cehennemdere) and 0.224 (in Çamlıyayla) in Bolkar Mountains. When all populations were considered, the highest expected heterozygosity was in Asar (0.296) and the lowest was in Cehennemdere (0.181). Mean heterozygosity for all populations was 0.191 for observed and 0.248 for expected. There was a difference between observed and expected heterozygosities which suggests that a considerable amount of inbreeding may be taking place in Anatolian black pine populations (*Table 1*).

In Kazdağı populations of Anatolian black pine, there were less heterozygotes than expected in almost all 21 loci studied. In Bolkar Mountain populations, deviations of observed heterozygosities from expected were not significant in most of 21 loci (data not given here).

Genetic identity and distance estimates between studied populations revealed that Eybekli-Asar (in Kazdağı Region) and Çamlıyayla-Gülekdere (in Bolkar Mountains) populations were the most genetically related populations (genetic identities being 0.996 and 0.997, respectively). The genetically least related population pairs were Katrandağ-Ulukışla and Kapıdağ-Gülekdere (genetic identity = 0.514). Estimated genetic distance indicated that populations from two geographically distant regions were also genetically quite different (genetic distance=0.574; data available on request). To show the differentiation and grouping of studied populations graphically, a cluster analysis was performed using *Unweighted Pair Group Method* with Arithmetic Means (Nei, 1978). In cluster analysis, two main branches were observed (*Figure 2*). The first main branch covered the Kazdağı Region, including Eybekli, Asar, Kalkım and Gürgendağ in one branch and Katrandağ, Kapıdağ and Mıhlıdere in the other branch. The second main branch covered the Bolkar Mountains (*Figure 2*).

Table 1. Mean number of alleles per locus, proportion of polymorphic loci, mean observed and expected heterozygosities in the Anatolian black pine populations.

Populations	Sample size	Mean Number of Alleles (A)	Poly-morphic Loci (P)	Mean Heterozygosity (H)	
Kazdağı Region				Observed (H_{obs})	Expected (H_e)
Eybekli	45	1.7 ± 0.1	61.9 %	0.161 ± 0.041	0.292 ± 0.053
Asar	45	1.7 ± 0.1	61.9	0.184 ± 0.048	0.296 ± 0.053
Katrandağ	45	1.6 ± 0.1	57.1	0.184 ± 0.051	0.259 ± 0.053
Kalkım	45	1.6 ± 0.1	57.1	0.131 ± 0.037	0.269 ± 0.054
Gürgendağ	45	1.7 ± 0.1	61.9	0.119 ± 0.037	0.269 ± 0.053
Kapıdağ	45	1.6 ± 0.1	57.1	0.149 ± 0.042	0.243 ± 0.051
Mıhlıdere	45	1.7 ± 0.1	61.9	0.144 ± 0.043	0.265 ± 0.052
Mean		*1.66*	*59.84*	*0.153*	*0.270*
Bolkar Mountains					
Çamlıyayla	48	1.5 ± 0.1	42.9	0.287 ± 0.084	0.224 ± 0.058
Cehennemdere	42	1.6 ± 0.1	47.6	0.212 ± 0.070	0.181 ± 0.049
Ulukışla	50	1.6 ± 0.1	52.4	0.258 ± 0.069	0.212 ± 0.052
Gülekdere	50	1.5 ± 0.1	42.9	0.274 ± 0.083	0.213 ± 0.056
Mean		*1.55*	*46.45*	*0.258*	*0.208*
Overall Mean		***1.62***	***54.97***	***0.191***	***0.248***

F_{ST} , equal to Nei's G_{ST} , was 0.0592 for Kazdağı and 0.0358 for Bolkar Mountains, meaning that 5.92% and 3.58% of total variation were between populations (*Table 2*).

Discussion

Within Population Diversity

The results of this study indicated that high genetic diversity within Bolkar Mountain populations and relatively less genetic diversity in Kazdağı populations of Anatolian black pine were present. Nevertheless, mean expected heterozygosities were considerably higher than the average value reported pines (Hamrick *et al.* 1992). The mean number of alleles per locus and the percent of polymorphic loci did not vary significantly among the populations studied. These findings can be explained by the fact that the populations within regions are located within gene flow distance. The amount of gene flow between populations should vary depending on prevailing wind direction during pollen shed, timing of pollen dispersal and the position of the population in the species range in the Kazdağı Region and Bolkar Mountains (Tolun *et al* 2000, Çengel *et al.* 2001). Nicolić and Tucić (1983) and Scaltsoyiannes *et al.* (1994) however, reported relatively higher polymorphism values than ours (54.97%), that is 66.1% and 70% polymorphism, respectively.

This discrepancy could be explained by the differences in the enzyme systems used in their studies and ours. Since the percent of polymorphic loci is dependent on the enzyme systems studied, it is not a reliable measure for genetic diversity. In fact, Scaltsoyiannes *et al.* (1994) studied 10 enzyme systems while Nicolić and Tucić (1983) studied only four enzyme systems, compared to 14 enzyme systems used in this study. Both the pattern and magnitude of genetic diversity and determination of deviations from Hardy-Weinberg expectations are also important. The segregation tests for heterozygous mother trees of both Bolkar and Kazdağı Region populations showed significant deviations from Hardy–Weinberg genotypic expected frequencies for most loci. Deviations from expectations were also apparent from Wright's F-Statistics.

Deviations were observed for all of the loci for Kazdağı Region and Bolkar Mountains, separately. For all loci considered, the fixation index within subpopulations (F_{IS}) was 0.4244 for Kazdağı Region, meaning that homozygotes were 42.44% higher than expected. For Bolkar Mountains, F_{IS} was estimated as –0.2592, meaning 25.92% excess of heterozygotes than expected. Deviations from the Hardy-Weinberg expectations suggest that there were more homozygous genotypes than heterozygous ones in Kazdağı while there were more heterozygous genotypes than the homozygous ones in Bolkar Mountain populations. These deviations indicate that there may be some evolutionary forces favoring heterozygous in Bolkar Mountain populations while inbreeding may be reducing the heterozygous in Kazdağı populations. Kazdağı populations are located in a relatively small area (75000 ha) and most of these populations form mixed-forests with broad-leaved species such as chestnut, beech and fir. These factors may have favored the mating between relatives, resulting in observation of high rate of homozygotes in Kazdağı populations (Çengel *et al.* 2001). Bolkar Mountain populations are distantly located and form pure black pine forests in diverse microhabitats which may be the reason for observation of high rate heterozygosity in these populations (Tolun *et al.* 2000).

The mean F_{ST} value was very low (for Kazdağı Region, 0.0592 and for Bolkar Mountains, 0.0358) indicating that there was little genetic differentiation among populations. This was also apparent from the low-estimated genetic distances between populations of the regions when the regions were considered separately. But the genetic distance between regions was quite high indicating high regional differentiation of populations. Cluster analysis (dendrogram) based on Nei's genetic distances also showed that Kazdağı Region and Bolkar Mountains were located in separate clusters. The Mıhlıdere population was the most distantly clustered one compared to other populations of Kazdağı Region.

Table 2. F-Statistics for the Kazdağı Region and Bolkar Mountains.

	Kazdağı		
	F_{IS}	**F_{IT}**	**F_{ST}**
Mean	*0.4244*	*0.4585*	*0.0592*
	Bolkar Mountains		
	F_{IS}	**F_{IT}**	**F_{ST}**
Mean	*-0.2592*	*-0.2141*	*0.0358*

F_{IS}: The fixation index within subpopulation; F_{IT}: Fixation index over total population; F_{ST}: the reduction in fixation index due to differences among subpopulations.

These findings are not surprising since Mıhlıdere population is located in extreme habitats (e.g., isolated, low elevation, dry environment and, shallow soil), at the western edge of the species range in the Kazdağı Region. On the other hand, Eybekli and Asar populations, which are in the same cluster, are located at the eastern edge in the Kazdağı Region. Considerable amount of gene flow between these populations probably occurs considering the topography of the area, prevailing wind direction and physical distance between these populations. In Bolkar Mountains, Cehennemdere was strongly separated from the other three populations. Çamlıyayla and Gülekdere were the most closely related ones.

Geographic Diversity and in situ Gene conservation

Protection of all studied Anatolian black pine populations in the Kazdağı Region and Bolkar Mountains may not be economically feasible. The number and size of GMZs as an *in situ* reserve could not be determined by the genetic data presented here alone since there will be other biological, economical and administrative issues which should be considered during decision making. However, based on present genetic data, we could point out some potential populations as GMZ sites in both Kazdağı and Bolkar Regions to help forest managers and conservation biologists in their decision making.

In Kazdağı Region, considering the genetic distances between populations and clustering pattern and genetic diversity statistics of populations (heterozygosity and polymorphic loci), Asar (or Eybekli), Mıhlıdere and Gürgendağ populations of Anatolian black pine could be suggested as potential GMZ populations. These populations are the representatives of the cluster groups formed in cluster analysis. Asar is an extreme population for black pine due to its altitude (300 meters); and it has the highest heterozygosity and percent polymorphic loci (Çengel *et al.*2001). The Eybekli, Mıhlıdere and Gürgendağ population sites are also rich in plant biodiversity, not only tree species, but also other plant species (Personal Communication, Nihal Özel, Aegean Forest Research Institute, June 1998). In Bolkar Regions, again considering small genetic distances between populations and genetic diversity statistics, Gülekdere and Ulukışla populations could be taken into consideration as *in situ* reserves. Ulukışla population has high polymorphic loc and located at the edge of the Bolkar Mountains Tolun *et al.* 2000). Gene flow from the other populations is probably low. Therefore, it is reasonable that Ulukışla can be considered as one of the populations for *in situ* conservation area in the Bolkar Mountain Region.

However, for the final decision on Gene Management Zone selection in the Kazdağı Region and Bolkar Mountains as an *in situ* reserve, genetic diversity of a species is not the unique criterion. Other factors such as presence of other economically important species and degree of biological diversity in the site, accessibility and suitability of the area for *in situ* preservation and protection should be also taken into consideration

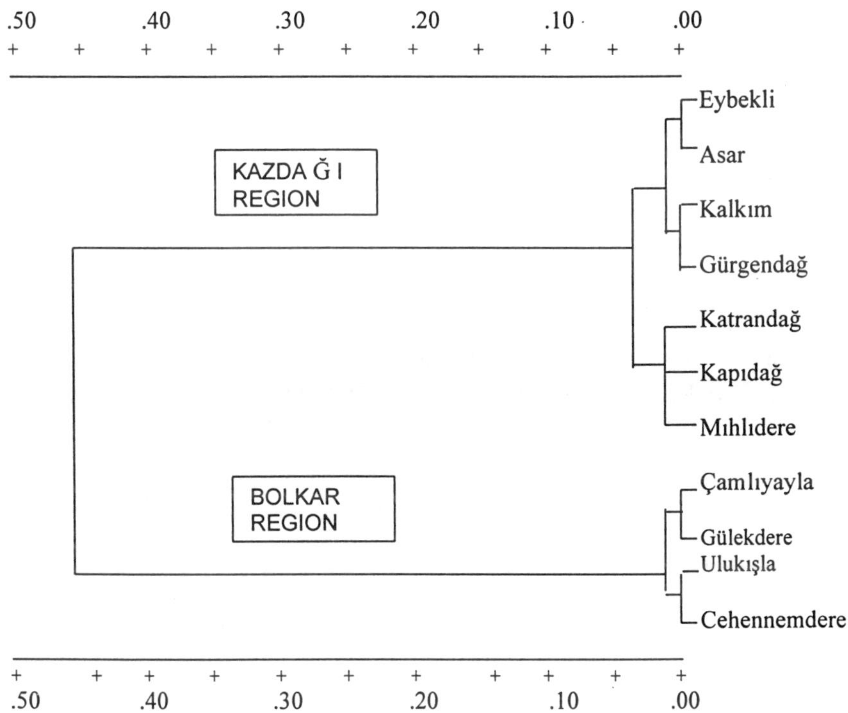

Figure 2. Dendrogram, constructed using Nei's (1978) Genetic Distance Statistics, showing the clusters of Anatolian black pine populations in Turkey

Acknowledgements

This study was supported by the Ministry of Forestry, The Republic of Turkey as a part of "*In situ* Conservation of Plant Genetic Diversity" project which was funded by the World Bank (Project #: GEF-28632-TU). We are grateful to Drs. D. J. Raynal and two unknown reviewers for their critical reviews and comments on the paper.

References

Adams, W. T. and Joly, R. J. 1980a: Genetics of allozyme variants in loblolly pine. The Journal of Heredity 71: 33-40

Conkle, M. T., Hodgskiss, P. D., Nunnally, L. B. and Hunter, S. C. 1982: Starch gel electrophoresis of conifer seeds: a laboratory manual. Pacific Southwest Forest and Range Experiment Station, California

Çengel , Burcu (Nazlıer), E. Velioglu, A.A. Tolun and Kaya, Z. 2001: Pattern and Magnitude of Genetic Diversity in *Pinus nigra* Arnold subspecies *pallasiana* Populations from Kazdağı (in press) Silvae Genetica.

Gemici, Y. 1994: Bolkar dağlarının(Orta Toroslar)flora ve vejetasyonu üzerine genel bilgiler: Turkish Journal of Botany 18(2):81-90 (in Turkish)

Hamrick, J. L., Godt, M. W. and Sherman-Broyles, L. 1992: Factors influencing levels of genetic diversity in woody plant species. New Forests 6: 95-124

Kaya, Z., Kün E. and Güner A. 1997: National plan for *in-situ* conservation of plant genetic diversity in Turkey. Milli Eğitim Basımevi, İstanbul

Kaya, Z. and Temerit, A. 1994: Genetic structure of marginally located *Pinus nigra* var *pallasiana* populations in central Turkey. Silvae Genetica 43: 272-276

Koski, V. and Antola, J. 1993: National Tree Breeding and Seed Production Program for Turkey. 1994-2003

Ledig, F. T. 1988: The conservation of diversity in forest trees. Bioscience 38: 471-478

Ledig, F. T. 1998: Genetic diversity in tree species: With special reference to conservation in Turkey and the eastern Mediterranean. In: The Proceedings of International Symposium on *In-situ* Conservation of Plant Genetic Diversity, edited by Zencirci, N., Kaya, Z., Anikster, Y., Adams, W. T. Central Research Institute for Field Crops, Turkey 231-247

Meffe, G. K. and Carroll, C. R.1994: Principles of Conservation Biology. Sinauer Associates, Inc.

Nazlier, B. 1998: The pattern and magnitude of genetic variation in *Pinus nigra* subsp. *pallasiana* populations sampled in Kazdağı. . M. Sci. Thesis, Middle East Technical University, Department of Biology, Ankara

Nei, M. 1972: Genetic distance between populations. Amer. Natur. 106: 83-292

Nei, M. 1978: Estimation of average heterozygosity and genetic distance from a small number of individuals. Genetics 89: 583-590

Nei, M. 1987: Molecular Evolutionary Genetics. Columbia University Press, New York.

Nicolić, D. and Tucić N. 1983: Isoenzyme variation within and among populations of European Black pine (*Pinus nigra* Arnold). Silvae Genetica 32: 80-89

Scaltsoyiannes, A., Rohr, R., Panetsos, K. P. and Tsaktsira, M. 1994: Allozyme frequency distributions in five European populations of Black pine (*Pinus nigra* Arnold). Silvae Genetica 43: 20-30

Swofford, D. L. and Selander, R. B. 1989: BIOSYS-1, A computer program for the analysis for allelic variation in population genetics and biochemical systematics Release 1.7

Tolun, A.A., Velioglu, E., Cengel, B. (Nazlıer) and Kaya. Z. 2000: Genetic Structure of Black pine (*Pinus nigra* Arnold subspecies *pallasiana*) Populations Sampled from the Bolkar Mountains. Silvae Genetica 49 (3) : 113-119

Tolun, A. A. 1998: Genetic structure of black pine (*Pinus nigra* Arn subsp. *pallasiana*) populations sampled from the Bolkar Mountains. M. Sci. Thesis, Middle East Technical University, Department of Biology, Ankara

Wendel, J. F. and Weeden, N. F. 1989: Visualization and Interpretation of Plant Isozymes. In Isozymes in Plant Biology edited by Soltis, D.E. and Soltis, P.S. Advances in Plant Scienecs Series Vol. 4, Dioscorides Press, Portland, Oregon

Yeh, F. C., Rongcai, Y. and Boyle, T. 1997: Population Genetic Analysis POPGENE Version 1.21 (32-Bit): A Quick User's Guide. A Joint Project Development by University Alberta, Canada and Center for International Forestry Research, USA

Genetic variation of silver fir (*Abies alba*) in unevenaged forests („Plenter“ forest) in comparison with evenaged forests (Altersklassenwald)

M. KONNERT and E. HUSSENDÖRFER

[1]*Bavarian Institute for Forest Seeding and Planting, D-83317 Teisendorf, Germany;* [2]*Forest Research Station Baden-Württemberg, Department of Botany and Site Ecology, D-79100 Freiburg, Germany*

Key words: genetic variation, *Abies alba*, isozymes, silviculture, sustainable forest management

Abstract: In order to elucidate the effects of management systems on the existing potential of biodiversity the genetic variation of silver fir (*Abies alba*) in unevenaged forests ("Plenter" forests) and evenaged forests (Altersklassenwald) was investigated by means of isozyme gene-markers at 14 to 16 gene loci. In total about 3300 adult silver firs in 9 „Plenter" forests and 16 evenaged forests in Switzerland, Baden-Württemberg (Germany) and Bavaria (Germany) were examined. In two of the „Plenter" forests complete inventories of all fir trees were carried out; in the other stands a partial sampling was performed. In four stands also individuals from the natural regeneration were studied genetically.The genetic diversity and heterozygosity was, with only few exceptions, lower in the unevenaged forests than in the evenaged forests. Possible reasons for these results are the uniform adaptive processes under the comparatively homogeneous site conditions of the „Plenter" forests caused by almost continuous shading effects in time and space. According to this hypothesis, silver firs of „Plenter" forests can be regarded to be well adapted to specific environmental conditions of this forest type. However the genetic adaptability of silver fir in unevenaged forests seems to be lower as in evenaged forests. In respect to the genetic multiplicity, diversity and heterozygosity there was little difference between the natural regeneration and the adult trees in the „Plenter" forest as well as in the evenaged forest. However, significant differences in genetic structures were found in „Plenter" forests at some gene loci. These findings are supposed to be mainly dependent on particularities of the reproduction system and adaptive processes in this stands. The consequences of these results are discussed in respect to genetically sustainable forest management practices.

Introduction

The risk of restricting or even losing the biological resources in forest ecosystems led to the expansion of the term „sustained yield forestry" to include the dimension of „ecological sustainability". The expanded term „sustainable forest development" implies a prospective development of forest ecosystems at all levels from the genetic level to the ecosystem level. Important aspects in these regards are the protection, the maintenance and the improvement of the biodiversity in forest ecosystems (Anonymus 1993, 1998; Müller-Starck 1996). In order to carry this out one must assess the effects of management systems on the existing potential of biodiversity. In this paper are presented the results of investigations concerning the above mentioned aspects in respect to two forest management systems: the unevenaged forest („Plenter" forest) and the evenaged forest („Altersklassenwald"). Unevenaged forest management is a production system distinguished by its principles of perennation and sustained yield of the forest over small areas. Because of the continual renewal of the stand over time it is generally assumed that this type of silvicultural management system would be most suitable in maintaining the genetic variation and adaptability. In the age class management scheme the regeneration occurs only in a limited period of the production process (discontinuous), which, depending upon the type of regeneration can be of varying length. In the light of population genetic a slow regeneration in small patches is considered most advantageous (Hattemer *et al.* 1993, chapter 18.2; Müller 1990a, 1990b, Geburek and Thurner 1993, Rotach 1994).

To verify these hypothesis silver fir (*Abies alba)* was selected and the genetic structures were compared between silver firs of „Plenter" forest and evenaged forests. In addition, genetic structures of adult trees and their natural regeneration were investigated.

Silver fir is a shade tolerant tree species, „feeling well only in unevenaged forest structures" as described by Waldherr (1996), and showing high vitality and volume yield there. Compared to other tree species, silver fir shows less genetic variation, smaller degrees of heterozygosity, less gene variants, more gene loci with minorpolimorphism or fixation (e.g. Bergmann *et al.* 1990, Bergmann 1991). Therefore it seems more important not only to stop the decline of silver fir in forest stands by an adapted silviculture to this species, but also to aim at the preservation of the whole genetic potential (e.g. Hussendörfer 1997).

Material and Methods

Material

The genetic investigations were carried out in 18 stands in Switzerland, 6 stands in Baden-Württemberg and two stands in Bavaria. Included were 9 „Plenter" forests and 17 evenaged forests („Altersklassenwald"). Description of the stands and number of trees investigated are given in Table 1. In most of the stands the selection of the sample trees was carried out using a systematic grid-sampling (generally 100 trees per stand). In three stands (Nr. 21, 22, 23) the number of trees examined per stand was increased. This was done to obtain a better proximation of the genetic variation, especially of rare variants.

Table 1. Survey of site and stand characteristics of silver fir (*Abies alba*) populations. P = „Plenter“ forest, A = evenaged forest

Nr.	Population	Sample size	Forest region	Percentage of silver fir (vol %)	Forest type
1	Le Brassus	100	Jura (CH)	40	P
2	Bretonniéres	100	Jura (CH)	50	A
3	Couvet-South*	100	Jura (CH)	30	P
4	Couvet-North	100	Jura (CH)	70	P
5	Neuenburg	100	Jura (CH)	40	A
6	Martigny	100	Alps (CH)	60	A
7	Gstaad	100	Foothills of the Alps (CH)	40	A
8	Ochsenboden*	100	Alps (CH)	60	A
9	Leuk	100	Alps (CH)	40	A
10	Signau	100	Foothills of the Alps (CH)	60	P
11	Zofingen	100	Middle land (CH)	50	A
12	Schwarzenberg	100	Foothills of the Alps (CH)	60	P
13	Vergeletto	100	Southern Alps (CH)	50	P
14	Prato	100	Southern Alps (CH)	40	A
15	Cavagnano	100	Southern Alps (CH)	30	A
16	Präz	100	Alps (CH)	40	A
17	Bondo	100	Alps (CH)	50	A
18	Tschlin	100	Alps (CH)	25	A
19	Spänplatz	100	Southern Black Forest (BW)	35	A
20	Schwarzahalden	100	Southern Black Forest (BW)	35	P
21	Bad Herrenalb	200	Northern Black Forest (BW)	55	A
22	Welzheim	200	Swabian-Frankonian Forest (BW)	40	A
23	Pforzheim	200	Northern Black Forest (BW)	35	A
24	PLW-16	366	Northern Black Forest (BW)	30	P
25	Kreuzberg 1*	326	Bavarian Forest (Bay)	36	P
26	Kreuzberg 2*	43	Bavarian Forest (Bay)	20	A

*additional natural regeneration

In the evenaged forest Nr. 26 („Kreuzberg 2“) all old trees of silver fir within a game fence of approximately 2.5 ha fenced as well as some individuals outside the fence in the imidiate vicinity (all together 43 old silver firs) were investigated. 183 randomly selected individuals from the oncoming natural regeneration were additionally sampled. In two of the nine „Plenter" forests nearly complete inventories of all fir trees with D.b.h. > 7 cm were carried out: In the stand Nr. 24 „PLW-16" 366 out of 430 silver fir trees were sampled, in the stand Nr. 25 „Kreuzberg 1" 326 trees out of 350. This two stands are long-term research plots of the Forest Research Station (FVA) Baden-Württemberg (PLW-16) and the Institute for Forest Growth of the Forestry Faculty at the Technical University of Munich („Kreuzberg 1"). Thus there was detailed data on the growth and development of the stands available. In the Plenter forest Nr. 25 „Kreuzberg 1" the natural regeneration was also sampled (192 randomly selected individuals with D.b.h.< 7 cm).

Of the stands registered as being „Plenter" forests (P in Table 1), the stands Nr. 13 „Vergeletto" and Nr. 20 „Schwarzahalden" are actually out of management. It is assumed that the „plenterlike" structure of these stands is the result of anthropogenic factors in the past as well as of specific site conditions (rock slide), which might have favored the natural development of a „plenterlike" structure in the stands (Schütz 1989).

Methods

Isoenzymes were used as gene-markers in determining the genetic structures of the silver fir trees. Enzymes were extracted from dormant buds and separated by horizontal starch-gel zone-electrophoresis. The following enzyme systems were investigated: aminopeptidase (AP, EC 3.4.11.1), glutamate oxalacetate transaminase (AAT, EC 2.6.1.1), isocitrate dehydrogenase (IDH, EC 1.1.1.42), malate dehydrogenase (MDH, EC 1.1.1.37), menadione reductase (MNR, EC 1.6.99.3), NADH-dehydrogenase (NADH, EC 1.6.99.3), 6-phosphogluconate dehydrogenase (6-PGDH, EC 1.1.1.44), phosphoglucomutase (PGM, EC 2.7.5.1) and shikimate dehydrogenase (SKDH, EC 1.1.1.25). For stands Nr. 1 – Nr. 24 fourteen gene loci were analyzed, for stands Nr. 25 and Nr. 26 sixteen loci. The gene loci investigated were: AP-A, AP-B, AP-C, AP-D, AAT-A, AAT-B, AAT-C, IDH-A, IDH-B, MDH-A, MNR-B, NADH-A, 6PGDH-A, 6PGDH-B und PGM-A, PGM-B and SKDH-A. Methodical details can be found under Hussendörfer *et al.* (1995) and Konnert and Maurer (1995).

To quantify the genetic variation within the different populations (e.g. adult trees, natural regeneration) the following parameters were determined: genetic multiplicity (mean number of alleles per locus, A_L), genetic diversity (v), gametic multilocus-diversity (v_{gam}), actual heterozygosity (H_a). The pairwise comparison of the genetic structures between collectives was done by computing the Gregorius genetic distances (Gregorius 1974). A detailed description of these parameters can be found under Müller-Starck and Gregorius (1986) and Hattemer *et al.* (1993). The hypothesis in respect to homogeneity of the observed frequencies of alleles (e.g. genotypes) were tested by means of G-Test and contingency table (Lozán 1992), with significance levels of $\alpha = 0.1$ (*), $\alpha = 0.01$ (**) and $\alpha = 0.001$ (***). The computation of the different parameters and the statistical analysis was performed with the GSED-program (Gillet, Univ. of Göttingen, 1994) and the program „MacGEN-TM Populationsgenetik with SAS" (Stauber and Hertel, BFH Waldsieversdorf, 1995).

Results

Comparison of „Plenter" forests and evenaged forests from different regions

In Table 2 are listed the values of genetic multiplicity, diversity and heterozygosity for the populations from Switzerland (Nr. 1 to Nr. 18) and Baden-Württemberg (Nr. 19 to Nr. 24) analyzed in the laboratory of Freiburg.

The genetic multiplicity (mean number of alleles per locus, A_L) range from 2.00 to 2.57, with an average value for all population of 2.19. A_L-values above or under the average are found in „Plenter" forests as well as in evenaged forests. Taking into account only alleles with frequencies higher than 5 %, the genetic multiplicity A_{L5} vary between 1.50 and 1.93, with an average value of 1.69. A_{L5}-values above or under the average are found also in populations managed in both ways. The lowest A_{L5}-value (1.50) was registrated in two „Plenter" forests, the highest value (1.86) in the evenaged forests. In the populations Nr. 4 and Nr. 24 („Plenter" forests) one third of all alleles have frequencies under 5 % ($A_L = 2.29$; $A_{L5} = 1.5$). Which means, that there is a high risk for one third of all alleles to get lost in genetic drift processes.

The mean values of observed heterozygosity, H_a, vary from 15.9 % (population Nr. 20) to 22.2 % (population Nr. 5); which means 6.3 percent total. The average value for all populations amounts to 19.2 %. Only one of the 8 „Plenter" forests owns a observed heterozygosity–value above the average, all the others have even significant lower values (e.g. populations Nr. 20, 10, 13, 24). Among the 16 evenaged forests 11 have H_a-values above the average. For some of them the values are even significantly higher (e.g. populations Nr. 5, 14, 15, 8).

The genetic diversity, v (effective number of alleles per locus), range from 1.20 (populations Nr. 20) to 1.29 (populations Nr. 8, 14, 17); this provides a ratio of 1:1.58. The average for all populations is 1.26. Only two of the eight „Plenter" forests (Nr. 2 and Nr. 4) have diversity values above the average. For the other six the values of v are partly significantly lower as the average. At the other side the evenaged forests Nr. 8, 14 and 17 have significant higher values.

The gametic multilocus-diversity, $v_{gam,}$ vary between 18.54 (population Nr. 20) and 65.96 (population Nr. 14); this provides a ratio of 1:3.56. The mean

Table 2. Levels of genetic variation of silver fir (*Abies alba*) within the „Plenter" forests and evenaged forests. P = „Plenter" forest, A = evenaged forest

Stand	Multiplicity		Heterozygosity	Diversity		Forest
Nr.	A_L	A_{L5}	$H_{a\,(\%)}$	v	v_{gam}	type
1	**2.21**	**1.71**	*17.2*	1.22	22.13	P
2	**2.21**	*1.64*	*18.8*	1.25	33.39	A
3	**2.21**	***1.71***	*18.8*	1.26	35.74	P
4	**2.29**	*1.50*	**19.6**	1.26	34.80	P
5	*2.14*	**1.71**	**22.2**	1.27	45.34	A
6	*2.07*	*1.57*	**19.6**	1.27	41.49	A
7	*2.00*	*1.64*	**19.9**	1.26	40.91	A
8	**2.29**	*1.57*	**21.4**	1,29	56.79	A
9	*2.14*	**1.71**	*18.2*	1.24	30.29	A
10	**2.21**	*1.64*	*16.3*	1.22	23.98	P
11	*2.00*	**1.71**	**19.7**	1.26	37.58	A
12	*2.14*	**1.71**	*18.6*	1.24	28.43	P
13	*2.14*	*1.57*	*16.7*	1.23	28.90	P
14	*2.14*	*1.64*	**21.8**	1.29	65.96	A
15	*2.07*	**1.79**	**21.7**	1.28	63.48	A
16	**2.36**	**1.71**	*18.6*	1.28	55.67	A
17	**2.21**	**1.79**	**19.9**	1.29	59,05	A
18	*2.00*	**1.71**	*18.6*	1.25	48.14	A
19	*2.29*	**1.79**	*19.1*	1.26	38.53	A
20	*2.14*	**1.71**	*15.9*	1.20	18.45	P
21	*2.14*	*1.64*	**19.5**	1.27	45.80	A
22	**2.57**	**1.86**	**19.8**	1.27	42.69	A
23	**2.36**	**1.93**	**21.0**	1.28	51.55	A
24	**2.29**	*1.50*	*16.7*	1.22	21.56	P
Mean	***2.19***	***1.69***	***19.2***	***1.26***	***40.44***	

value for all populations amounts to 40.44. Among the four „Plenter" forests there is none with a v_{gam} value above the average. In contrary there are 12 evenaged forests with gametic diversity values higher then this mean value.

Comparison of the genetic variation in a „Plenter" forest and an evenaged forest within the same ecodistrict

The „Plenter" forest Nr. 25 „Kreuzberg 1" and the evenaged forest Nr. 26 „Kreuzberg 2" are both located in the communal forest „Kreuzberg" within the Forest District Freyung, ecodistrict or growing district 11.3 "Interior Bavarian Forest", in the near neighbourhood to each other. Very similar site conditions can be assumed. In both stands silver fir is considered autochthonous. A comparison of the genetic structures (type and frequency of gene variants) of both stands at the investigated gene loci with the variation patterns known for silver fir (i.e. Schroeder 1988; Breitenbach-Dorfer *et al.* 1992; Bergmann, 1993; Konnert 1993; Konnert and Bergmann 1995) clearly indicates their autochthony. A few years ago a group selection cutting was carried out in the evenaged stand "Kreuzberg 2". The natural regeneration is well established.

A comparison of the variation parameter values contained in Table 3 show that the genetic multiplicity (A_L-values) taking into account all alleles is considerably higher in the „Plenter" forest "Kreuzberg 1" than in the evenaged forest "Kreuzberg 2". The number of alleles with frequencies over 5 % (A_{L5}-values) is almost alike in both stands. That means that in the „Plenter" forest there are more rare alleles with frequencies less than 5 %. These alleles are exposed to a high risk of loss if drift processes occurs.

Table 3. Levels of genetic variation of silver fir (*Abies alba*) within the „Plenter" forest and the evenaged forest located in the same ecodistrict

Genetic parameter	Communal forest "Kreuzberg"	
	"Plenter" forest	evenaged forest
A_L	2.35	1.88
A_{L5}	1.59	1.65
H_a (%)	18.4	18.7
v	1.23	1.25
v_{gam}	70.0	88.0

The mean values of observed heterozygosity in the two stands are very similar. In respect to the diversity this comparison repeats the tendency indicated under 3.1 as well: slightly smaller values for the „Plenter" forest as compared to the evenaged forest.

Genetic variation in the natural regeneration of „Plenter" forests and evenaged forests in comparison with the adult population

In the populations Nr. 3 „Couvet-South" (P), Nr. 8 „Ochsenboden" (A), Nr. 25 „Kreuzberg 1" (P) and Nr. 26 „Kreuzberg 2" (A) besides the older silver fir trees also individuals from the natural regeneration (in „Plenter" forests individuals with D.b.h.< 7 cm) distributed over the entire area were studied genetically. In Table 4 the values for genetic multiplicity, diversity and heterozygosity are recorded for the regeneration and the collective of old firs in evenaged forests respectively older firs (lower, middle and upper layer) in the „Plenter" forests. It shows that the differences for all of the four parameters of variation in all stands are low.

Table 4. Genetic variation in the adult trees and the natural regeneration in 2 „Plenter" forests and 2 evenaged forests

Stand designation	Developmental stage	Genetic multiplicity A_L	Diversity v	v_{gam}	Heterozygosity H_a
Kreuzberg 1	adult	2.35	1.23	70.4	18.4
P	regeneration	2.35	1.25	94.8	18.8
Kreuzberg 2	adult	1.88	1.25	88.0	18.7
A	regeneration	1.94	1.24	79.1	18.2
Couvet-South	adult	2.21	1.26	35.7	18.8
P	regeneration	2.21	1.25	35.1	18.9
Ochsenboden	adult	2.28	1.29	56.8	21.4
A	regeneration	2.21	1.28	55.9	21.4

When studying the variation parameter values in the two neighbouring stands from the communal forest „Kreuzberg", however some interesting statements can be made:

In the evenaged forest the allelic multiplicity, (that means the number of different alleles), is in the natural regeneration slightly higher (A_L=1,94) than in the collective of the old firs (A_L=1,88). For the diversity the opposite is true.

Although in the „Plenter" forest the number of alleles in the natural regeneration is identical with that one of the older firs, not the same rare alleles can be found coherently within both collectives. The alleles AAT-A1, MNR-B1 and SKDH-A3 were only found in the regeneration whereas the allels MNR-A1, PGM-B1 and NADH-A1 only in the older trees. The diversity (v resp. v_{gam}) is a little higher in the natural regeneration than in the collective of the older firs. This higher multilocus - value results from the clearly higher single locus values at the gene loci AAT-B, AAT-C, IDH-A and IDH-B. In other words these gene loci exhibit a more balanced distribution of gene variants in the natural regeneration than in the collective of the older firs. This is opposite to the situation at the gene loci AP-A and AP-C were the natural regeneration shows a smaller diversity.

In the two evenaged forests „Kreuzberg 2“ and „Ochsenboden“ the differences between the genetic structures (distribution of genotypes and alleles) of natural regeneration and old stand are low. Even at gene loci with comparatively higher values for the genetic distance (e.g. AP-A, AP-B, AP-C, 6-PGDH-A) the differences are statistically not significant. In the two „Plenter“ forests „Kreuzberg 1“ and „Couvet-South“ the differences between the genetic structures of natural regeneration and older firs are higher and for some gene loci (AP-A, AAT-B, AAT-C and IDH-A in „Kreuzberg 1“ resp. AP-D and AAT-A in „Couvet-South“) statistically significant. The mean genetic distances between the natural regeneration and the older firs are nearly 1.5 times higher in the two „Plenter“ forests than in the two evenaged forests (Tab. 5).

Table 5. Genetic distances at polymorphic loci between natural regeneration and adult populations of silver fir (*Abies alba*) in 2 „Plenter“ forests and 2 evenaged forests. Significant differences between allele frequencies are marked by asterisk for different significance levels: $\alpha = 0.1$ % (*); $\alpha = 0.001$ % (***)

Gene locus	Genic distance			
	"Plenter" forest		evenaged forest	
	Kreuzberg 1	Couvet-South	Kreuzberg 2	Ochsenboden
AP-A	5.2*	1.5	1.3	6.0
AP-C	5.5	9.5	6.8	5.0
AP-D	not calculated	5.5*	not calculated	1.5
AAT-A	1.8	3.5*	0.0	1.5
AAT-B	7.0***	6.5	0.5	1.5
AAT-C	7.5*	1.5	3.5	1.5
IDH-A	7.8***	4.0	4.2	2.5
IDH-B	4.7	0.5	0.8	4.0
MNR-B	0.5	4.0	0.0	1.0
6PGDH-A	0.6	0.1	7.6	1.5
6PGDH-B	0.8	1.5	0.6	1.0
PGM-A	2.7	0.0	0.1	0.5
MDH-A	0.1	8.5	2.0	0.5
NADH-A	1.2	0.5	0.0	0.0
Gene pool-distance	*3.3*	*3.6*	*2.1*	*2.1*

Discussion

The comparative genetic investigations in unevenaged forests („Plenter“ forests) and evenaged forests show the advantages and disadvantages of these two management systems for the *in situ* conservation of genetic ressources of silver fir (*Abies alba*).

„Plenter“ forest obviously is better adapted than evenaged stands to the conservation of the genetic multiplicity because more rare gene variants are preserved. This was shown by the A_{L5} values of all investigated stands and by the direct comparison of the two neighbouring stands in the Bavarian Forest, which, with high probability, originate from the same gene pool and which are treated differently. In this two stands almost a complete inventory of all older trees within an area of 1.5 ha for the „Plenter“ forest and 2.5 ha for the evenaged forest respectively was carried out. Thus, the calculated genetic values reflect precisely the actual extent of the presently existing genetic variation in that defined area. The number of silver firs varies between the respective areas significantly: 326 trees in the „Plenter“ forest (trees with a D.b.h. > 7 cm) and 43 adult trees in the evenaged forest. Although the volume related percentage of silver fir in the „Plenter“ forest is only little higher (see Table 1) this forest type exhibits considerably more stems on smaller area. This is caused by the middle and lower layer usually containing many stems with low volume in the „Plenter“ forest.

The fact that in „Plenter“ forests the number of trees per area is by far larger than in the evenaged forests in higher age classes is of great importance for the *in situ*-conservation of the genetic variation. The more rare alleles a population contains, the bigger is its „latent genetic potential“ (Bergmann *et al.* 1990) and its so called evolutionary adaptibility. The ability of a population "to store" rare gene variants is directly related to its number of individuals. On the other hand the risk of loosing genetic variants by drift processes increases with decreasing population size. Theoretically a „Plenter“ forest with a large and constant number of trees over generations may contain more (rare) genes per given area than an evenaged forest with a lower number of trees in the higher age classes. At the same time in a „Plenter“ forest the composition of the reproductive population changes continuously with the gradual extraction of the present upper layer and the ingrowth of the middle layer. This dynamics in time which governs the reproductive system distinguishes this management form from evenaged forest management and reduces the loss in genetic multiplicity.

The diversity of silver fir populations in „Plenter“ forests is mostly smaller than in evenaged forest stands. That means that the distribution of alleles in the „Plenter“ forest is less balanced than in evenaged forests. On the one hand, „Plenter“ forests are mostly found in the ecological optimum of the silver fir

(*Abieti-Fagetum*). On the other hand, the environmental conditions in „Plenter" forests are comparatively homogeneous and relatively constant in time (e.g. shading conditions or frost hazard). Both factors can cause the specialisation to those „optimum" conditions, leading to a diminution of genetic diversity.

The higher degrees of homozygosity in „Plenter" forests compared to the evenaged stands plead also for a high degree of specialisation and adaptation. Small degrees of heterozygosity are mainly caused by inbreeding and/or selection-advantages of homozygote individuals. Although inbreeding is frequently observed in silver fir (reasons are little dispersion of pollen and seeds; low self incompatibility), it can be excluded as the only reason for small heterozygosity in the investigated material because excess of homozygotes is not observable at every gene locus. Thereby it can be supposed that inbreeding is not the only and not the most important reason for smaller degrees of heterozygosity in „Plenter" forests, but more the relative enrichment of homozygote genotypes. Causative could be once again the better adaption of those genotypes to the comparatively homogeneous conditions in „Plenter" forests, as a consequence of long shaded terms and the relatively balanced climate inside the stand (Schütz 1989).

Thus the high degree of adaption of silver fir in the „Plenter" forest to the environmental conditions is obvious. Under constant environmental conditions this means most a higher fitness of the population and higher production yields. This situation stays optimum as long as the conditions won´t change and the plenter system won´t be interrupted. During the constant reproduction the existing genetic composition will be handed over to the next generation, which itself underlies the selection to the specific conditions in „Plenter" forests very early. But if these conditions change, there is high risk for the width of physiological reactions and adaptation, being not wide enough to react in the best way to those changes. Especially under complex stress situations a high degree of heterozygosity is considered advantageous for the physiological fitness of an individual, a low degree of heterozygosity correspondingly desadvantageous (Mitton and Grant 1984; Koehn and Bayne 1989; Ziehe *et al.* 1990; Müller-Starck 1993). Silver fir in „Plenter" forests reacts, compared to spruce, much more sensitive to extreme climatic events (frost) and changed environmental conditions, e.g. after strong thinnings. These observations made by Spiecker (1991) possibly can be extended and can be seen in connection with genetic factors.

Hattemer *et al.* (1993) recommend to give some attention to every observable gene variant for possible adaption in the future, because the expected changing environmental conditions can not be forecasted, neither their direction nor their intensity. After this estimation a management system, as it can be found in „Plenter" forests, seems to be the best option for the conservation of genetic variation, because it guarantees the conservation of very many genetic variants over a long time period. But it can not be said, facing the drastical changes in the environment, if this advantage is big enough to equalize the disadvantage of reduced physiological adaptability. Under this point of view evenaged forests show clear advantages if the stand area is large, the relative amount of silver fir in the stand is big enough and browsing by game does not cause high losses. Facing these conclusions, „Plenter" forests and evenaged forests should exist and be managed at the same time. Compared to evenaged forests the „Plenter" forest obviously is advantageous in respect to the conservation of genetic variation in the case when we are dealing with very small stands and/or with stands of low percentage of silver fir. In larger stands or in stands with high portion of silver fir this advantage is not given necessarily.

In evenaged forests slow regeneration in small patches should be the target. The results show that there is no reduction in the genetic variation of the natural regeneration in an evenaged forest, if the number of members of the stand which effectively participate in the reproduction process is sufficiently large. Investigations in evenaged forests in Baden-Württemberg (Konnert 1992), Switzerland (Hussendörfer 1997) and Thüringen (Bergmann 1996) showed similar results. Different results were obtained, however, in fir relictes in Saxonia in which individual trees were severely isolated. Here all stands showed statistically significant differences between natural regeneration and old stand as well as pronounced reduction of the genetic variation (multiplicity and diversity) in the natural regeneration (Llamas-Gómez 1998). The same was found in a stand in Thüringen in which the firs were represented only in the middle layer having suppressed crowns. Here it was found, that only 27 % of the tested old trees could be considered as parents of the actual natural regeneration (Bergmann 1996).

The results and conclusions presented are valid only for silver fir (*Abies alba*). From a genetic point of view, more conclusions can finally be drawn, if further investigations in „Plenter" forests for other species e.g spruce (*Picea abies*) and beech (*Fagus sylvatica*) were carried out.

Acknowledgments

This study was supported in part by the Bundesamt für Umwelt, Wald und Landschaft (BUWAL) Switzerland and by the Ministerium Ländlicher Raum Baden-Württemberg (Programm UFO 55-95-6). The technical assistance of S. Nowak, S. Fuchs (LSP Teisendorf), M. Müller (WSL Birmensdorf), M. Dägele and S.Thumm (FVA Freiburg) is greatly appreciated.

References

Anonymus 1993: General Guidelines for the Conservation of the Biodiversity of European Forests (Resolution H2). Ministerial Conference on the Protection of Forests in Europe, Helsinki 1993

Anonymus 1998: Pan-European Criteria, Indicators and Operational Level Guidelines for Sustainable Forest Management (Resolution L2). 3rd Ministerial Conference on the Protection of Forests in Europe, Lissabon 1998

Bergmann, F. 1991: Causes and consequences of Species-specific Genetic Variation Patterns in European Forest Tree Species: Examples with Norway Spruce and Silver Fir. Pp 192 –205 in: Müller-Starck and Ziehe (Eds.): Genetic Variation in European Populations of Forest Trees. J.D. Sauerländer's Verlag, Frankfurt am Main

Bergmann, F. 1993: Die genetische Struktur in Weisstannen-Populationen Mittel-und Südeuropas. pp. 97-105 in: Wolf, H. (Ed.): Weisstannenherkünfte. Ecomed-Verlagsgesellschaft, Landsberg am Lech

Bergmann, F. 1996: Die genetische Struktur der Naturverjüngung bei der Weisstanne in Abhängigkeit vom Altbestand. AFZ/Der Wald 51: 1046-1047

Bergmann, F., Gregorius, H.-R. and Larsen, J.B. 1990: Levels of genetic variation in European Silver Fir (*Abies alba*). Are they related to the species' decline? Genetica 82: 1-10

Breitenbach-Dorfer, M., Pinsker, W., Hacker R. and Müller, F. 1992: Clone identification and clinal allozyme variation in populations of *Abies alba* from the Eastern Alps (Austria). Pl. Syst.Evol. 181: 109-120

Geburek, T. and Thurner, G. 1993: Verändert sich der Genpool von Waldbeständen durch forstwirtschaftliche Massnahmen? Centralbl. gesamte Forstwes. 110: 49-62

Gregorius, H.-R. 1974: Genetischer Abstand zwischen Populationen. I. Zur Konzeption der genetischen Abstandsmessung. Silvae Genetica 23: 22-27

Hattemer, H.H., Bergmann, F. and Ziehe, M. 1993: Einführung in die Genetik für Studierende der Forstwissenschaft. Zweite Auflage, Sauerländer's Verlag, Frankfurt/M.

Hussendörfer, E. 1997: Untersuchungen über die genetische Variation der Weisstanne (*Abies alba* Mill.) unter dem Aspekt der *in situ* Erhaltung genetischer Ressourcen in der Schweiz. Beih. Z. Schweiz. Forstver. 83: 151

Hussendörfer, E., Konnert, M. and Bergmann, F. 1995: Inheritance and linkage of isozyme variants of Silver Fir (Abies alba Mill.). Forest Genetics 2: 29-40

Koehn, R.K. and Bayne, B.L. 1989: Towards a Physiological and Genetical Understanding of the Energetics of the Stress Response. Bio. J. Lin. Soc. 37: 157-171

Konnert, M. 1992: Genetische Untersuchungen in geschädigten Weisstannenbeständen *(Abies alba* Mill.) Südwestdeutschlands. Mitt. der Forstl. Versuchs-u. Forschungsanstalt Baden-Württemberg, Freiburg 167: 119 pp

Konnert, M. 1993: Untersuchungen über die genetische Variation der Weisstanne (*Abies alba* Mill.) in Bayern. Allg. Forst-u. Jagdztg. 162: 100-106

Konnert, M. and Bergmann, F. 1995: The geographical distribution of genetic variation of silver fir (*Abies alba, Pinaceae*) in relation to its migration history. Pl. Syst. Evol. 196: 19-30

Konnert, M. and Maurer, W. 1995: Isozymic Investigations on Norway Spruce (*Picea abies* (L.) Karst.) and European Silver Fir (*Abies alba* Mill.): A Practical Guide to Separation Methods and Zymogram Evaluation. Bund-Länder-Arbeitsgruppe "Erhaltung genetischer Ressourcen", Bayerische Landesanstalt für forstliche Saat- und Pflanzenzucht, Teisendorf

Llamas-Gómez, L. 1998: Populationsgenetische Untersuchungen bei der Weisstanne (*Abies alba* MILL.) in Sachsen. Dissertation, Fakultät f. Forst-, Geo- und Hydrowissenschaften der TU Dresden, 136 pp

Lozán, J.L. 1992: Angewandte Statistik für Naturwissenschaftler. Verlag Paul Parey, Berlin

Mitton, J.B. and Grant, C. 1984: Associations among protein heterozygosity, growth rate and developmental homeostasis. Ann.rev.Eco.Syst. 15: 479-499

Müller, F. 1990a: Die wichtigsten Naturverjüngungsverfahren und Betriebsarten. Österreichische Forstzeitung, 12: 57

Müller, F. 1990b: Naturverjüngung und genetische Vielfalt. Österreichische Forstzeitung, 12: 58-60

Müller-Starck, G. 1993: Auswirkungen von Umweltbelastungen auf genetische Strukturen von Waldbeständen am Beispiel der Buche (*Fagus sylvatica* L.). Schr. Forstl. Fak. Univ. Gött. Niedersächs. forstl. Versuchsanst, Band 112, Sauerländer's Verlag, Frankfurt am Main

Müller-Starck, G. 1996: Beiträge der Forstgenetik zur nachhaltigen Waldbewirtschaftung. Pp. 259-283 in: Müller-Starck, G. (Ed.): Biodiversität und nachhaltige Forstwirtschaft. Ecomed Verlagsgesellschaft, Landsberg am Lech

Müller-Starck, G. and Gregorius, H.-R. 1986: Monitoring genetic variation in forest tree populations. Proc. 18th IUFRO W. Congr., Div. 2, Vol. II: 589-599

Rotach, P. 1994: Genetische Vielfalt und praktische forstliche Tätigkeit: Probleme und Handlungsbedarf. Schweiz. Z. Forstwes. 145: 999-1020

Schroeder, S. 1988: Die Isoenzym-Variation der Weisstanne (*Abies alba* Mill.) 16 europäischer Provenienzen. Mitt. Ver.f. Forstl. Standortskunde u. Forstpflanzenzüchtung 32: 81-85

Schütz, J.-P. 1989: Der Plenterbetrieb. Fachbereich Waldbau, ETH, Zürich

Spiecker, H. 1991: Zur Dynamik des Wachstums von Tannen und Fichten auf Plenterwald-Versuchsflächen im Schwarzwald. AFZ/Der Wald 46: 1076-1080

Waldherr; M. 1996: Weißtannen-Verjüngungs- und Pflegeverfahren im Bayerischen Wald. Mitteilungen der Landesanstalt für Wald und Forstwirtschaft Gotha, Heft 11: 121-135

Ziehe, M., Hattemer, H.H. and Müller-Starck, G. 1990: Zur Bedeutung der Heterozygotie für die dynamische Genkonservierung. pp. 46-57 in: Hattemer, H.H. (Ed.): Schr. Forstl. Fak. Univ. Gött. Niedersächs. forstl. Versuchsanst., Band 98: Erhaltung forstlicher Genressourcen. Sauerländer's Verlag, Frankfurt am Main

PART 4: TREE BREEDING FOR UNCERTAIN FUTURE ENVIRONMENTS

Multiple population breeding for uncertain climatic futures with *Alnus rubra*: ecological genetics and selection experiments

A. HAMANN, G. NAMKOONG AND M. P. KOSHY

University of British Columbia, Department of Forest Science, Forest Science Center, Vancouver, B. C., Canada, V6T 1Z4

Key words: red alder, ecological genetics, phenology, tree breeding

Abstract: In a case study with red alder (*Alnus rubra* Bong.) we develop multiple breeding populations to cope with uncertainty in the response of genotypes to environmental changes. Genetic parameters and geographic variation in adaptive traits were studied to tailor breeding populations to specific climatic environments. Orthogonal clines were identified in bud break and growth cessation with spatial statistical analysis, implying that it is not possible to cope with effects of climatic warming by simply shifting provenances to new locations. Instead climatic warming may require the development of genotypes with novel combinations in adaptive traits. Selection targets for bud break and leaf abscission for six breeding populations were derived that cover current and possible future climatic conditions. Genetic parameters for adaptive traits were estimated, and it was confirmed that selection targets for the proposed breeding populations could be met in one generation. Heritabilities for the date of bud break and leaf abscission were moderately high (0.41 to 0.54) and genetic correlations implied that selection for advanced bud break and delayed growth cessation will positively influence growth traits.

Introduction

Timing of phenological events in forest trees such as bud burst, onset of dormancy, and frost hardening are considered essential adaptations in temperate climates. If the phenology is not well synchronized with climatic conditions, trees may suffer frost damage or fail to fully utilize favorable growing conditions at the site. Global climate change could cause such a lack of synchronization. The global temperature is expected to rise in future decades, and warming has been predicted to be especially pronounced during autumn, winter, and spring in regions located at high latitudes, such as British Columbia (Bilger 1992). The predicted warming could have various impacts on the synchronization of phenological events in forest trees and climatic events at the site. First, the dates marking the beginning and the end of favorable growing conditions may shift; second, environmental cues that control

Müller-Starck, G. and Schubert, R. (eds.). Genetic Response of Forest Systems to Changing Environmental Conditions. Vol. 70 (For. Sci.), 321-330

phenological events may disappear or shift to different dates; and third, probabilities of climatic events (e. g. frost) subsequent to an environmental cue may change. In addition to lack of synchronization, there could be a lack of adaptation to conditions such as a higher moisture deficit during summer.

The response of phenotypes to climate change is likely to differ among provenances of wide ranging species (Wang and Perry 1958; Perry and Wang 1960; Sharik and Barnes 1976). In regions with mild winters, for example, Cannell and Smith (1986) and Kramer (1994) demonstrated that climatic warming regularly causes a chilling deficit and therefore a delayed bud burst, resulting in failure to utilize favorable growing conditions in earlier springs. In other regions, where chilling requirements are far exceeded by winter temperatures, Murray *et al.* (1989) and Hänninen (1991) showed that climatic warming would force bud burst at an earlier date and could increase the risk of frost injuries in spring.

Designing a breeding strategy for a species that takes effects of climate change into account requires knowledge of clinal variation in adaptive traits, genetic parameters for adaptive traits, and the response in phenological traits to warming for a sample of populations from the region of interest. Such data is partly available for red alder in British Columbia (Hamann *et al.* 1998, 2000). Since there is a limited understanding of response of genotypes to changes in climate, and potentially a large error in predicted climate change itself, a breeding strategy must incorporate elements of uncertainty. Multiple population breeding has been suggested for such a situation (Namkoong *et al.*, 1988; Eriksson *et al.* 1993). Coverage of future climatic conditions may be accomplished by simply assigning new deployment targets for multiple populations, or it may require the development of several populations with novel combinations in adaptive and growth traits.

In this report we develop a multiple population breeding strategy intended to cope with possible effects of climatic warming, given existing provenance data and incomplete information on phenotypic response to climate change. We derive which trait combinations are most likely required for current and possible future climates in British Columbia, and test if selection targets could be met within one generation of breeding.

Methods

Phenology

Data were collected in a provenance trial of red alder covering its natural range in British Columbia including some sources from Oregon and Alaska. (Hamann 1999). Monthly height measurements from 100 trees for each of the 65 provenances over the period of one year were evaluated to examine differences in growth patterns among provenances during the second growing season. Abscission of the top five leaves and bud break of the top five buds was recorded in weekly intervals at the end of the first growing season and the beginning of the second growing season. The average day of bud break and the average date of leaf abscission were calculated as weighted averages for individual trees from these repeated measurements. The experimental design of the provenance trial is described in the next section.

A model of geographic variation in bud break and leaf abscission was developed with the kriging method, which yields smooth varying surfaces of best linear unbiased predictions of provenance performance over the geographic range of the species. The program GSTAT by Pebesma and Wesseling (1998) and the GIS software IDRISI (Clarke Labs 1999) was used to fit variogram models for kriging and to generate maps of predicted values. Methodological details are described in Hamann *et al.* (2000).

Genetic testing

Genetic parameters were estimated based on half-sib families of a provenance trial, and full-sib families from a selection experiment. The provenance trial was laid out as a split plot designs and analyzed according to the following model:

$$Y_{ijkl} = \mu + B_i + P_j + P{\times}B_{ji} + F(P)_{k(j)} + F(P){\times}B_{k(ij)} + \varepsilon_{l(ijk)}$$

where Y is the measurement of seedling l in block i from provenance j and family k; μ is the overall mean; B, P, and $F(P)$ are the effects of block, provenance, and family within provenance, respectively. Individual heritabilities and genetic correlations were calculated based on family within provenance variance components following standard procedures.

Families for the second experiment were generated from the above provenance trial using a restricted index selection (Kempthorne and Nordskog 1959), where one or two traits were advanced while keeping the others constant. Three populations were generated by advancing exclusively seedling height, date of leaf abscission or date of bud break. In three more populations every combination of two traits was advanced simultaneously. Each population consisted of eight male and eight female trees mated in a disconnected four parent factorial design to generate 32 families per population. Offspring was planted in a randomized complete block design with each family represented by one offspring per block using eight replications. Because of variable success in obtaining offspring from controlled crosses (420 offspring of 1536 required) the unbalanced data were evaluated with the mixed model procedure of the SAS statistical software package (SAS Institute 1997) according to the following model:

$$Y_{ijklm} = \mu + B_i + P_j + P{\times}B_{ji} + F(P)_{k(j)} + F(P){\times}B_{k(ij)} + M(P)_{l(j)} + M(P){\times}B_{l(ij)} + M(P){\times}F(P)_{kl(ij)} + \varepsilon_{m(ijkl)}$$

where Y is the measurement of seedling m in block i from population j, female parent k and male parent l; μ is the overall mean; B, P, $M(P)$ and $F(P)$ are the effects of block, population, and male and female within populations, respectively. Individual heritabilities were calculated based on the mean of male and female family variance components. Standard errors of heritabilities were estimated according to the delta-method (Lynch and Walsh 1998, Appendix 1).

Results and discussion

Ecological genetics

Analysis of geographic patterns in bud break shows that coastal provenances in the southern part of British Columbia break bud later than inland sources (Fig. 1) conforming to typical maritime patterns (Campbell and Sugarno 1979, Myking and Heide 1995). In southern coastal British Columbia, winter days warm enough to encourage growth are interspersed with periods of freezing, and trees are exposed to forcing temperatures throughout the winter. Further inland, further north and at higher elevations, temperature fluctuations between growing and freezing conditions are restricted to shorter periods in fall and spring. Therefore, high dormancy stability to prevent premature bud burst is of greater importance in southern, coastal and low elevation populations.

Leaf abscission occurs earlier in northern provenances and thus at longer critical photoperiods than southern sources (Fig. 2). As in most temperate trees, growth cessation in alder is primarily controlled by the shortening of the photoperiod, since in colder northern environments favorable growing conditions end at longer photoperiods than in southern environments (Perry 1971; Morgenstern 1996). Leaf abscission has been shown to be a reliable indicator for growth cessation in red alder, being highly correlated with growth late in the season (Hamann 1999). The reason is that alder exhibits indeterminate growth patterns and young leaves that are formed late in the season will not drop readily.

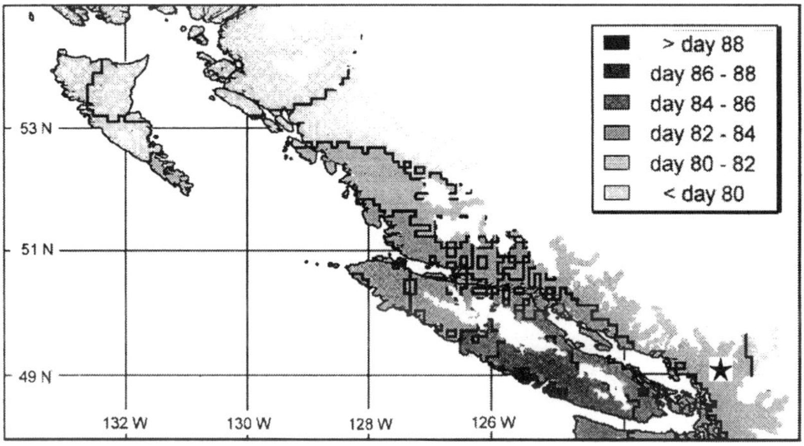

Figure 1. Date of bud break predicted for red alder provenances when grown at a southeastern location (★). Contours are drawn at intervals of approximately 0.25 standard deviations.

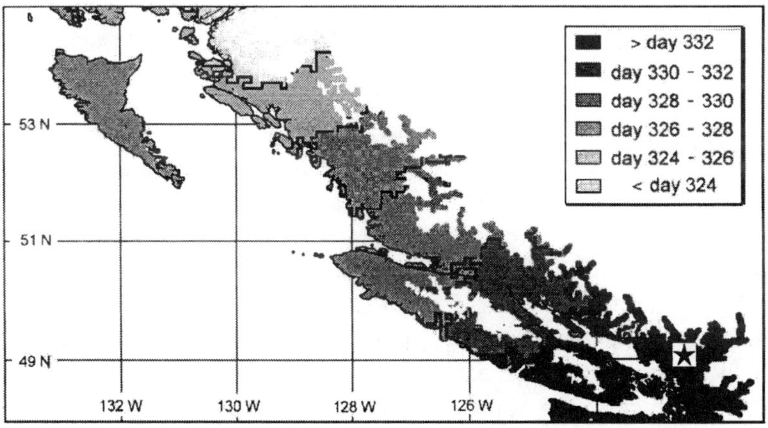

Figure 2. Date of leaf abscission predicted for red alder provenances when grown at a southeastern location (★). Contours are drawn at intervals of approximately 0.25 standard deviations.

Analysis of growth rates throughout the season also indicates that synchronization of growth initiation and growth cessation with the available growing season has important effects on growth. When northern provenances are transferred to a southern site with favorable environments in fall, premature growth cessation appears to be a disadvantage compared to local sources (Fig. 3). In addition, provenances sampled from a very far northern location showed reduced growth during the summer months (Fig. 3, Alaska).

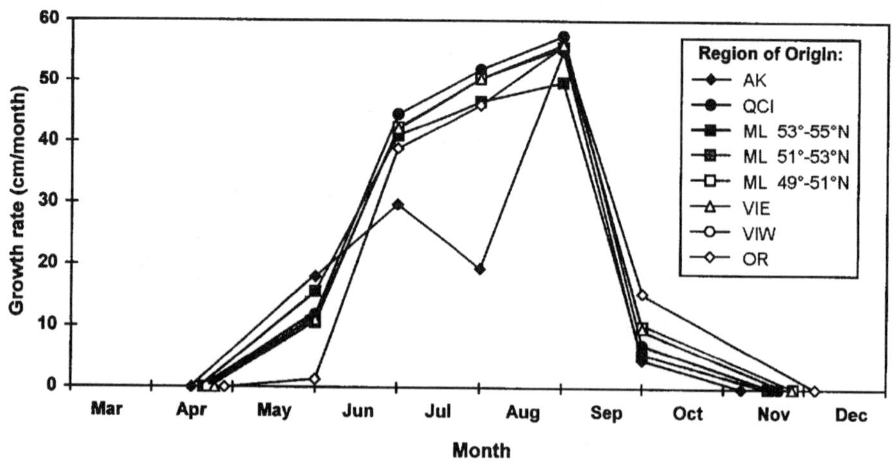

Figure 3. Growth rates during the second growing season for provenances summarized by region of origin (AK, Alaska; ML, mainland British Columbia, QCI, Queen Charlotte Islands; VIE Vancouver Island East; VIW, Vancouver Island West; OR, Oregon).

The adaptive value of timely growth cessation in harsher environments is more questionable. It could be argued that damage by fall frost should not have serious consequences, since it only prevents the recovery of nutrients from leaves, which is not an important process in red alder. However, provenances that cease growth later than local sources (Fig. 3, Oregon) exhibit severely retarded growth rate during the first 30 days after bud break. Active growth late in the season may expend resources that would otherwise be available in spring to sustain high growth rates early in the season. Hence, premature as well as delayed growth cessation relative to local climates in target areas of breeding populations should be avoided.

Early leaf abscission as well as late bud break were also shown to be correlated with reduced height and biomass at the end of the growing season (Hamann *et al.* 1998). These observations are good circumstantial evidence that the traits bud break and leaf abscission are relevant indicators for adaptive traits in red alder, and can be used for tailoring breeding populations to specific climatic environments.

Breeding populations

For current climatic conditions we delineated three breeding zones for British Columbia, which ensures that transfer of seed does not result in a reduction of growth traits larger than 10% relative to a local source (Hamann *et al.* 2000). In red alder a 10% change in growth or survival is associated with approximately 8 days change in the date of growth cessation and 4 days in the date of bud break. (Hamann *et al.* 1998). Assuming that either shift in phenology could cause the 10% reduction in growth, these values were used as conservative thresholds, where the use of a new breeding population over another would be justified under a scenario of uniform climatic warming.

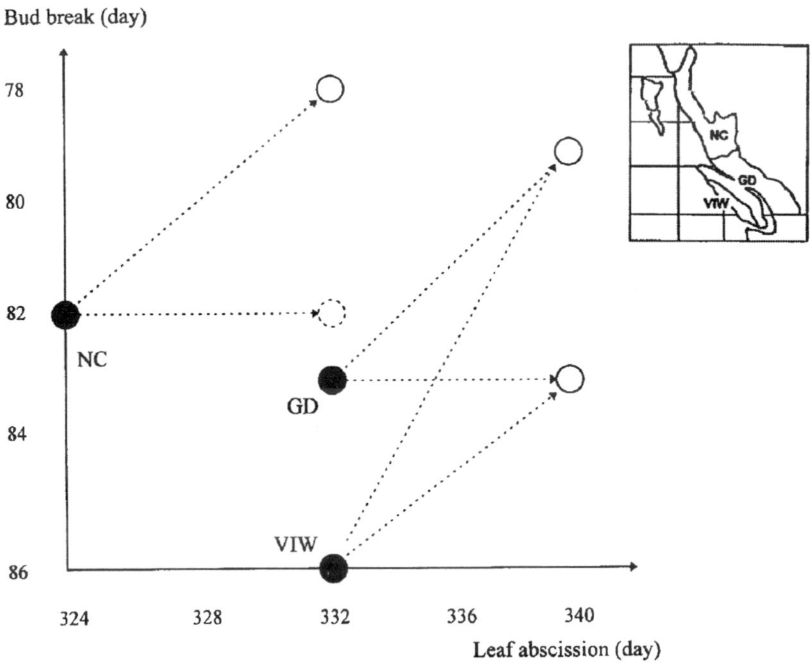

Figure 4. Proposed multiple breeding populations for current (dark circles) and possible future (light circles) climatic conditions (NC, British Columbia north of 51°N latitude; GD, Georgia Depression; VIW, Vancouver Island Westcoast).

Since there is a simple linear cline over latitude and the environmental cue for growth cessation is independent of climate change itself, identification of selection targets for growth cessation is straightforward. A delayed date of bud break by 8 days for all populations would allow that longer favorable growing conditions in fall can be utilized (Fig. 4).

The situation for the date of bud break is more ambiguous, since the exact effect of climatic warming on growth initiation is unknown. The sources from southern coastal British Columbia showed the highest dormancy stability and are most likely to suffer delayed bud break due to a chilling deficit. Selecting alternatively for 4 and 8 days of advanced bud burst in the VIW-breeding population could compensate for this effect. For the other populations with less dormancy stability, the effect of climate warming is likely to be neutral or less pronounced. Alternatively, bud break is not advanced or it is advanced by only 4 days (Fig. 4).

Genetic testing

Heritabilities for bud break and leaf abscission based on half-sib as well as full-sib progeny tests suggests that adaptive traits and seedling height are under moderate genetic control (Tab. 1 and 2).

Table 1. Estimated individual heritabilities with standard errors for traits based on half-sib offspring in a provenance trial

	Height	Bud break	Leaf abscission
Heritability (h^2)	0.35 (0.04)	0.43 (0.04)	0.54 (0.04)

Table 2. Estimated genetic variances and individual heritabilities with standard errors for traits based on full-sib offspring of controlled crosses

	Height	Bud break	Leaf abscission
Additive Variance V_A	249 (27.1)	8.66 (1.01)	20.22 (2.81)
Dominance Variance V_D	0.64 (1.12)	1.16 (2.34)	10.16 (5.35)
Heritability h^2	0.46 (0.21)	0.47 (0.22)	0.41 (0.21)

Poor success in generating offspring in the selection experiment did not allow precise estimates of realized gains from selection (reported in Hamann 1999). However, delaying leaf abscission by 8 days would require a selection intensity of approximately 1.9. Similarly, advancing the date of bud break by 4 days would require a selection intensity of approximately 1.3. Both could be achieved in current test plantations. Advancing the date of bud break by 8 days would require two breeding cycles.

Table 3. Genetic correlations (upper right) and phenotypic correlations (lower left) estimated from half-sib offspring of a provenance trial.

	Height	Bud break	Leaf abscission
Height	-	-0.38	0.42
Bud Break	0.02	-	0.08
Leaf abscission	-0.21	0.32	-

As expected, growth traits will be positively influenced by selecting for an advanced date of bud break or a delayed date of leaf abscission, presumably due to a longer growing season. Further, the date of leaf abscission and the date of bud break are genetically uncorrelated (Tab. 3).

Summary and recommendations

Red alder showed a typical latitudinal trend in growth cessation and conforms to maritime patterns in bud burst phenology. Southern and coastal sources exhibited the highest dormancy stability, implying that these populations are most likely to experience a chilling deficit due to climatic warming. Under this hypothesis, a greater risk of frost damage due to climatic warming is not expected for red alder. In fact, unusual hardiness of coastal species and provenances has been observed after an exceptionally warm winter (van der Kamp and Worrall 1990). Allowing an advanced date of bud break and a delayed date of growth cessation to varying degrees in breeding populations is expected to enhance growth rates of genotypes by utilizing a longer available growing season under the scenario of climate warming.

References

Bilger, B. 1992: Global warming. Chelsea House Publishers, New York

Campbell, R.K. and Sugano, A. 1979: Genecology of bud burst phenology in Douglas-fir: response to flushing temperature and chilling. Bot. Gaz. 140: 223-231

Cannell, M.G.R. and Smith, R.I. 1986: Climatic warming, spring bud burst and frost damage on trees. J. Appl. Ecol. 23: 177-191

Clarke Labs 1999: Idrisi GIS and Image Processing. Release 3.2. Worchester, Massachusetts

Eriksson, G., Namkoong, G. and Roberts, J.H. 1993: Dynamic gene conservation for uncertain climatic futures. For. Ecol. Manage. 62: 15-37

Hamann, A. 1999: Utilization and Management of red alder genetic resources in British Columbia. Ph. D. Dissertation, University of British Columbia, Vancouver, B. C.

Hamann, A., Koshy, M., Namkoong, G., and Ying, C.C. 2000: Genotype × environment interactions in *Alnus rubra*: developing seed zones and seed transfer guidelines with spatial statistics and GIS. For. Ecol. Manage. 136: 107-119

Hamann, A., El-Kassaby, Y.A., Koshy, M.P. and Namkoong, G. 1998: Multivariate analysis of allozymic and quantitative trait variation in *Alnus rubra*: geographic patterns and evolutionary implications. Can. J. For. Res., 28: 1557-1565

Hänninen, H. 1991: Does climatic warming increase the risk of frost damage in northern trees? Plant Cell Environ. 14: 449-454

Kempthorne, O. and Nordskog, A.W. 1959: Restricted selection index. Biometrics 15: 10-19

Kramer K. 1994: A modeling analysis of the effects of climatic warming on the probability of spring frost damage to tree species in the Netherlands and Germany. Plant Cell Environ. 17: 367-377

Lynch, M. and Walsh, B. 1998: Genetics and analysis of quantitative traits. Sinauer Associates, Inc., Sunderland, Massachusetts, pp. 807-821

Morgenstern, E.K. 1996: Geographic Variation in Forest Trees: Genetic Basis and Applications of Knowledge in Silviculture. UBC Press, Vancouver

Murrey, M.B., Cannell, M.G.R. and Smith, R.I. 1989: Date of bud burst of fifteen tree species in Britain following climatic warming. J. Appl. Ecol. 26: 693-700

Myking, T. and Heide, O.M. 1995: Dormancy release and chilling requirements of buds of latitudinal ecotypes of *Betula pendula* and *B. pubescens*. Tree Physiol. 15: 697-704

Namkoong, G., Kang, H.C. and Brouard, J.S. 1988: Tree Breeding: Principles and Strategies. Springer Verlag, New York

Pebesma, E.J. and Wesseling, C.G. 1998: Gstat, a program for geostatistical modeling, prediction and simulation. Computers Geosci. 24: 17-31

Perry, T.O. 1971: Dormancy of trees in winter. Science 171: 29-36

Perry, T.O. and Wang, C.W. 1960: Genetic variation in winter chilling requirement for date of dormancy break for *Acer rubrum*. Ecology 41: 785

SAS Institute. 1997: SAS/STAT user's guide. Release 6.12, Cary, NC

Sharik, T.L. and Barnes, B.V. 1976: Phenology of shoot growth among divers populations of yellow birch (*Betula alleghaniensis*) and sweet birch (*B. lenta*). Can. J. Bot. 54: 2122-2129

van der Kamp B.J. and Worrall, J. 1990: An unususual case of winter bud damage in British Columbia interior conifers. Can. J. For. Res. 20: 1640-1647

Wang, C.W. and Perry, T.O. 1958: The ecotypic variation of dormancy, chilling requirement, and photoperiodic response in *Betula* species. Volume 2. Proceedings of the 10th International Congress of Genetics. McGill University, Montreal

Early growth characteristics of some *Abies* hybrids

A.KORMUŤAK and B. VOOKOVÁ

Institute of Plant Genetics and Biotechnology, Slovak Academy of Sciences, SK-950 07 Nitra, Slovak Republic

Key words: *Abies*, hybrids, height growth

Abstract: Height growth performance of 5-year old seedlings originating from selfing, controlled intraspecific and interspecific crossings of *Abies alba, A. nordmanniana, A.cephalonica, A. numidica, A. pinsapo, A.concolor* and *A. grandis* was compared. Some of the interspecific crosses attempted were found to exhibit somatic heterosis outgrowing the progenies of the respective species from self- and cross-pollination. A few intraspecific crosses of *A. alba* have also reached the height growth parameters comparable with those of interspecific hybrids. Based on these characteristics, practical suggestions have been made concerning utilization of *Abies* hybrids in reforestation of ecologically disturbed areas.

Introduction

Species hybridization has been offered as an alternative to the extensive die-back of silver fir in central Europe (Vincent and Kantor 1971). Low adaptability of the species to changing climatic conditions is believed to be caused by a narrow genetic variability of silver fir in the region (Larsen 1988). Among suggestions concerning the extension of silver fir variability and increasing of its adaptive potential, the hybridization of the species with exotic firs is recommended (Kobliha 1994; Greguss 1996). As an example, the vitality of the forests in Greece is believed to result from introgression between *A. alba* and *A. cephalonica* (Mitsopoulos and Panetsos 1987; Fady and Conkle 1993). According to Kaňák (1988), the highly variable and heterozygous progenies of both intra- and interspecific hybridization of forest trees can adapt themselves to the new ecological niches much faster than pure species. In the light of expected global warming and subsequent changes in other factors, the plasticity of individuals together with sufficient intrapopulation variation in adaptive traits are considered to be a cornerstone of breeding strategies in the future (Koski 1996). With special reference to *Abies*, Rohmeder and Eisenhut (1961) reported of a cross between *A. veitchii* x *A. alba* sharing a higher degree of frost resistance than *A. alba*. The same is true of *A. nordmanniana* x *A. veitchii*, *A. concolor* x *A. procera* and *A. veitchii* x *A. alba* hybrids which are postulated to

Müller-Starck, G. and Schubert, R. (eds.). Genetic Response of Forest Systems to Changing Environmental Conditions. Vol. 70 (For. Sci.), 331-338

be resistant against emissions and fungal diseases (Müller 1989). In addition to increased resistance, an overwhelming majority of interspecific crosses of firs were reported to exhibit also somatic heterosis (Mergen and Gregoire 1988). The phenomenon refers to a higher degree of viability or productivity of the offspring than the average characteristics of the kind in the corresponding parents (Dubovský and Maršálek 1968).

In this study we provide evidence for hybrid superiority to be concomitant feature but not a general rule of *Abies* interspecific hybridization.

Material and methods

Two groups of seedlings were involved in our study. The first group originated from artificial hybridization experiments in 1991 using maternal trees of *Abies alba* Mill., *A. numidica* De Lann., *A. pinsapo* Boiss., *A. nordmanniana* (Stev.) Spach, *A. cephalonica* Loud., *A. concolor* (Gord et Glend) Lindl. and *A. grandis* (Dougl.) Lindl. The second group consisted of seedlings which were obtained in the same way in 1992 when additional individuals of *A. alba* Mill. and *A. nordmanniana* (Stev.) Spach. were used as mother trees. Except for the interspecific combinations, the variants with selfing, controlled intraspecific crossing and open pollination were taken into account serving as a control. Five trees of silver fir (*A. alba* Mill.) growing in the natural stand of the species in Jedľové Kostolany (west Slovakia) and one or two individuals of the exotic species introduced in the arboretums in Mlyňany and Kysihýbel were used as parent trees. The species were crossed in both intra- and interspecific combinations, the list of which is given in Table 1. The seeds collected in October of each year were sown in the nursery Kamenec pod Vtáčnikom in the autumn of the same year. Height growth characteristics of individual progenies were measured on 5-year old seedlings.

Results

Height growth potential of different progeny classes revealed differences between the mating types performed within a given maternal tree as well as between the crosses of different species.

The data presented in Table 1 indicate that progenies from self-pollination exhibited the worst performance in a prevailing majority of mother trees. In particular, it is true of selfed progenies of *A. alba* no. 5, *A. alba* no. 6, *A. concolor* no. 1 and *A. concolor* no. 2, all of which differed significantly from the corresponding intraspecific crosses of the respective mother trees reaching a higher level of their height growth. The only exception was self-pollination variant of *A. alba* no. 7 sharing comparable potential for height growth as progenies from the two controlled intraspecific crossings and from open pollination of the mother tree. Comparable with selfed progenies were also the height growth characteristics of the open pollination progenies in *A. alba* no. 6 and *A. concolor* no. 2 suggesting a high proportion of natural selfing in both these trees. Of the species tested so far, the species *A. pinsapo* was exceptional exhibiting a better performance by its self-pollination progeny than in the interspecific cross *A. pinsapo* x *A. numidica.*

Contrary to self-pollination, the outcrossing positively affected growth potential of the respective progenies. This effect was especially conspicuous in both mother trees of *A. concolor* as well as in *A. alba* no. 5 and *A. alba* no. 6 whose progenies from controlled intraspecific crossings had outgrown significantly the corresponding control variants from selfing. Some of the intraspecific crosses obtained within a given mother tree were difficult to distinguish mutually (*A. alba* no. 5, *A. alba* no. 7), whereas the others differed

Table 1. Height growth performance of 5-year old hybrids obtained within maternal trees of *A. alba, A. numidica, A. pinsapo, A. nordmanniana, A. cephalonica, A. concolor* and *A. grandis* in 1991 and 1992

No	Crosses	n	Average (cm)	Tukey test[1]		
			1991			
1	*A.alba* no.1 x *A.alba* no.3	42	30.8 ± 5.04			X
2	*A.alba* no.1 x *A.alba* no.4	13	23.4 ± 3.57		X	
3	*A.alba* no.1 – open pollination	39	24.8 ± 5.56		X	
4	*A.alba* no.2 x *A.alba* no.3	29	21.1 ± 4.71	X		
5	*A.alba* no.2 x *A.alba* no.4	25	30.1 ± 4.71			X
6	*A.alba* no.2 – open pollination	122	24.7 ± 6.09		X	
7	*A.numidica* – open pollination	132	27.4 ± 4.66			X
8	*A.numidica* x *A.alba*	12	20.1 ± 3.26	X		
9	*A.numidica* x *A.pinsapo*	15	21.3 ± 3.03	X		
10	*A.pinsapo* - selfing	39	30.7 ± 5.04			X
11	*A.pinsapo* x *A.numidica*	43	24.9 ± 3.10		X	
12	*A.nordmanniana* x *A.alba*	38	22.7 ± 4.75		X	
13	*A.nordmann.* x *A.cephalonica*	44	29.5 ± 5.36			X
14	*A.cephalonica* x *A.alba*	13	26.9 ± 4.31		X	

Continuation of Table 1.

15	*A.cephalonica* x *A.nordmann.*	12	24.0 ± 5.15	X							
16	*A.concolor* no.1 - selfing	15	35.7 ± 11.0				X				
17	*A.concolor* n.1 x *A.concolor* 2	66	43.6 ± 7.46					X			
18	*A.concolor* no.1 x *A.grandis*	186	48.8 ± 8.53						X		
19	*A.concolor* no.2 - selfing	23	36.2 ± 6.89				X				
20	*A.concolor* no.2 – open pollin.	104	37.6 ± 11.12				X				
21	*A.concolor* 2 x *A.concolor* n.1	59	46.3 ± 5.92					X			
22	*A.concolor* no.2 x *A.grandis*	105	56.3 ± 8.89							X	
23	*A.grandis* – open pollination	38	61.6 ± 13.20								X
24	*A.grandis* x *A.concolor* no.1	62	50.3 ± 6.73						X		
25	*A.grandis* x *A.concolor* no.2	69	51.6 ± 8.18						X		
			1 9 9 2								
26	*A.alba* no.5 - selfing	11	25.4 ± 7.25	X							
27	*A.alba* no.5 x *A.alba* no.4	37	34.6 ± 4.30			X					
28	*A.alba* no.5 x *A.alba* no.7	35	35.2 ± 4.09			X					
29	*A.alba* no.5 x *A.nordmanniana*	110	34.5 ± 3.62			X					
30	*A.alba* no.6 - selfing	24	29.0 ± 3.78		X						
31	*A.alba* no.6 x *A.alba* no.4	99	36.1 ± 3.71			X					
32	*A.alba* no.6 x *A.alba* no.7	100	38.2 ± 4.98				X				
33	*A.alba* no.6 – open pollination	13	31.0 ± 6.74		X						
34	*A.alba* no.6 x *A.nordmanniana*	135	37.1 ± 3.70			X					
35	*A.alba* no.6 x *A.cilicica*	92	39.7 ± 3.63				X				
36	*A.alba* no.7 - selfing	24	29.6 ± 4.12		X						
37	*A.alba* no.7 x *A.alba* no.4	91	33.8 ± 5.33		X						
38	*A.alba* no.7 x *A.alba* no.5	76	32.0 ± 4.36		X						
39	*A.alba* no.7 – open pollination	16	33.7 ± 5.96		X						
40	*A.alba* no.7 x *A.cilicica*	72	39.2 ± 4.48				X				
41	*A.nordmann.* no.1 – open poll.	52	31.5 ± 4.59		X						
42	*A.nordman.*1 x *A.nordman.* 2	43	25.7 ± 5.58	X							
43	*A.nordmanniana* no.1 x *A.alba*	88	32.8 ± 4.46		X						
44	*A.nordmann.*1 x *A.cephalonica*	95	34.4 ± 4.38			X					
45	*A.nordmann.* no.1 x *A.cilicica*	99	35.9 ± 4.56			X					
46	*A.nordmann.* n.1 x *A.numidica*	50	32.4 ± 4.54		X						

1) Means of the progenies of a given mother tree with crosses in the same column do not differ significantly

significantly in their height growth characteristics (*A. alba* no. 1, *A. alba* no. 2, *A. alba* no. 6). It is worth to mention in this connection that intraspecific crosses of both *A. alba* no. 5 and *A. alba* no. 6 mother trees shared similar or higher potential for height growth than interspecific hybrids *A. alba* no. 5 x *A.nordmanniana* and *A. alba* no. 6 x *A. nordmanniana.*

As to other interspecific crosses, only some of them exhibited more or less distinct symptoms of somatic heterosis overgrowing control variants from selfing, outcrossing and open pollination of the mother trees. In silver fir, it has primarily concerned the crosses *A. alba* no. 6 x *A. cilicica* and *A. alba* no. 7 x *A. cilicica* both of which achieved the highest height growth among the crosses in which *A. alba* was used as mother tree. The extent of height growth prevalence of these hybrids over controls from intraspecific crossings reached 10% in the former and 22% in the latter. The same is true of the interspecific crosses of *A. nordmanniana* no. 1 with *A. alba, A. cephalonica, A. cilicica* and *A. numidica,* all of which dominated the intraspecific control *A. nordmanniana* no. 1 x *A. nordmanniana* no. 2. Unfortunately, owing to the absence of control variants, it was not possible to confirm or deny the heterotic growth of the hybrid combinations attempted within mother trees of *A. nordmanniana* and *A. cephalonica* in 1991 (Table 1). Contrary to the interspecific hybrids given above, the combinations *A. numidica* x *A. alba* and *A. numidica* x *A. pinsapo* were characterized by the slowest growth lagging behind the open pollination progeny of the maternal tree. However, owing to the fact that open pollination progeny of any exotic species growing under arboretum conditions cannot be accepted as a control in artificial hybridization experiments, no conclusion about the negative heterosis of these combinations can be made. It is reasonable to suppose that open pollination control mentioned above might also contain the hybrids.

The differential potential for height growth of the progenies from selfing, intraspecific and interspecific crossings was demonstrated in *A. concolor* crosses. The inbred nature of selfed progenies obtained within mother trees *A. concolor* no. 1 and no. 2 has as a rule contrasted with the intermediate growth characteristics of *A. concolor* no. 1 x *A. concolor* no. 2 and reciprocal as well as with the highest parameters of the kind in the interspecific hybrid *A. concolor* x *A. grandis* derived from the two mother trees. The advantage of *A. concolor* x *A. grandis* seedlings over progenies from intraspecific control of both mother trees reached accordingly the levels of 12% and 21%. However, it is necessary to point out that both *A. concolor* x *A. grandis* and reciprocal combination exhibit heterotic growth with regard to the control variants of *A. concolor* no. 1 and no. 2 only but not with regard to the open pollination control of *A. grandis* which was found to perform best under nursery conditions (Table 1).

Discussion

Somatic heterosis is generally considered to be a characteristic feature in majority of *Abies* interspecific hybrids (Gathy 1957; Rohmeder 1960; Mergen and Gregoire 1988; Greguss 1992). The first artificial hybrid of *A. pinsapo* x *A. cephalonica* obtained by Vilmorin in 1867 has already been reported to exhibit heterotic growth (Sargent 1898). The same type of performance observed in 7-year old saplings of *A. veitchii* x *A. alba*, *A. concolor* x *A. procera* and *A. nordmanniana* x *A. veitchii* (Rohmeder and Eisenhut 1961) was found to persist also in 17-year old trees (Müller 1989). Comparing the height growth performance of the interspecific crosses of *A. alba*, *A. cephalonica*, *A. cilicica*, *A. nordmannina* and *A. numidica,* all of which were obtained according to the incomplete diallelic model, Greguss (1992) postulated a better performance for the interspecific hybrids than for control from open pollination of the parental species. According to the author, heterosis of the hybrid progenies was higher in 4-year old than in 1- and 10-year old seedlings. For example, the observed prevalence of the 4-year old seedlings of *A. alba* x *A. cilicica* combination over open pollination progeny of the maternal tree has reached 16% in the trials mentioned above. A comparable degree of the interspecific hybrid prevalence over controlled intraspecific crossings of the mother trees was also found in the 5-year old crosses of *A. alba* no. 6 and *A. alba* no. 7 with *A. cilicica* in our study. However, it is worth of mentioning that the same effect as with interspecific hybridization may be achieved probably also with intraspecific crossing, as evidenced by the corresponding crosses of the maternal trees *A. alba* no. 5, *A. alba* no. 6 and *A. alba* no.7. This is in accord with recommendation by Vincent and Kantor (1971) who call for „blood refreshment“ in withering stands of silver fir in former Czechoslovakia by means of involving into intraspecific crossing the trees from Eastern Great Carpathian Arc where silver fir exhibits vigorous growth and vitality. In support of this approach, the finding by Moulalis (1986) may be mentioned concerning a better performance of 6-year old seedlings from controlled intraspecific crossing of *A. alba* than was the height growth performance of the seedlings from either self- or open pollination.

The North American firs *A. concolor* and *A. grandis* are known to intercross spontaneously in zones of their sympatry in southwestern part of Oregon and in northwestern California, producing offsprings with clearly expressed symptoms of somatic heterosis (Duffield and Snyder 1958; Frederick 1977). The artificial hybrids of these species were also reported to exhibit such growth during the juvenile stages of their development and in mature trees (Schlepitz 1956; Rohmeder 1960; Gautlitz 1985). Taking into account the results of presented study, we may add that this conclusion is valid on the seedling level of the hybrid as well. It is obvious that *A. grandis* genes improve height growth in *A. concolor*. The same kind of evidence we have provided also for the interspecific crosses obtained by Greguss (1992). In particular, it concerns the heterotic nature of *A. nordmanniana* crosses with *A. cephalonica*, *A. cilicica* and *A. numidica*, all of which are characterized by the higher survival rates at age of 10 and 15 years than intraspecific control of *A. nordmanniana.*

We have observed only partial correlation between the growth intensity of the interspecific hybrids and geographic distribution of the respective parental species as postulated for the hybrid trial in Connecticut by Mergen and Gregoire (1988). The species combinations *A. nordmanniana* x *A. cephalonica*, *A. nordmanniana* x *A. cilicica* and *A. cephalonica* x *A. alba* with parents sharing neighbouring geographic ranges have as a rule grown more vigorously than crosses *A. nordmanniana* x *A. alba*, *A. nordmanniana* x *A. numidica* and *A. cephalonica* x *A. nordmanniana* whose parents are geographically more separated. On the contrary, the hybrid combination of geographically very allied species *A. numidica* x *A. pinsapo* from southern part of Europe was difficult to distinguish from the interspecific hybrid *A. numidica* x *A. alba* with the male parent occurring naturally in central Europe.

According to Kobliha (1994), even more important than somatic heterosis of *Abies* interspecific hybrids is their adaptive heterosis conferring a higher tolerance of the hybrids to changing ecological conditions. This property of the hybrids is of paramount importance in reforestation of heavily polluted areas where silver fir cannot withstand the emission pressure. The Mediterranean firs and their hybrids have been offered in this connection as a perspective solution to the global warming problems in the region of central Europe (Kobliha 1994). Experimentally this concept is supported by the finding of higher survival rates in *Abies* interspecific hybrids than in pure species (Greguss 1996). However, this aspect of hybrid performance was not the subject of our study. Therefore, the next step in testing the superiority of *Abies* interspecific hybrids over intraspecific combinations will be oriented towards establishment of the permanent test plot in heavily polluted region of the Upper Nitra using seedling material which has been analyzed in present study.

Acknowledgement

This study was financially supported by the VEGA grant agency, project no. 2/4162/99.

References

Dubovský, J. and Maršálek, L. 1968: Plant Genetics. (In Slovak). Slovenské vydavateľstvo pôdohospodárskej literatúry, Bratislava, pp. 411

Duffield, J.W. and Snyder, E.B. 1958: Benefits from hybridizing American forest tree species. Jour. For. 56: 809-815

Fady, B. and Conkle, M.T. 1993: Allozyme variation and possible phylogenetic implications in *Abies cephalonica* Loudon and some related Eastern Mediterranean firs. Silvae Genetica 42: 351-359

Frederick, D.J. 1977: An integrated population of *Abies grandis* x *A. concolor* in Central Idaho and its relation to decay. Silvae Genetica 26: 8-10

Gaudlitz, G. von 1985: Merkmale und Eigenschaften luxurierender Tannenbastarde. pp. 222- 238 in: Kramer, W. (Ed.) Ergebnisse des 4. Tannen-Symposiums. Band 80. Schriften aus der Forstlichen Fakultät der Universität Göttingen. J.D. Sauerländer's Verlag frankfurt am Main

Gathy, P.P. 1957: A propos de l'hybride natural *Abies concolor* (Gord.) Engelmann x *A. grandis* (Link.). Silvae Genetica 6: 186-190

Greguss, L. 1992: Assessment of the initial growth of inter-specific fir hybrids on the example of the permanent experimental plot Drieňová. (In Slovak). Lesn. Čas. - Forestry Journal 38: 223-238

Greguss, L. 1996: Survival of inter-specific fir hybrids until the age of 15 years in three test plots. (In Slovak). Lesn. Čas. – Forestry Journal 42: 363-370

Kaňák, K. 1988: Contribution to maintaining continuity of the Norway spruce in the Ore Mountains. Folia Musei Rerum Naturalium Bohemiae Occidentalis. Plzeň. Botanica 27: 1-62

Kobliha, J. 1994: Hybridization within the genus *Abies* with special emphasis on obtaining F_2 generation. (In Czech). Lesnictví-Forestry 40: 513-518

Koski, V. 1996: Breeding plans in case of global warming. Euphytica 92: 235-239

Larsen, J.B. 1988: Waldbauliche Probleme und Genökologie der Weisstanne (*Abies alba* Mill.) pp. 9-19 in: Paule, l. and Korpeľ, Š. (Eds.) 5. IUFRO-TANNENSYMPOSIUM, Hochschule für Forstwirtschaft und Holztechnologie, Zvolen

Mergen, F. and Gregoire, T.G. 1988: Growth of hybrid fir trees in Connecticut. Silvae Genetica 37: 118-124

Mitsopoulos, D.J. and Panetsos, C.P. 1987: Origin of variation of fir forests of Greece. Silvae Genetica 38: 1-15

Moulalis, D. von. 1986: Selbstinkompatibilität und Inzucht bei der Weisstanne (*Abies alba* Mill.). Forstw. Cbl. 105: 487-494

Müller, K.W. 1989: Manifestieren wir den Reichtum der Natur. Deutsche Baumschule 1: 32-34

Rohmeder, E. 1960: Bastardierung in der Gattung *Abies*. Silvae Genetica 9: 136-137

Rohmeder, E. and Eisenhut, G. 1961: Bastardierungsversuche an der Gattung *Abies*. Allg Forstzeitschrift 34: 495-497

Sargent, C.S. 1898: The Silva of North America. Vol. 12 (Coniferae). Houghtan and Mifflin and Co., Boston, pp.95-139

Schlepitz, B. 1956: Über einen natür lichen *Abies*-Bastard. Morphologische und Holztechnologische Untersuchung an Artbastarden von *Abies concolor* x *A. grandis.* Z. Forstgen. Forstpfl. 5: 71-79

Vincent, G. and Kantor, J. 1971: Das frühzeitige Tannensterben, seine Ursachen und Vorbeugung. Clb. ges. Forstwesen. 88:101-115

Field testing of model populations of Norway spruce (*Picea abies* L. KARST.) with different genetic structures for reforestation in air polluted regions

H. HERTEL[1] AND M. PAUL[2]

[1]*Institute for Forest Genetics and Forest Tree Breeding, Federal Research Centre for Forestry and Forest Products, D-15377 Waldsieversdorf, Germany;* [2] *Department of Genetics and Breeding, Saxon State Institute for Forestry, D-01827 Graupa, Germany*

Key words: *Picea abies*, population genetics, gene conservation, breeding strategies, reforestation, crossing, progeny test, isozyme gene markers, heterozygosity

Abstract: The presented paper refers to special importance of long-term programs in the field of gene conservation and forest tree breeding for the Saxon Ore Mountains as well as accompanying genetic investigation. The situation required taking the necessary steps, because many Norway spruce stands have died back in part or completely in the Ore Mountains in the eastern part of Germany due to several decades of heavy SO_2 impact. Conservation of spruce trees which can survive these conditions was started 25 years ago using several methods for evacuation. In our days, reforestation in the Ore Mountains needs suitable material that can be obtained from these evacuated plants. The use of this material is described for the example of a clonal plantation. All clones of the plantation were characterized by 25 polymorphic isozyme gene markers prior to choosing pairs of parents to form three categories of offspring families: low level of heterozygosity with low number of rare alleles (model population 1), high level of heterozygosity with low number of rare alleles (model population 2), high level of heterozygosity with high number of rare alleles (model population 3). The objective of genetic investigations in this program is to determine whether population genetic parameters (heterozygosity, number of rare alleles) are suitable tools to support the selection of Norway spruce plant material for reforestations. Associations between population genetic parameters and phenotypic traits were examined in a progeny test. Preliminary results demonstrate that the genetic basis of progenies seems to be adequate to contribute to the restoration of Norway spruce populations adapted to the local conditions of Saxon Ore Mountains.

Introduction

Situation in damaged forests and natural conditions in Ore Mountains

Since the 70's and the 80's of the last century, the burning of large quantities of brown coal in power stations resulted in heavy SO_2-immissions that were loaded on the Saxon highlands in the Ore Mountains (Erzgebirge). The

Müller-Starck, G. and Schubert, R. (eds.). Genetic Response of Forest Systems to Changing Environmental Conditions. Vol. 70 (For. Sci.) 339-352

continuous pollutant deposition led to catastrophic damages, especially in spruce ecosystems. At present, two thirds of the total forested area of Saxony are damaged by air pollution. That means 6.000 ha with very severe damages, 32.000 ha with severe damages, 78.000 ha with medium damages, and 166.000 ha with light damages (Liebold and Drechsler 1991). The problem is still an acute one as has been demonstrated by the winter 1995/96. After special weather conditions (inversion situation) in connection with a high level of SO_2, about 50.000 ha spruce stands were damaged, among them 20.000 ha with very severe damages and 3.000 ha are dead (Anonymus 1996).

The Ore Mountains are characterized by rough natural conditions in the higher elevations between 650 and 1200 m above sea level. The annual mean temperature amounts to 4 to 6 °C, and the annual precipitation is 900 to 1000 mm. The combination of snow, fog and mean wind speeds of 3 to 7 m/s often results in ice-break. This environment with extreme climatic and site conditions led to the evolution of an adapted spruce race in the natural forest community in the high elevations of the Ore Mountains. The actual stocking however was mainly artificially established, often with unsuitable progenies (Nestler et. al 1957; Paul 1993). In addition to the immission damages, this fact led to increasing ecological instability (break and insect calamities).

Aims, strategies and measures

The present task of forestry in Saxony is the realization of a sustained multifunctional forest (Braun 1995). Reforestation has to create mixed stands with predominate Norway spruce to keep or restore all functions of the forest. That means the combination of timber production, landscape protection, and recreation purposes at the same area. It was observed that autochthonous or site adapted stands have survived for a longer time in the SO_2-polluted area than non-adapted provenances from former artificial reforestations. The reconstruction of the natural vegetation has to consider the genetic constitution of the populations (Mejnartowicz 1983; Braun and Kohlstock 1990).

With the beginning of the extensive die-back of stands, foresters and forest scientists have searched for possibilities to take measures against the damages (Piesnack and Flöhr 1988; Mauersberger *et al.* 1991). The first step of all breeding programs is the gene conservation to rescue the material that had survived (Braun *et al.* 1987; Weiß *et al.* 1990; Paul 1998a, b). The strategy includes measures of gene conservation combined with different breeding programs and short-term reactions as a crisis management by using untested material (Fig. 1, Tzschacksch 1983b; 1998; Kohlstock 1985).

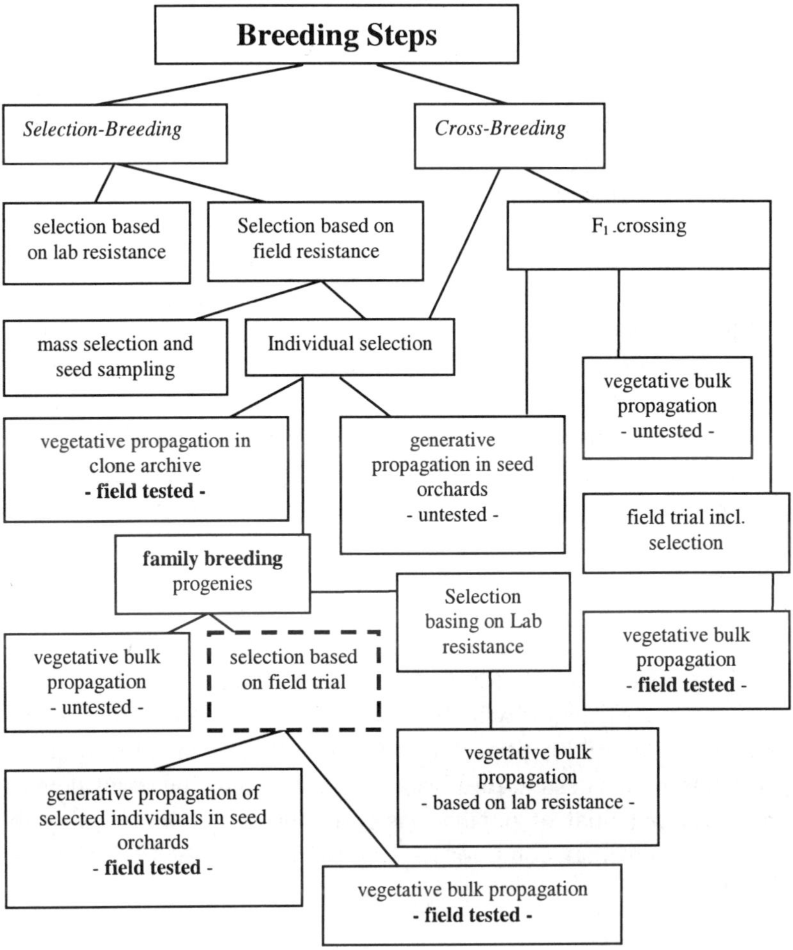

Figure 1. Original breeding strategy cited from a research report by Tzschacksch (1983a).

Experimental example for the use of evacuated material

Several ways exist to use *in situ* and *ex situ* sources. The use of *in situ* sources was started more than 25 years ago with the harvest of seeds from the remaining adult individuals, but the number of seed crop stands in higher elevations was not sufficient. Also natural regeneration was limited because it was possible only in a few cases. The introduction of suitable provenances from other regions of the natural area of Norway spruce can be recommended for the cultivation in

Saxony based on the results of long-term provenance trials (Weiß 1994). Selected juvenile and adult trees from the damaged area were propagated by cuttings or grafts in order to establish clone collections, seed orchards, and mother tree plantations in areas with a lower level of air pollution. It is assumed that autochthonous populations were among the saved material at a very high probability. Now they are the only remainders of these spruce stands, which can be integrated into reforestation.

In 1998, a field trial was established to test reintroduced populations in according to the general strategy set up for the reforestation of the Ore Mountains. The attempt was made to combine this reforestation trial in polluted areas with a model experiment for more scientific questions on adaptation processes and genetic structures. The hypotheses on relationships between genetic structure of populations and their stability in adaptation processes should be mentioned that played a role for the choice of experimental material.

There are numerous indications for a positive association between heterozygosity and fitness-related or adaptive traits (as growth, survival and fecundity) in populations of forest trees and other organisms (actual reviews including arguments against by David 1998, Deng and Fu 1998). Therefore, different levels of mean heterozygosity were regarded.

In this study, besides heterozygosity the presence of rare alleles was included into consideration. Rare alleles were described to have a negative influence on survival or growth (Bush and Smouse 1991, Simon *et al.* 1995, Strauss and Conkle 1986; Strauss and Libby 1987, v. Wühlisch and Krusche 1991). Despite their possible disadvantages, rare alleles are considered to be a latent adaptive potential of populations and a precondition for adaptability to changing environmental conditions (Bergmann *et al.* 1990, Finkeldey 1993, Turok 1996).

Material and methods

Origin of plant material, genetic characterization and formation of model populations

Norway spruce (*Picea abies* (L.) Karst.) clones originate from the forest districts Seiffen, Oberlochmühle and Deutscheinsiedel in a region strongly influenced by SO_2 in the upper part of the Saxon Ore Mountains (Fig. 2). They were selected in the early 70's according to the main criteria of lack of needle damage and frost hardiness at the age of 40 to 50 years (Tzschacksch 1983a), and they were propagated by cuttings in the following years. A plantation was established near Waldsieversdorf (Brandenburg) in 1979 (Weiser and Schachler

1988) using rooted cuttings of 58 selected clones with an average number of 34 ramets (3...40) per clone.

The genetic characterization of spruce clones and seedlings of offspring families by isozyme markers was carried out on bud and megagametophyte tissue. The electrophoretic separation of enzyme proteins in starch and polyacrylamide gels, staining solutions and the interpretation of banding patterns are described by Konnert & Maurer (1995) and Hertel & Ewald (1995). Leucine aminopeptidase (LAP-A), 6-Phosphogluconate dehydrogenase (PGDH-A), and Shikimate dehydrogenase (SKDH-A) are among the 25 polymorphic gene loci, which were evaluated. The genetic characterization showed that these clones do not differ significantly from other spruce material from Ore Mountains. Rare alleles were defined as alleles with frequencies below 0.05 in the plantation. The same alleles are rare or absent in other populations from this region (unpublished results).

All clones were characterized by isozyme gene markers prior to choosing pairs of parents to form three categories of offspring families. A crossing program for model populations with defined genetic structures was created: low level of heterozygosity with low number of rare alleles (model population 1), high level of heterozygosity with low number of rare alleles (model population 2), and high level of heterozygosity with high number of rare alleles (model population 3). The genetic structures of offspring families were calculated from the parents' multilocus genotype under the condition of 1:1-segregation of alleles of heterozygous loci, free recombination and absence of selection.

30 clones of the total of 58 were included into the crossing program. Crossings were carried out in spring 1993, and cones were harvested in autumn 1993. 48 crossings were found to be successful out of 55 combinations. All offspring families are outcrossed; no crossing of relations is expected.

Observation of phenotypic traits and establishment of field trials

Raw seeds were sown into open ground in spring 1994 including the provenance Kleingießhübel as standard. The data on number of seedlings and mean height per family after the first and second vegetation period were collected, and the survival and increment rates 1994-1995 were calculated.

Offspring families with numbers of 150 seedlings or more were kept separately. Offspring families with numbers lower than 150 were pooled according to their model population to generate progenies with numbers sufficient for field trials in block design. The seedlings were planted into the nursery in spring 1996. The number of plants and mean height per progeny after the third and fourth vegetation period was measured and survival and increment rates 1996-1997 were calculated. In addition, the frost damage after winter 1996/1997 was scored.

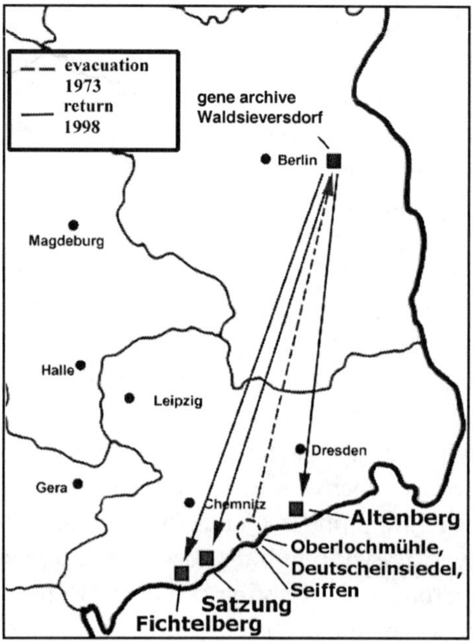

Figure 2. Origin of survived Norway spruce, place of evacuated clones and location of field trials

In 1998, a progeny test with 3 experimental plots was established in the damaged area to observe the model populations in a field trial (Fig. 2, Tab. 1). A further standard provenance (Carlsfeld) was added.

Table 1. Site conditions at three experimental locations in the Ore Mountains

Field trial	Altenberg	Satzung	Fichtelberg
Elevation (m a.s.l.)	830	880	1.170
Precipitation (mm)	940 - 1.000	1.000	1.000 – 1.200
Annual mean temperature	4.5 – 5.5°C	< 5°C	< 4.5°C
Soil	poor, moist	poor, moist	poor, dry
Natural Vegetation (PNV)	Piceetum	Piceetum	Piceetum
Tree species (PNV)	Spruce, Beech	Spruce	Spruce

Results

Genetic structures and phenotypic traits of model populations

The model populations are characterized by varying levels of heterozygosity and the numbers of rare alleles as it was designed for this experiment (Fig. 3).

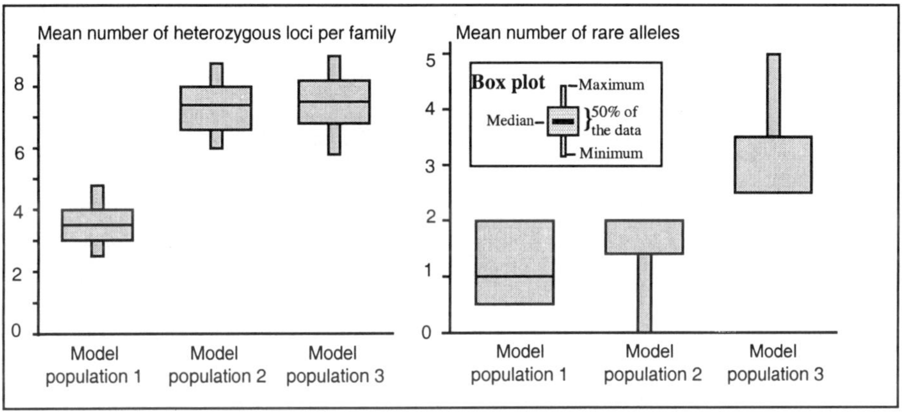

Figure 3. Genetic characteristics of three model populations (Box plots for the range of number of heterozygous loci and for the range of number of rare alleles)

The model populations 2 and 3 have the same high level of heterozygosity whereas their mean number of rare alleles differ remarkably. The model population 1 however shows a low level for both parameters.

The comparison between model populations for several phenotypic traits depending on plant age is given in Table 2. More detailed results for the seedling stage of development are given in Hertel and Kohlstock 1996. The height growth and the annual increment rates are statistically significant greater for the model population 3 for the whole time of experiment. The lowest values were observed for model population 1. The survival and frost damage rates show a tendency for the same relationship.

Genetic structures and phenotypic traits of single offsprings

In addition to the comparison between the three model populations, the data evaluation was carried out for all 48 offspring families. Figure 4 illustrates the relationship between the height growth of 1-year-old seedling families and the expected family heterozygosity. The Pearson correlation coefficient amounts to +0.52 ($p<0.001$, $n=48$).

Table 2. Comparison between model populations (Statistics: Wilcoxon rank test, the same letters indicate no significant differences between model populations, p<0.05)

	Model population 1	Model population 2	Model population 3	Provenance Kleingießhübel	Provenance Carlsfeld
Seedling stage 1994-1995	13 families	16 families	19 families		
Height 1994 (cm)	3.2 a	4.1 b	4.7 b	3.8	
Height 1995 (cm)	14.0 a	14.5 a	18.0 b	14.3	
Increment 94-95 (cm)	10.8 a	10.3 a	13.3 b	10.5	
Survival Oct94 – Oct95 (%)	78.5 a	75.6 a	88.0 a	93.0	
Nursery stage 1996-1997	3 families + 2 pooled progenies	5 families + 3 pooled progenies	9 families + 2 pooled progenies		
Height 1997 (cm)	43.2 a	48.2 a	54.5 b	52.2	
Increment 96-97 (cm)	22.4 a	28.7 ab	31.6 b	35.8	
Survival Apr96 – Sept97 (%)	76.5 a	77.6 a	79.8 a	79.2	
Frost damage 96/97 (%)	14.4 a	8.5 a	7.2 a	10.6	
Field trials since 1998 (3 sites)	9	15	25	3	3
Height 1999 (cm)	49.7 a	54.7 b	56.7 b	53.3	51.4
Survival Apr98 – Sept99 (%)	95.6 a	96.9 a	97.4 a	94.7	93.3

Up to now, observed data on genetic structures are available for only 3 offspring families, one out of each model population. Genetic investigation was carried out after the second vegetation period for almost all isozyme loci with 2 or 4 different genotypes in the offspring. We observed 3 cases of significant deviation (out of 18 tests) between the observed and the expected genotypic structure in the offspring families. In 2 cases, a deficiency of a rare allele was detected (Tab. 3).

Discussion

Besides the practical aims of the field trials to get information on the suitability of evacuated material for reforestation, the scientific objective was to investigate, whether the genetic parameters based on isozyme markers could support practical measures. The adaptivity of single isozyme loci under special

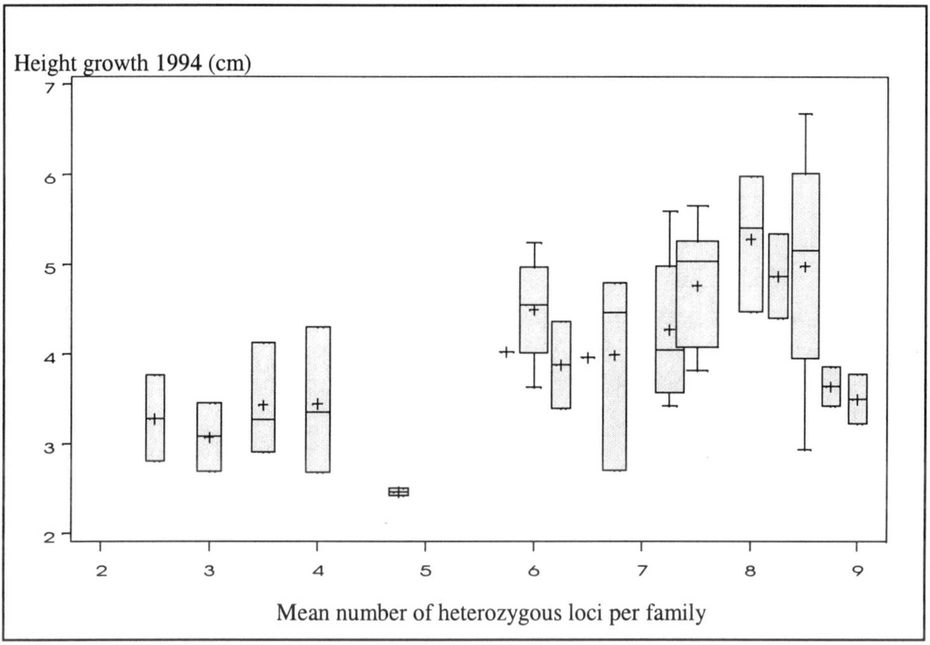

Figure 4. Box plot of height growth of 1-year-old seedlings with different levels of family heterozygosity (box width varies with subgroup size 1...7)

Table 3. Deviation between expected and observed allele frequencies in offspring families (Chi-squared test * p<0.05)

Offspring family / model population	Allele	Frequency in the plantation	Number of tested seedlings	Ratio in the offspring (rare : common)
32 (304 x 325) / M2	SKDH-A1	0.054	62	39 : 23 *
36 (304 x 292) / M3	LAP-B1	0.036	48	17 : 31 *
36 (304 x 292) / M3	PGDH-A3	0.027	45	13 : 32 *

environmental conditions or in certain life stages was described in numerous examples. Viability selection during the reduction of population size was detected by a shift of genetic structures at single loci (beech: Kim 1985, Müller-Starck 1993, Norway spruce: Geburek 1987, Konnert 1991, silver fir: Konnert 1993, sessile oak: Kim 1999). The adaptive significance of isozyme marker loci may become obviously if individual genotypes and respective phenotypes are considered as in the examples above, and it could be hidden for population means of genetic parameters and phenotypic traits because of the larger portion of variance within populations.

It is known from the investigations about genetic consequences of air pollution impact that groups of tolerant trees are often more heterozygous than groups of sensitive trees. This was found for mean heterozygosity of populations of several tree species, for example Norway spruce (Bergmann and Scholz 1989, Raddi *et al.* 1994, Prus-Glowacki and Godzik 1995), Scots pine (Geburek *et al.* 1987, Bakhtiyarova and Yanbaev 1996, Prus-Glowacki *et al.* 1998), beech (Müller-Starck 1985, Hertel and Zander 1991), and oak (Hertel and Zaspel 1996). Sometimes results were published that are describing advantages for more homozygous genotypes (Bergmann and Scholz 1989, Norway spruce clone mixtures in fumigation experiments, Mejnartowicz 1983, Prus-Glowacki and Godzik 1991, selection processes in Scots pine stands under influence of special air pollutants) or no relationship between heterozygosity and vitality (Zaspel and Hertel 1999, phase of partial recovery from oak decline in long-term observation plots). It seems to be a clearer relationship between heterozygosity and fitness in case of enhanced stress than under normal conditions. Independent from air pollution or environmental stress, the reports on relationships between heterozygosity and growth or adaptive traits are inconsistent. With the exception of Bongarten *et al.* 1985, no distinction was made for major and minor polymorphisms. The idea of an optimum heterozygosity was represented by Bongarten *et al.* 1985, Altukhov *et al.* 1986, Dukharev *et al.* 1987 and Strauss and Libby 1987. It could be explained by the overlay of two effects: the heterozygosity advantage (balanced selection for major polymorphisms) and the negative influence of some rare alleles.

In our example of model populations, the model population 1 (low number of heterozygous loci) has the minimum growth, or regarding 48 full-sib families, the linear correlation coefficient between number of heterozygous loci and seedling growth for family means amounted to +0.52. All parental clones are non-related and were considered to be members of one population because of the restricted area of their origin. These preconditions do not allow assuming inbreeding depression as a cause for the lower growth of families in model population 1. However, it seems that pairs of individuals within a population with larger genetic distances (measured as number of different alleles at the tested isozyme loci) generate offsprings with a higher growth and survival. These results support the hypothesis of heterozygote advantage.

Nevertheless, as demonstrated in genetic analyses of random samples of 3 offspring families, the reduction of family size was coupled with selection processes in 2 of the 3 cases. In the family belonging to model population 3, the selection against 2 rare alleles resulted in a lower level of heterozygosity as expected. Further genetic analyses of offspring families are planned to investigate, whether this is an exception or not.

The preliminary results demonstrate that offsprings derived from a limited number of parental clones develop at least as well as the two reference provenances. The genetic basis seems to be adequate for reconstruction of stable populations despite the strongly selected material after air pollution and vegetative propagation. Thus, it is allowed to conclude that progenies of survived and evacuated clones are a valuable complement for the restoration of populations adapted to the local conditions and adaptable to changing environmental conditions. Long-term observations of field trials and further marker-aided genetic analyses to obtain knowledge about population genetic structures and the inheritance of phenotypic traits (adaptability, morphology, resistance, growth) will support the forest practice for breeding and establishing seed orchards and clonal mixtures that are based on field-tested material.

Acknowledgement

We would like to thank Otto Tzschacksch (resistance breeding programs), Martin Weiß (provenance research of Norway spruce), Kurt Nestler (forester), Dietrich Ewald (co-operation during crossings), Mrs. Elke Ewald and Mrs. Marianne Schneck (laboratory assistance), Roberto Böhme (nature reservation officer), Peter Brantz (district forest officer) and Olaf Nowatzki (assistant) who contributed their work to these results, and two anonymous reviewers for their helpful critical remarks.

References

Altukhov, Y. P., Gafarov, N. J., Krutovskii, K. V. and Dukharev, V. A. 1986: Allozyme polymorphism in a natural spruce population III. Correlation between individual heterozygosity and relative portion of non-viable seeds. Genetika 22: 2825-2830 (in Russian)

Anonymus 1996: Waldschadensbericht 1996. Sächsisches Staatsministerium für Landwirtschaft, Ernährung und Forsten, Dresden

Bakhtiyarova, R. M. and Yanbaev, Y. A. 1996: Genetic variation in Scotch pine (*Pinus sylvestris* L.) under industrial pollution in the southern Urals: Comparison of tree groups differing in life state. Genetika 32: 1135-1141 (in Russian)

Bergmann, F. and Scholz, F. 1989: Selection effects of air pollution in Norway spruce (*Picea abies*) populations. pp. 121-128 in: Scholz, Gregorius, Rudin (Eds.): Genetic effects of air pollutants in forest tree populations. Springer-Verlag Berlin Heidelberg

Bergmann, F., Gregorius, H.-R. and Larsen, J. B. 1990: Levels of genetic variation in European silver fir (*Abies alba*) – are they related to the species decline? Genetica 82: 1-10

Bongarten, B. C., Wheeler, N. C. and Jech, K. S. 1985: Isoenzyme heterozygosity as a selection criterion for yield improvement in Douglas-fir. pp. 121-128 in: New ways in forest genetics. Proceedings of the Improvement Association

Braun, H. 1995: Strategien des Waldumbaus in Sachsen. Der Wald 45: 46-50

Braun, H. and Kohlstock N. 1990: Aufgaben und Ergebnisse der Forstpflanzenzüchtung speziell für die Immissionsschadgebiete der Mittelgebirge der DDR. Allgemeine Forstzeitschrift München 45: 868-873

Braun, H., Weiß, M. and Kohlstock, N. 1987: Erhaltung des genetischen Potentials der Fichte. Sozialistische Forstwirtschaft Berlin 37: 149-152

Bush, R.M. and Smouse, P. E. 1991: The impact of electrophoretic genotype on life history traits in *Pinus taeda*. Evolution 45: 481-498

David P. 1998: Heterozygosity-fitness correlations: new perspectives on old problems. Heredity 80: 531-537

Deng H.-W. and Fu Y.-X. 1998: Conditions for positive and negative correlations between fitness and heterozygosity in equilibrium populations. Genetics 148: 1333-1340

Dukharev, V. A., Romanovskii, M. G. and Ryabokon, S. M. 1987: Heterozygosity and seed production in Scots pine. Lesovedenye 2: 87-89 (in Russian)

Finkeldey, R. 1993: Die Bedeutung allelischer Profile für die Konservierung genetischer Ressourcen bei Waldbäumen. Göttinger Forstgenetische Berichte 14

Geburek, T., Scholz, F., Knabe, W. and Vornweg, A. 1987: Genetic studies by isozyme gene loci on tolerance and sensitivity in an air polluted *Pinus sylvestris* field trial. Silvae Genetica 36: 49-53

Hertel, H. and Zander , M. 1991: Genetische Unterschiede zwischen gesunden und geschädigten Buchen eines belasteten Bestandes. pp. 227-229 in: Berichte des Forschungszentrums Waldökosysteme, Reihe B, Bd. 22

Hertel, H. and Zaspel, I. 1996: Investigations on vitality and genetic structure in oak stands. Annales des Sciences Forestieres 53: 761-773

Hertel, H. and Ewald, C. 1995: Conservation of genetic diversity in SO_2-polluted stands of Norway spruce. pp. 413-420 in: Baradat, Adams, Müller-Starck (Eds.) Population genetics and genetic conservation of forest trees. SPB Academic Publishing Amsterdam, The Netherlands

Hertel, H. and Kohlstock, N. 1996: Forstpflanzenzüchtung und Biodiversität: Ein Modellversuch zur Untersuchung des Zusammenhangs zwischen genetischer Variation und Stabilität am Beispiel der Fichte (*Picea abies* L. Karst.). pp. 93-104 in: Müller-Starck (Ed.) Biodiversität und nachhaltige Forstwirtschaft. ecomed Landsberg

Kim, T.-S. 1999: Einfluß von Vermehrungsmethoden auf die genetische Struktur sowie phänotypische und phänologische Eigenschaften von drei Populationen der Traubeneiche (*Quercus petraea* (Matt.) Liebl.). Dissertation, Universität Hannover

Kim, Z.-S. 1985: Viability selection on an allozyme locus during development in European Beech (*Fagus sylvatica* L.). Silvae Genetica 34: 181-186

Kohlstock, N. 1985: Stand und Perspektiven der Rauchtoleranz-Züchtung und Möglichkeiten der Überführung. Beiträge für die Forstwirtschaft 19: 153-157

Konnert, M. 1991: Vergleich der genetische Struktur verschiedener Generationen zweier natürlich verjüngter Fichtenbestände des Schwarzwaldes. Silvae Genetica 40: 60-65

Konnert, M. 1993: Untersuchungen zum Einfluß genetischer Faktoren auf die Schädigung der Weißtanne. Forstwissenschaftliches Centralblatt 112: 20-26

Konnert, M. and Maurer W. 1995: Isozymic Investigations on Norway Spruce (*Picea abies* (L.) Karst.) and European Silver Fir (*Abies alba* Mill.) – A Practical Guide to Separation Methods and Zymogram Evaluation. 74 p., Bayerische Landesanstalt für forstliche Saat- und Pflanzenzucht, Teisendorf

Liebold, E. and Drechsler, M. 1991: Schadenszustand und -entwicklung in den SO_2-geschädigten Fichtengebieten Sachsens. Allgemeine Forstzeitschrift München 46: 492-494

Mauersberger U., Gärtner, Richter H. and Liebold, E. 1991: Gegenmaßnahmen der Forstwirtschaft in den sächsischen SO_2-Schadgebieten. Allgemeine Forstzeitschrift München 46: 496-499

Mejnartowicz, L. 1983: Changes in genetic structure of Scots pine population affected by industrial emission of fluoride and sulphur dioxide. Genetica Polonica 24: 41-50

Müller-Starck, G. 1985: Genetic differences between "tolerant" and "sensitive" beeches (*Fagus sylvatica* L.) in an environmentally stressed adult forest stand. Silvae Genetica 34: 241-247

Müller-Starck, G. 1993: Auswirkungen von Umweltbelastungen auf genetische Strukturen von Waldbeständen am Beispiel der Buche (Fagus sylvatica L.). Schriften aus der Forstl. Fak. d. Univ. Göttingen u. d. Niedersächs. Forstl. Versuchsanst., Bd. 112

Nestler, K., Mann, E. and Riess, G. 1957: Die Nebelfrostablagerungen im Fichtelberggebiet. Fachschule für Forstwirtschaft Schwarzburg, Abschlußarbeit

Paul, M. 1998a: Die Nutzung immissionsresistenter Fichten - Picea abies (L.) Karst.- für die forstliche Generhaltung. pp. 69-72 in: Forstpflanzenzüchtung für Immsionsschadgebiete. Schriftenreihe der Sächsischen Landesanstalt für Forsten 13/1994

Paul, M. 1998b: The importance of forest tree breeding trials and collections for the forest gene-conservation of the Norway spruce. In: Skroppa, Paule, Gömöry (Eds.): Genetics and Breeding of Norway Spruce. Arbora Publishers, Zvolen

Paul, M.. 1993: Forstpflanzenzüchtung und naturnaher Waldbau. Der Wald 43: 296-299

Piesnack J. and Flöhr W. 1988: Waldbauliche Methoden in immissionsbelasteten Waldgebieten der DDR. Sozialistische Forstwirtschaft 38: 213-215

Prus-Glowacki, W. and Godzik, S. 1991: Changes induced by zinc smelter pollution in the genetic structure of pine (*Pinus sylvestris* L.) seedling populations. Silvae Genetica 40: 184-188

Prus-Glowacki, W. and Godzik, S. 1995: Genetic structure of *Picea abies* trees tolerant and sensitive to industrial pollution. Silvae Genetica 44: 62-65

Prus-Glowacki, W.; Oleksyn, J. and Reich, P.B. 1998: Relation between genetic structure and susceptibility to air pollution of European *Pinus sylvestris* populations from a IUFRO-1982 provenance experiment. Chemosphere 36: 813-818

Raddi, S., Stefanini, F. M.; Camussi, A. and Giannini, R. 1994: Forest decline index and genetic variability in *Picea abies* (L.) Karst. Forest Genetics 1: 33-40

Simon, J.-P., Payette, Y. and Longpre, M.-H. 1995: Comparative analysis of the genetic composition of canopy and juvenile sugar maple individuals in an old-growth forest in southern Quebec as related to anthropogenic disturbance. Can J. For. Res. 25: 743-752

Strauss, S.H. and Conkle, M. T. 1986: Segregation, linkage, and diversity of allozymes in knobcone pine. Theoretical and Applied Genetics 72: 483-493

Strauss, S.H. and Libby, W.J. 1987: Allozyme heterosis in radiata pine is poorly explained by overdominace. Am. Nat. 130: 879-890

Turok, J. 1996: Genetische Untersuchungen bei der Buche – Genetischen Anpassungsprozesse und die Erhaltung von Genressourcen in Buchenwäldern (*Fagus sylvatica* L.). Dissertation, Universität Göttingen

Tzschacksch, O. 1983a: Immissionsresistente Fichten-Herkunftssorte und immissionsresistente Fichten-Mehrklonsorte. Institut für Forstwissenschaften Eberswalde, Forschungsbericht, 27 p.

Tzschacksch, O. 1983b: Möglichkeiten der Begründung von Fichtenbeständen verminderter Anfälligkeit gegenüber phytotoxischen Immissionen durch Hochpfropfungen. Sozialistische Forstwirtschaft Berlin 33: 339-341

Tzschacksch, O. 1998: Wege und Irrwege der Immissionsresistenzzüchtung. pp. 12-23 in: Forstpflanzenzüchtung für Immsionsschadgebiete. Schriftenreihe der Sächsischen Landesanstalt für Forsten 13/1998

v. Wühlisch, G. and Krusche, D. 1991: Single and multilocus genetic effects on diameter growth in *Picea abies* (L.) Karst.. pp. 77-86 in: Fineschi, Malvolti, Cannata, Hattemer (Eds.) Biochemical markers in the population genetics of forest trees, SPB Academic Publishing bv, The Hague, The Netherlands

Weiser, F. and Schachler, G. 1988: Aufbau sowie erste Ergebnisse zur Entwicklung und Nutzung eines Stecklings-Mutterquartiers mit Fichtenklonen verminderter Anfälligkeit gegenüber SO_2. Beiträge für die Forstwirtschaft 22: 55-61

Weiß, M. 1994: 45 Jahre Forstpflanzenzüchtung in Graupa. pp. 13-32 in: Forstpflanzenzüchtung Quo vadis? Schriftenreihe der Sächsischen Landesanstalt für Forsten 1/1994

Weiß, M., Paul, M. and Braun, H. 1990: Das genetische Potential der Fichte. Der Wald 40: 332 334

Zaspel, I. and Hertel, H. 1999: Genetische Aspekte bei der Untersuchung geschädigter Stiel- und Trauben-Eichen. pp. 77-83 in: Berichte über Landwirtschaft, Bd. 77, Landwirtschaftsverlag GmbH Münster-Hiltrup

Provenance-environment interactions of Norway spruce (*Picea abies* [L.] Karst.) on German and Hungarian test sites

M. LIESEBACH[1], A. O. KÖNIG[2] AND E. UJVÁRI-JÁRMAY[3]

[1]*Federal Research Centre for Forestry and Forest Products (BFH), Institute for Forest Genetics and Forest Tree Breeding, D-15377 Waldsieversdorf, Germany;* [2]*D-22927 Grosshansdorf, Germany;* [3]*Forest Research Institute, ERTI, Experiment Station, H-3232 Mátrafüred, Hungary*

Key words: Norway spruce (*Picea abies*), provenance-environment interactions, climatic change

Abstract: The present paper attempts to analyse the response of 100 seed sources of one block of the IUFRO 1964/68 Norway spruce provenance experiment to varying site conditions. Data from three German and one Hungarian test sites were used. These trial site can be seen as a model to investigate the prognosticated change of climate. Between the German sites the annual mean temperature increase corresponds to a warming scenario of +0.7 °C and +1.8 °C, respectively. At the Hungarian site the temperature extremes are more distinct than at the German sites. In general a wide range with regard to survival and diameter growth could be observed on each test site. A few provenances are well adapted to all sites, whereas most of them perform well only under specific conditions.

Introduction

Norway spruce (*Picea abies* [L.] Karst.) has a wide natural distribution area, extending from Central Europe into Asia. It is considered to be one of the morphologically most variable coniferous species, and as well a very plastic species when transferred to other geographical regions (Schmidt-Vogt 1987). Provenance variation has been studied in numerous experiments. They have revealed valuable information on the response of different geographical origins when planted under different growing conditions (Schmidt-Vogt 1987, Dietrichson 1979, Giertych 1979, 1985, Weisgerber 1979, Krutzsch 1992). Scandinavian trials have shown that Norway spruce originating from more southern regions will often grow faster than the local provenance (Langlet 1960, Kiellander 1956, Dietrichson 1979). However, the general conclusion that southern provenances should be preferred in silviculture, can not be drawn, as there is a large variation in growth performance within central and south-eastern European regions. The climatic conditions of the planting site are of great importance (Walberg 1980 after Fottland and Skrøppa 1989). On good sites continental spruce

Müller-Starck, G. and Schubert, R. (eds.). Genetic Response of Forest Systems to Changing Environmental Conditions. Vol. 70 (For. Sci.), 353-363 © *2001 Kluwer Academic Publishers, Dordrecht, Boston, London*

frequently shows a large annual ring width variation, delayed lignification, and low latewood percentage, what may result in stem cracks on standing trees (Dietrichson *et al.* 1985).

Many of the old provenance experiments comprise only a limited number of provenances and are usually planted on a few test sites. The international IUFRO provenance experiment of 1964/68 however permits a thorough investigation of the overall geographic variation in the species and is a good base for the study of uniformity within and between provenance regions (Krutzsch 1975).

The main aim of the present paper is to study the effect of an increased annual mean temperature on the performance of Norway spruce provenances in relation to the predicted warming.

Material and methods

For the present study one block of the international Norway spruce experiment IUFRO 1964/68 was evaluated. The entire provenance trial comprises 1100 seed sources, which in the majority can be considered as collections from defined provenances; additionally samples from provenance regions or single tree lots were included. The seed was sown at the nursery of the Institute for Forest Genetics and Forest Tree Breeding in Schmalenbeck, Germany, in 1964. After transplanting, the plants were distributed to interested partner institutions in Europe and Canada. The experiment was established as series of field trials on 20 sites in 13 countries in 1968 (details see Krutzsch 1975). For experimental reasons the total number of seed sources was subdivided into eleven groups (blocks) of 100 seed sources each. Each block comprises provenances from all provenance regions and can therefore be considered as an independent provenance experiment. In most locations the field design consisted of single tree plots with 20 or 25 replications, respectively. In Germany three complete sets of provenances, and in Hungary one set were planted. For the present study data of block 7 from Germany and Hungary were evaluated. Block 7 was chosen because of the rather uniform soil conditions over the four locations (different types of brown soil), and its varying annual mean temperatures over the German test locations (Buchert and Hoppe 1986; Szönyi and Ujvári 1970). For the test sites and the locations, where the 100 seed sources had been collected long standing temperature and precipitation data are available. The climatic data were interpolated between European climate stations with regard to the location and elevation of the test and collection sites, respectively. In Table 1 the main geographic and climatic data are summarized.

Table 1. Main information on the test sites

Test site	Laasphe	Radolfshausen	Hinterweidenthal	Nyírjes
Country	Germany	Germany	Germany	Hungary
Internation. trial no.	15	14	13	20
Geogr. latitude	51° 00' N	51° 40' N	49° 11' N	47° 56' N
Geogr. longitude	08° 30' E	10°07 ' E	07° 41' E	19° 58' E
Altitude a.s.l.	400 m	250 m	305 m	560 m
Temperature:				
Year	7.5 °C	8.2 °C	9.3 °C	7.6 °C
May-Sept.	13.9 °C	14.7 °C	16.0 °C	15.8 °C
Range (Jan.-July)	17.1 °C	17.5 °C	17.9 °C	22.7 °C
No. of days >10 °C	146	159	173	159
Precipitation:				
Year	990 mm	730 mm	800 mm	780 mm
May-Sept.	390 mm	350 mm	370 mm	400 mm

The mean annual temperature of the site at Radolfshausen is +0.7 °C higher than that of Laasphe, and the increase to Hinterweidenthal is +1.8 °C. The block at Nyírjes was chosen because an additional increase of the mean annual temperature was expected, but this is only partly true for the mean temperature from May to September. A more continental influence can be seen from the higher difference between summer and winter temperature. On the other hand the precipitation during the vegetation period is in Hungary slightly higher than on the German sites.

Height and diameter measurements have been carried out at more or less regular intervals. The corresponding plant ages are listed in Table 2. They are unfortunately not equal for Germany and Hungary, with the exception of the last diameter measurement at age 29. Therefore main weight will be put on this age.

Table 2. Plant ages when height and diameter were measured

Test site	Laasphe	Radolfshausen	Hinterweidenthal	Nyírjes
Height	5, 7, 10, 13, 16	7, 10, 13, 16	5, 7, 10, 13, 16	5, 9, 14, 20
Diameter	16, 21, 29	16, 21, 29	16, 21, 29	14, 20, 29

For the listed ages survival rates were calculated. With the exception of the trial Hinterweidenthal all experiments were thinned once before the last measurement.

Dietrichson and Skrøppa (1977) proposed 20 zones, which are considered more or less homogeneous with regard to geography and climate. From the zone "Foothills of the Alps" six German provenances were studied in detail. They are listed in Table 3.

Table 3. Information on the six German provenances from the zone "Foothills of the Alpes"

Provenance	Königs-bronn	Tussen-hausen	Burg-hausen	Ipsheim	Landsberg	München-Süd
Prov. no.	0701	0715	0721	0771	0780	0782
Geo. latitude	48° 44' N	48° 06' N	48° 10' N	49° 32' N	48° 02' N	48° 05' N
Geo. longitude	10° 08' E	10° 35' E	12° 49' E	10° 30' E	10° 53' E	11° 35' E
Altitude a.s.l.	500 m	500 m	500 m	400 m	800 m	580 m
Temperature:						
Year	7.6 °C	7.8 °C	7.8 °C	8.1 °C	6.7 °C	7.6 °C
May-Sept.	14.6 °C	15.0 °C	15.3 °C	15.2 °C	13.5 °C	14.8 °C
Range (Jan. July)	18.9 °C	19.3 °C	20.2 °C	19.1 °C	18.3 °C	19.4 °C
No. of days >10 °C	153	156	159	158	139	154
Precipitation:						
Year	850 mm	880 mm	1040 mm	690 mm	1080 mm	1040 mm
May-Sept.	420 mm	530 mm	580 mm	350 mm	620 mm	610 mm

Statistical evaluation was done with help of the SAS software package (SAS 1989, 1990). To provide greater details on the distribution of the growth variables means and quantiles were calculated and these results presented in box plots separately by trial sites, zones and provenances. Then analyses of variance were run and variance components were estimated. Furthermore correlation coefficients between the test sites and between growth and survival rates were estimated.

Results

Survival and growth development of all provenances of block 7

The development of the experimental means for survival, height and diameter growth are shown in Figure 1. At age 29 the survival amounted to 92 % for the trial Nyírjes, and varied between 78 % (Laasphe) and 85 % (Radolfshausen and Hinterweidenthal) at the German sites. Early overall height was better on the Hungarian than on the German sites. This is confirmed by the experimental mean diameters at age 29: Nyírjes: 17.0 cm, Laasphe 15.6 cm and Radolfshausen 13.0 cm. The poorest diameter growth was observed at Hinterweidenthal with 10.6 cm.

From the curves in Figure 1 it can be seen that the increment rate decreased between the last two measurements as compared to the years before, especially in the trials at Radolfshausen and Hinterweidenthal.

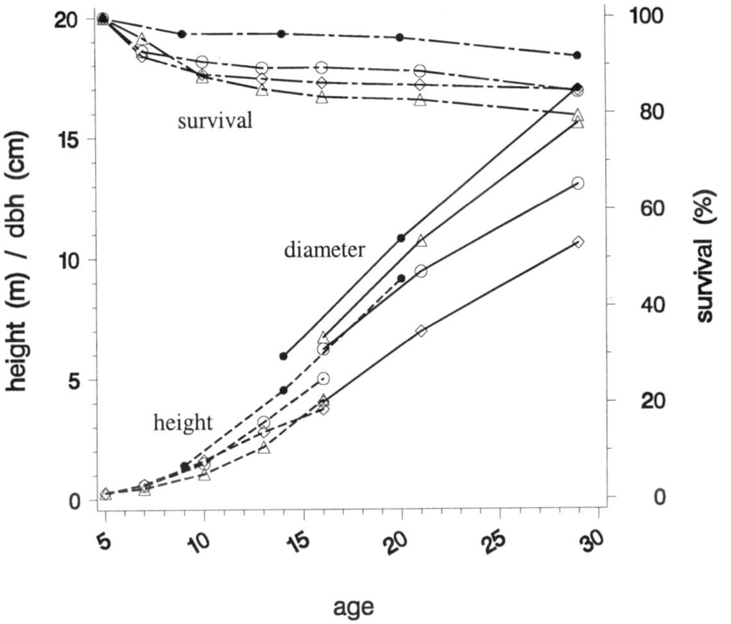

Figure 1. Development of survival (lines at the top with changing dots), height growth (dotted lines) and diameter at 1.3 m (solid lines) in the field trials Laasphe (triangle), Radolfshausen (circle), Hinterweidenthal (rhomb) and Nyirjes (dot).

For the trials Hinterweidenthal and Radolfshausen the overall coefficient of variance is around 32 %, whereas in Laasphe it is 24 % and in Nyírjes 20 %.

At age 29 the correlation between survival of the provenances among test sites is very weak (between r_P= 0.31 and r_P= 0.45). A stronger correlation was found for the mean dbh of provenances among the different trials (between r_P= 0.66 and r_P= 0.74). Also diameter growth and survival is weakly correlated (between r_P= 0.36 at Radolfshausen and r_P= 0.53 at Nyírjes).

Effects of provenance transfers

To quantify the effects of sites, provenances and interaction an analysis of variance was run for diameter growth at age 29. The results are listed in Table 4.

38 % of the total variance can be explained by differences between locations (site conditions), 9 % by effects of the provenance and about 1 % by the site-provenance interaction.

The climatic differences between the location of origin and the test site play an important role with regard to adaptability of provenances. In Figure 2 these differences are put in relation to the diameter growth in the trial of Laasphe.

The x-axis presents the differences in temperature, whereas the y-axis is the percentage of precipitation of the original site as compared to the test site. Right of the vertical line in Figure 2 the provenances which come from colder locations than the test site are plotted and below the horizontal line are those originating from dryer sites. It can be seen that for majority of the provenances the planting at Laasphe implies a transfer to a warmer climate, and about half of the provenances come from dryer sites.

Table 4. Analysis of variance and relative components of variance for diameter growth at age 29

Source of variation	DF	MS	F-value	Components of variance
test site	3	10 405	884.06 ***	38 %
provenance	99	118	9.99 ***	9 %
site * provenance	297	16	1.32 ***	1 %
error	5 554	12		52 %

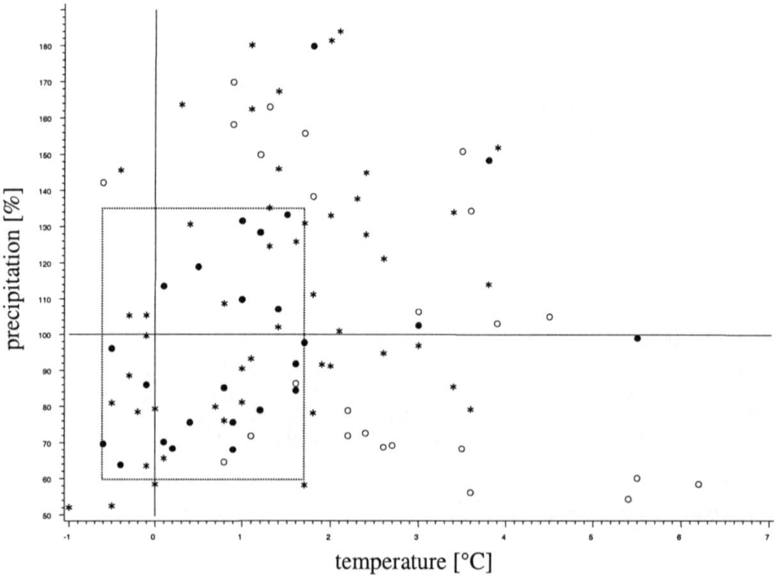

Figure 2. Graph of differences in temperature between the test site Laasphe and the location of the provenances and the precipitation ratio of test site to location of the provenances. The provenances are assigned according to their diameter growth at age 29 (• = 25 % best growing provenances; * = medium range; o = 25 % poorest growing ones).

In order to establish the relation to growth, the 25 % best growing provenances were assigned by a dot, the 25 % poorest growing by a circle and the medium range by a star. With regard to temperature most of the better growing provenances are those from locations with a temperature difference to the test site from between -0.6 to +1.7 °C. Regarding precipitation the better growing provenances are from locations with 60 to 135 % of the precipitation at the test site. The four additional points (out site the spotted rectangle) represent provenances from elevations between 940 and 1410 m with equal or higher precipitation than the test site. The worse growing provenances are from locations with larger differences in temperature and in precipitation; the circles are concentrated in the right bottom and at the top of the graph. Among the worst growing provenances, there are a few Swedish ones with very low survival rates.

Performance of the selected provenances of the zone "Foothills of the Alps"

In block 7 the zone "Foothills of the Alps" is represented by six provenances. The original stands of these provenances are growing under more or less similar climatic conditions. Nevertheless, they have shown different response with regard to survival and diameter growth at age 29 at the trial sites. In Figure 3 box-plots are shown for the six provenances at the sites Laasphe and Radolfshausen. The upper and the lower horizontal lines of the box include half of the trees (medium range), the dotted vertical lines above and below the box show the range of the tallest and smallest quarter of trees, respectively.

These plots give an impression about the variation between and within provenances. In general a wide range of survival and diameter growth can be observed on each test site. Of special interest are the provenances no. 0721 (Burghausen) and 0771 (Ipsheim). At Laasphe there is nearly no overlapping between the medium range of these two provenances while at Radolfshausen the medium ranges are more or less similar. For survival similar patterns are observed.

In Figure 4 the different growth of the six selected provenances over test sites is shown. It can clearly be seen that in five of them diameter growth is decreasing with increasing temperatures. But the provenancc no. 0721, Burghausen has a different response: while it is the poorest at Laasphe and the second poorest at Hinterweidenthal, it is the best at Radolfshausen. The range between the smallest and biggest mean diameter is at Radolfshausen 4.1 cm whereas it is at Nyírjes only 1.9 cm.

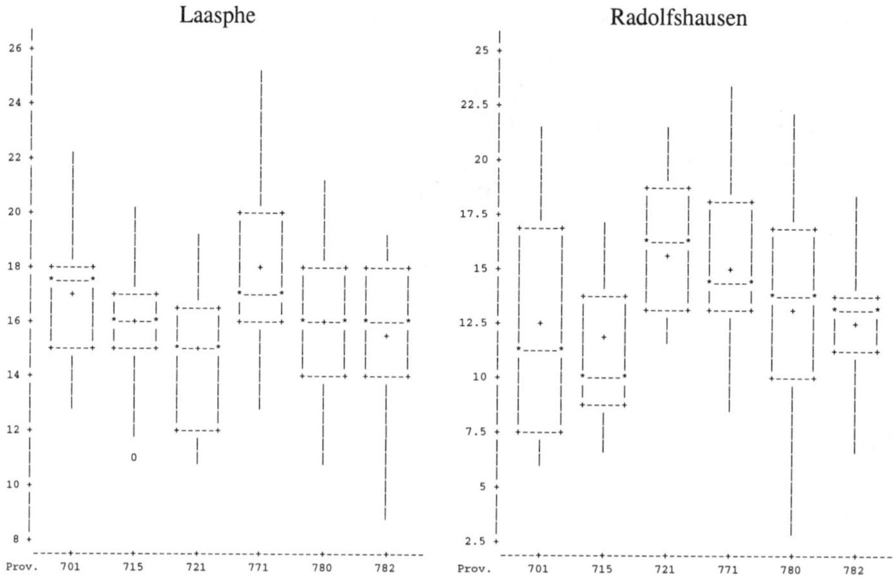

Figure 3. Boxplots showing mean diameters and variation parameters from six provenances of the zone "Foothills of the Alps" in the trials Laasphe (left) and Radolfshausen (right). The upper and lower horizontal lines are the 75 % and 25 % percentile, respectively. The line in the middle of each boxplot is the median and the plus sign (+) the arithmetic mean. The zero (0) stands for extreme values (here thin trees).

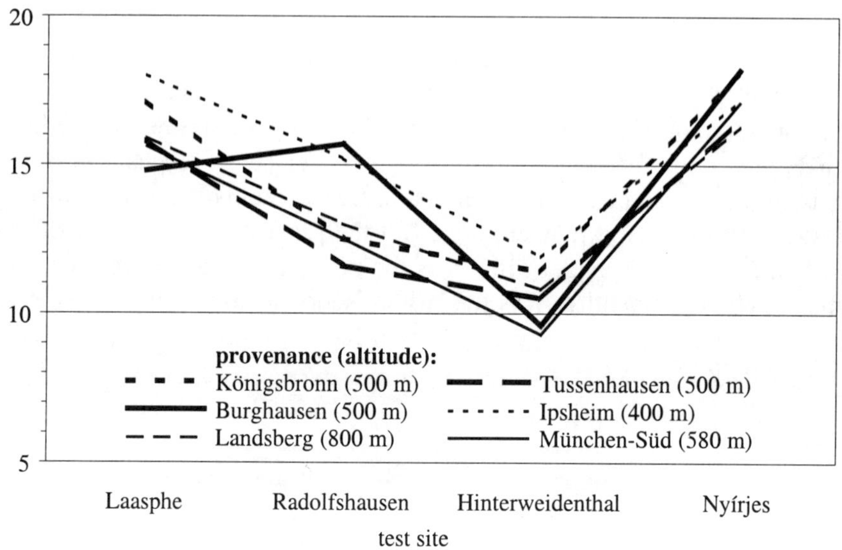

Figure 4. Rank changes of mean diameters between the four test sites

Discussion and Conclusions

On all sites, but especially on the German ones, a big differentiation between and within provenances is observed even if the provenances are from the same geographical zone with similar altitudes and climatic conditions. The variance component estimation showed, that the provenance effect is important and that provenance x environment interaction is generally low. But from Figures 3 and 4 it can be seen that certain provenances change ranks among test sites.

For five of the six more closely studied six provenances a decrease of diameter growth with increasing temperature on the test site could be observed. This is in accordance with models of Schmidtling (1994) who used data from the 1938 IUFRO Norway spruce provenance experiment and predicted a loss in height growth with increasing annual mean temperature. If only the mean annual temperature is considered, also the Hungarian trial fits into that observation: Mean annual temperatures are at Laasphe and Nyírjes with 7.5 °C and 7.6 °C very similar, and the same is true for the mean diameters of the six studied provenances with 16.2 cm and 17.2 cm. On the other hand one would expect a greater influence of the mean temperature of the growing season which is at Nyírjes 1.9 °C higher than at Laasphe. At Nyírjes the higher temperature in summer is paired with a slightly higher amount of precipitation.

These results seem to be contradictory to those obtained in Figure 2, where the bigger part of good growing provenances is coming from colder sites. Beuker (1994) found, that provenances from northern Europe showed an increase in wood production when transferred southwards into a climate with an annual mean temperature sum close to that which is expected in northern areas as a result of the predicted climate change.

These findings make it difficult to draw a general conclusion. It is possible that the interpolations of the climatic data are not in good agreement with the real site conditions. Furthermore it is a fact that many of the investigated provenances are not autochthonous at the location where the seed was collected. So they respond possibly different than expected, because they had been transferred to other site conditions one generation before. König (1981) evaluated height growth in the IUFRO 1964/68 experiment at age 13 and discussed especially the results of six seed sources of the well known provenance district Westerhof (Foothills of the Harz mountains). Within four Central European sites these seed sources occupied in total ranks between 1 and 68, and even within the same seed sources rank changes of 58 (no. 0579) or 46 positions (no. 0550), respectively occurred. Liesebach (1994) could not find any correlation making the approach to detect relationships between isozyme-genloci and morphological, phenological and growth traits.

The results show that for growth model studies which take into consideration a global warming, Norway spruce cannot be taken as a homogeneous entity, but the different response of provenances has to be taken into consideration, especially when a provenance transfer is planned. If a warming is expected, provenances from warmer and dryer conditions offer a possibility to choose good growing provenance. But with regard to the limited number plotted in the left lower corner of Figure 2, not much possibilities for a selection exist. Furthermore the stability would have to be confirmed through further investigations.

The variability patterns are under the influence of several evolutionary forces interacting in rather complex ways (Skøppa and Johnsen 1999). The genetic diversity within a provenance constitutes an adaptation to its present environment, as well as a store of adaptive options for a changing environment (Kräuchi 1993).

Acknowledgements

The authors are thankful to the responsible persons of the forest districts, where the trials were planted, and for the technical assistance of C. Dühring. The study was supported by a grant of the Bundesministerium für Bildung und Forschung BMBF (contract: 01LK95314, Genetic adaptability of important tree species in Germany). The assessments of IUFRO provenance experiment in Hungary were supported by Hungarian Scientific Research Fund (OTKA 990 and T 025 752).

References

Beuker, E. 1994: Long-term effects of temperature on the wood production of *Pinus sylvestris* L. and *Picea abies* (L.) Karst. in old provenance experiments. Scand. J. For. Res. 9: 34-45

Buchert, H. and Hoppe, K. 1986: Standortskartierung der Versuchsflächen des internationalen Fichten-Provenienzversuchs IUFRO 1964/68. Internal report

Dietrichson, J. 1979: Norway spruce provenance trials in nordic countries. Proc. IUFRO joint meeting of WP on Norway spruce provenances and Norway spruce breeding. Bucharest 1979, Pub. by Lower Saxony research institute, Escherode: 3-14

Dietrichson, J., Rognerud, P.A., Haveraaen, O. and Skrøppa, T. 1985: Stem cracks in Norway spruce (*Picea abies* [L.] Karst.). Medd. Nor. inst. for skogforsk. 38 (21): 1-32

Dietrichson, J. and Skrøppa, T. 1977: Skadene på granskogen i vinter og valg av provenienser. Skogeieren (6-7): 10-12

Fottland, H. and Skrøppa, T. 1989: The IUFRO 1964/68 provenance experiment with Norway spruce (*Picea abies*) in Norway. Variation in mortality and height growth. Medd. Nor. inst. skogforsk 43: 1-30

Giertych, M. 1979: Norway spruce (*Picea abies* L. Karst.) provenance experiments in eastern Europe. Proc. IUFRO joint meeting of WP on Norway spruce provenances and Norway spruce breeding. Bucharest 1979, Pub. by Lower Saxony research institute, Escherode: 15-27

Giertych, M. 1985: Report on the IUFRO 1938 and 1939 provenance experiments on Norway spruce (*Picea abies* [L.] Karst.). Polish Academy of Sciences, Kornik: 179 pp.

Kiellander, C.L. 1956: En beaktansvärd granproveniens. Svenska Skog. Fören. Tidskr. 54: 275-189

König, A. 1981: Einige Ergebnisse aus dem IUFRO-Fichtenprovenienzversuch von 1964/68 in der Bundesrepublik Deutschland. Allg. Forstztg. Wien 92: 300-303

Kräuchi, N. 1993: Climate change and forest ecosystems – an overview. Pp. 53-76 in: Schlaepfer, R. (ed). Long-term implications of climate change and air pollution on forest ecosystems; Progress report of the IUFRO task force "Forest, climate change and air pollution". IUFRO World Series 4

Krutzsch, P. 1975: Die Pflanzschulenergebnisse eines inventarisierenden Fichtenherkunftsversuches *Picea abies* Karst. und *Picea obovata* Ledeb. Skogshögskolan, Institutionen för skogsgenetik, Rapporter och Uppsatser nr. 14: 64 + 40 pp.

Krutzsch, P. 1992: IUFRO's role in coniferous tree improvement: Norway spruce (*Picea abies* (L.) Karst.). Silvae Genetica 41: 143-150

Langlet, O. 1960: Mellaneuropeiska granprovenienser i svenskt skogsbruk. K. Skogs- og Lanbr. Akad. Tidskr. 99: 259-329

Liesebach, M. 1994: Untersuchungen an ausgewählten Herkünften des internationalen Fichtenprovenienzversuchs 1964/68 über den Zusammenhang zwischen Isoenzym-Merkmalen und morphologischen, phänologischen sowie Wachstums-Merkmalen. Dissertation, Universität Hamburg: 210 pp.

SAS Institute Inc. 1989: SAS/STAT User's Guide, Version 6, Fourth Edition, Volume 1, Cary, NC: SAS Institute Inc.: 943 pp. and Volume 2, Cary, NC: SAS Institute Inc.: 846 pp.

SAS Institute Inc. 1990: SAS Procedures Guide, Version 6, Third Edition, Cary, NC: SAS Institute Inc.: 705 pp.

Schmidt-Vogt, H. 1987: Die Fichte: Ein Handbuch in 2 Bänden. Band I. Taxonomie, Verbreitung, Morphologie, Ökologie, Waldgesellschaften, Parey, Hamburg, Berlin: 647 pp.

Schmidtling, R.C. 1994: Use of provenance tests to predict response to climatic change: loblolly pine and Norway spruce. Tree physiology 14: 805-817

Skrøppa, T. and Johnsen, Ø. 1999: Patterns of adaptive genetic variation in forest tree species; the reproductive environment as an evolutionary force in *Picea abies*. In: Mátyás, C. (Ed.) Forest Genetics and Sustainability. Kluwer Acad. Publ., Dortrecht

Szönyi, L. and Ujvári, F. 1970: International (IUFRO) Norway Spruce Provenance Trial. Erdészeti kutatások, Budapest, Vol. 66 (3): 47-59

Weisgerber, H. 1979: Norway spruce experiments in Central Europe. Proc. IUFRO joint meeting of WP on Norway spruce provenances and Norway spruce breeding. Bucharest 1979, Pub. by Lower Saxony research institute, Escherode: 57-73

FORESTRY SCIENCES

43. R.T. Prinsley (ed.): *The Role of Trees in Sustainable Agriculture.* 1993 ISBN 0-7923-2030-1
44. S.M. Jain, P.K. Gupta and R.J. Newton (eds.): *Somatic Embryogenesis in Woody Plants*, Volume 3: Gymnosperms. 1995 ISBN 0-7923-2938-4
45. S.M. Jain, P.K. Gupta and R.J. Newton (eds.): *Somatic Embryogenesis in Woody Plants*, Volume 1: History, Molecular and Biochemical Aspects, and Applications. 1995 ISBN 0-7923-3035-8
46. S.M. Jain, P.K Gupta and R.J. Newton (eds.): *Somatic Embryogenesis in Woody Plants*, Volume 2: Angiosperms. 1995 ISBN 0-7923-3070-6
    Set ISBN (Volumes 44–46) 0-7923-2939-2
47. F.L. Sinclair (ed.): *Agroforestry: Science, Policy and Practice.* Selected Papers from the Agroforestry Sessions of the IUFRO 20th World Congress (Tampere, Finland, 6–12 August 1995). 1995 ISBN 0-7923-3696-8
48. J.H. Goldammer and V.V. Furyaev (eds.): *Fire in Ecosystems of Boreal Eurasia.* 1996 ISBN 0-7923-4137-6
49. M.R. Ahuja, W. Boerjan and D.B. Neale (eds.): *Somatic Cell Genetics and Molecular Genetics of Trees.* 1996 ISBN 0-7923-4179-1
50. H.L. Gholz, K. Nakane and H. Shimoda (eds.): *The Use of Remote Sensing in the Modeling of Forest Productivity.* 1996 ISBN 0-7923-4278-X
51. P. Bachmann, M. Köhl and R. Päivinen (eds.): *Assessment of Biodiversity for Improved Forest Planning.* Proceedings of the Conference on Assessment of Biodiversity for Improved Planning (Monte Verità, Switzerland, 1996). 1998 ISBN 0-7923-4872-9
52. G.M.J. Mohren, K. Kramer and S. Sabaté (eds.): *Impacts of Global Change on Tree Physiology and Forest Ecosystems.* Proceedings of the International Conference on Impacts of Global Change on Tree Physiology and Forest Ecosystems (Wageningen, The Netherlands, 26–29 November 1996). 1997 ISBN 0-7923-4921-0
53. P.K.R. Nair and C.R. Latt (eds.): *Directions in Tropical Agroforestry Research.* Selected Papers from the Symposium on Tropical Agroforestry organized in connection with the annual meeting of USA (Indianapolis, U.S.A., 5 November 1996). 1998 ISBN 0-7923-5035-9
54. K. Sassa (ed.): *Environmental Forest Science.* Proceedings of the IUFRO Division 8 Conference Environmental Forest Science. (Kyoto, Japan, 19–23 October 1998). 1998 ISBN 0-7923-5280-7
55. S.M. Jain, P.K Gupta and R.J. Newton (eds.): *Somatic Embryogenesis in Woody Plants*, Volume 4. 1999 ISBN 0-7923-5340-4
56. J.R. Boyle, J.K. Winjum, K. Kavanagh, E.J. Jensen (eds.): *Planted Forests: Contribution to the Quest for Sustainable Societies.* 1999 ISBN 0-7923-5468-0
57. K. von Gadow and G. Hui: *Modelling Forest Development.* 1999 ISBN 0-7923-5488-5
58. J. Abildtrup, F. Helles, P. Holten-Andersen, J.F. Larsen and B.J. Thorsen (eds.): *Modern Time Series Analysis in Forest Products Markets.* 1999 ISBN 0-7923-5524-5
59. S.M. Jain, P.K. Gupta and R.J. Newton (eds.): *Somatic Embryogenesis in Woody Plants*, Volume 5. 1999 ISBN 0-7923-5553-9
60. D. Auclair and C. Dupraz (eds.): *Agroforestry for Sustainable Land-Use.* Fundamental Research and Modelling with Emphasis on Temperate and Mediterranean Applications. 1999 ISBN 0-7923-5799-X

FORESTRY SCIENCES

61. F. Helles, P. Holten-Andersen and L. Wichmann (eds.): *Multiple Use of Forests and Other Natural Resources.* Aspects of Theory and Application. 1999 ISBN 0-7923-5859-7
62. A. Yoshimoto and K. Yukutake (eds.): *Global Concerns for Forest Resource Utilization.* Sustainable Use and Management. 1999 ISBN 0-7923-5968-2
63. C. Mátyás (ed.): *Forest Genetics and Sustainability.* 1999 ISBN 0-7923-6011-7
64. S.M. Jain and S.C. Minocha (eds.): *Molecular Biology of Woody Plants*, Volume 1. 1999 ISBN 0-7923-6012-5
65. E. Mälkonen (eds.): *Forest Condition in a Changing Environment.* The Finnish Case. 2000 ISBN 0-7923-6228-4
66. S.M. Jain and S.C. Minocha (eds.): *Molecular Biology of Woody Plants*, Volume 2. 2000 ISBN 0-7923-6241-1
67. S.M. Jain, P.K. Gupta and R.J. Newton (eds.): *Somatic Embryogenesis in Woody Plants.* Volume 6. 2000 ISBN 0-7923-6419-8
68. T. Treue: *Politics and Economics of Tropical High Forest Management.* A case study of Ghana. 2001 ISBN 0-7923-6931-9
69. K.E. Linsenmair, A.J. Davis, B. Fiala and M.R. Speight (eds.): *Tropical Forest Canopies: Ecology and Management.* 2001 ISBN 0-7923-7049-X
70. G. Müller-Starck and R. Schubert (eds.): *Genetic Response of Forest Systems to Changing Environmental Conditions.* 2001 ISBN 1-4020-0236-X

KLUWER ACADEMIC PUBLISHERS – DORDRECHT / BOSTON / LONDON